# Lecture Notes in Computer Science 14522

Founding Editors

Gerhard Goos
Juris Hartmanis

## Editorial Board Members

The series Lecture Notes in Computer Science (LNCS), including its subseries Lecture Notes in Artificial Intelligence (LNAI) and Lecture Notes in Bioinformatics (LNBI), has established itself as a medium for the publication of new developments in computer science and information technology research, teaching, and education.

LNCS enjoys close cooperation with the computer science R & D community, the series counts many renowned academics among its volume editors and paper authors, and collaborates with prestigious societies. Its mission is to serve this international community by providing an invaluable service, mainly focused on the publication of conference and workshop proceedings and postproceedings. LNCS commenced publication in 1973.

Jean Baratgin · Baptiste Jacquet · Hiroshi Yama
Editors

# Human and Artificial Rationalities

Second International Conference, HAR 2023
Paris, France, September 19–22, 2023
Proceedings

 Springer

*Editors*
Jean Baratgin 🆔
Université Paris 8
Saint Denis, France

Baptiste Jacquet 🆔
Université Paris 8
Saint-Denis, France

Hiroshi Yama 🆔
Osaka City University
Osaka, Japan

ISSN 0302-9743         ISSN 1611-3349 (electronic)
Lecture Notes in Computer Science
ISBN 978-3-031-55244-1         ISBN 978-3-031-55245-8 (eBook)
https://doi.org/10.1007/978-3-031-55245-8

# Preface

The Human and Artificial Rationalities (HAR) conference series is focused on comparing human and artificial rationalities, investigating how they interact together in a practical sense, but also on the theoretical and ethical aspects behind rationality from three main perspectives: Philosophy, Psychology, and Computer Sciences. HAR aims at building bridges between these three fields of research.

This volume, entitled Advances in Reasoning, from Human to Artificial Rationalities, contains the papers presented at the 2nd International Conference on Human and Artificial Rationalities (HAR 2023) which was held in Paris (France), during September 19–22. This edition continued the theme initiated in the first conference of bridging three different fields of study, Philosophy, psychology, and computer sciences to share knowledge about intelligent systems, whether natural or artificial, to generate a multidisciplinary understanding of reasoning.

In this second year of the HAR conference, we received 39 submissions. For publication in the proceedings, we accepted 18 of them as full papers and 5 as short papers. All submissions were carefully reviewed by the program committee and additional reviewers in a multi-stage single-blind peer-review process. Three reviewers were assigned to each submission by the program chairs according to their area of expertise and more could be assigned if reviewers did not agree on the decision. On average, there were 8 papers reviewed per reviewer. Reviewers were assigned based on the balance between expertise in the topic of the paper and minimization of potential conflicts of interest. If the situation where a reviewer but would whose have personal knowledge of the authors of a paper but would whose expertise would also made them too invaluable to replace, we ensured that the two other reviewers did not have personal contacts with the authors. Final decision was made by the program committee chairs and general chair based on the score on the 6 criteria (Content quality, Significance, Originality, Relevance, Rigor, and Coherence with conference) and on the written reviews. Decisions could include acceptance for publication, major revisions, minor revisions or rejection. The accepted papers cover a wide array of topics pertaining to rationality, such as reasoning, conceptual thinking, belief revision, judgments, theory of mind, anthropomorphism, language and culture on reasoning, socio-cultural aspects of artificial agents, evaluation of artificial agents, acceptance and trust in artificial agents, use of artificial devices in life-span approaches, neuropsychology of reasoning, and thinking with disabilities.

In addition to the list of accepted papers, the conference greatly benefited from three invited lectures by important researchers on reasoning from the field of philosophy: Daniel Andler, professor emeritus at Université Paris-Sorbonne, France (Philosophie des sciences et théorie de la connaissance), honorary member of the Institut Universitaire de France, member of the Académie des sciences morales et politiques; from the field of linguistics: Daniel Lassiter, senior lecturer in Semantics, School of Philosophy, Psychology and Language Sciences, University of Edinburgh, United Kingdom; and from the field of human-robot interaction: Bernard N'Kaoua, professor at the Université de Bordeaux, France (cognitive sciences), head of the Society, Politics, and Public Health doctoral school, member of the Inserm Research Center Bordeaux Population Health Center (BPH) and of the INRIA Research Center Bordeaux Sud-Ouest. Among them, two contributed invited papers to this volume which were peer-reviewed before publication.

We would like to thank all the members of the Program Committees, as well as the additional reviewers, who devoted their time for the reviewing process. We thank all the authors of submitted papers, the invited speakers, and the participants for their scientific contributions to the conference. Finally, we would like to thank all the Organizing Committee members (with a special thanks to Emmanuel Brochier and Antoine Gazeaud, from IPC, Facultés Libres de Philosophie et de Psychologie) for their excellent local organization, which made the HAR conference a success.

The HAR conference could not have happened without the organizational support of partner institutions: The CHArt laboratory of Université Paris 8, Facultés Libres de Philosophie et de Psychologie (IPC), and the P-A-R-I-S Reasoning Association. We would also like to particularly thank the MAIF, the BONHEUR laboratory of CY University, and the EDUS4EL from the ERASMUS+ program for their financial support.

December 2023
Jean Baratgin<br>Baptiste Jacquet<br>Hiroshi Yama

# Organization

## General Chair

Jean Baratgin                 Université Paris 8, France

## Program Committee Chairs

Baptiste Jacquet            Université Paris 8, France
Hiroshi Yama              Osaka Metropolitan University, Japan

## Program Committee

| | |
|---|---|
| Emmanuel Brochier | IPC, Facultés Libres de Philosophie, France |
| Julien Bugmann | Haute École Pédagogique du Canton de Vaud, Switzerland |
| Charles El-Nouty | Université Sorbonne Paris Nord, France |
| Patrice Godin | Université de Nouvelle Calédonie, France |
| Hirofumi Hashimoto | Osaka Metropolitan University, Japan |
| Alain Jaillet | CY Cergy Paris Université, France |
| Frank Jamet | CY Cergy Paris Université, France |
| Vassilis Komis | University of Patras, Greece |
| Hélène Labat | CY Cergy Paris Université, France |
| Florian Laguens | Facultés Libres de Philosophie, France |
| Laura Macchi | Università degli Studi di Milano Bicocca, Italy |
| Federico Manzi | Università Cattolica del Sacro Cuore, Italy |
| Laura Martignon | Pädagogische Hochschule Ludwigsburg, Germany |
| Emile Thalabard | Facultés Libres de Philosophie, France |

## Additional Reviewers

Kevin Bague
Maxime Bourlier
Fabien Calonne
Marion Dubois-Sage
Darya Filatova

Antoine Gazeaud
Léa Lachaud
David Over
Véronique Salvano-Pardieu

Co-funded by the
Erasmus+ Programme
of the European Union

# Contents

## Neuropsychology and Interaction

## Artificial Agents and Interaction

**Applied Reasoning**

# Invited Papers

# The Crucial Role of Compositional Semantics in the Study of Reasoning

Daniel Lassiter[✉]

School of Philosophy, Psychology, and Language Sciences, University of Edinburgh,
Edinburgh, UK
dan.lassiter@ed.ac.uk

**Abstract.** Reasoning work in both psychology and computer science is often conceived in terms of drawing inferences over strings. This approach works reasonably well in some cases, but it is sometimes dramatically misleading due to the pervasive presence in natural language of interpretive factors that are not reflected transparently in surface form. The problem includes not only the well-known issues of ambiguity, polysemy, and sensitivity to pragmatic context, but also a much less-studied factor: compositional semantic interactions where one expression in a sentence influences the interpretation of another. In two case studies, I show that careful attention to natural language semantics is crucial in understanding valid and invalid patterns of reasoning, and in avoiding fallacies.

**Keywords:** Psychology of reasoning · Computational reasoning · Compositional semantics · Validity

Linguistics is extremely important in psychological and computational work on reasoning. This has been recognized for some time, especially in the domains of pragmatics and lexical semantics. But compositional semantics—the study of how the form of sentences relates to their semantic interpretation—has not been given sufficient attention, especially in psychology. This paper draws attention to two ways that attention to semantics can clarify core issues in the study of reasoning, and help us avoid fallacious arguments.

## 1 Counter-Examples to Bedrock Reasoning Principles

Psychological research on reasoning has traditionally been concerned with the validity or invalidity of various principles of inference, mostly drawn from philosophical logic. Conditional reasoning is by far the most intensively investigated

I'm very grateful to the organizers of the conference "Human and Artificial Rationalities" for their kind invitation to give a plenary talk. Thanks also to David Over for very helpful discussion, to three anonymous reviewers for this volume, and to Jean Baratgin, Maxime Boulier, and Baptiste Jacquet, whose experimental collaboration inspired a major portion of the thoughts below.

J. Baratgin et al. (Eds.): HAR 2023, LNCS 14522, pp. 3–18, 2024.
https://doi.org/10.1007/978-3-031-55245-8_1

domain, and Modus Ponens (MP) is surely the best-supported non-trivial principle of conditional reasoning, being almost unanimously endorsed in numerous studies [6]. Modus Tollens (MT) is the ugly twin of MP: while participants endorse it somewhat less enthusiastically, most researchers consider MT to be valid nonetheless, attributing reduced endorsement to the psychological difficulty of applying the principle.

(1)  Modus ponens:
  a. If $A$, $C$.
  b. $A$.
  c. Therefore, $C$.

(2)  Modus tollens:
  a. If $A$, $C$.
  b. Not $C$.
  c. Therefore, not $A$.

MP and MT sure do look like valid rules of inference. When we fill in the variables with simple English sentences, the reasoning seems rock-solid.

(3)  a. If it's raining, Mike is carrying an umbrella.
  b. It's raining.
  c. Therefore, Mike is carrying an umbrella.

(4)  a. If it's raining, Mike is carrying an umbrella.
  b. Mike is not carrying an umbrella.
  c. Therefore, it's not raining.

Constructive Dilemma (CD) is a more complex principle, but one that is widely considered valid. Here it is in the abstract:

(5)  a. If $A$, $C$.
  b. If $B$, $D$.
  c. $A$ or $B$.
  d. Therefore, $C$ or $D$.

Again, filling in the variables with simple English sentences gives us what looks like impeccable reasoning:

(6)  a. If it's raining, Mike is carrying an umbrella.
  b. If it's sunny, he's wearing sunglasses.
  c. It's either raining or it's sunny.
  d. Therefore, either Mike is carrying an umbrella or he's wearing sunglasses.

I have bad news: all three principles admit of obvious counter-examples. Here is one.

(7)  a. If it's raining, Mike always carries an umbrella.
  b. It's raining.
  c. Therefore, Mike always carries an umbrella.

This seems to be an instance of the MP template above in (1), substituting *It's raining* for *A* and *Mike always carries an umbrella* for *C*. It's also a terrible argument. If we assume that the generalization in (7a) holds, and that it is raining right now, we still cannot conclude (7c) unless we have some *additional* information—namely, that it always rains. So this is a counter-example to MP.

Here is a similar counter-example to MT:

(8)  a.  If it's raining, Mike always carries an umbrella.

  b.  It's not the case that Mike always carries an umbrella.

  c.  Therefore, it's not raining.

Suppose that Mike carries an umbrella always when it rains, and never carries one when it's sunny. Furthermore, we live in a place where it rains on half of the days, and is sunny on the others. Then both (8a) and (8b) are true. But this knowledge doesn't allow us to conclude anything about the *current* state of the weather in (8c)! What has gone wrong?

For good measure, here's the matched counter-example to CD.

(9)  a.  If it's raining, Mike always carries an umbrella.

  b.  If it's sunny, Mike always wears sunglasses.

  c.  It's either raining or it's sunny.

  d.  Therefore, either Mike always carries an umbrella, or he always wears sunglasses.

Since one counter-example is enough to show that a rule of inference is not valid, this would seem to imply that MP, MT, and CD are not valid! Have we just overturned the intuitions of generations of philosophers and logicians, and the hard-won results of decades of careful psychological research? The answer is, as we'll see below, is "It depends". There is definitely something fishy going on in the counter-examples, but we need help from research in compositional semantics to see what it is and how we can respond.[1]

## 2  Foibles of String-Based Reasoning

Semantic theory sheds important light on the puzzle just posed. Lewis [23] pointed out that conditional antecedents interact compositionally with adverbs of quantification (AQs) like *always, frequently, seldom,* etc. When an AQ appears in the immediate scope of an *if*-clause, the *if*-clause is usually interpreted as providing a restriction on the quantificational domain of the AQ. As a result, the item *always* is interpreted in a special way in (the most prominent reading of) the major premise of the arguments above.

---

[1] Yalcin [32] gives a counter-example to MT similar to (8), but does not point out that the problem affects MP equally. I will discuss Yalcin's argument further below. McGee [27] famously gives a purported counter-example to MP involving nested conditionals, but also contends that MP is valid in other contexts. In any case, many have found McGee's counter-example less than fully convincing (e.g., [2,28]).

(10)  If it's raining, Mike always carries an umbrella.

a. **Wrong**: If it's raining (now), Mike carries an umbrella *in all relevant situations.*

b. **Right**: *In all relevant situations in which it's raining*, Mike carries an umbrella.

All of the purported counter-examples given above rely on this kind of shift in the domain of an AQ. When *always* occurs in the consequent of a conditional—e.g., the major premise of our counter-examples to MP and MT—it quantifies over a restricted domain of situations that are contextually relevant *and* satisfy the conditional antecedent. When *always* occurs unembedded, it quantifies over a larger domain: the situations that are contextually relevant, full stop. The schematic interpretations given below make clear why these arguments are not convincing. Let $R$ be a variable standing for the contextually relevant situations. Lewis' analysis implies that the purported MP counterexample (11) is interpreted along the lines of (12), and the MT counterexample (13) as in (14).[2]

(11)  a. If $A$, always $C$.
b. $A$.
c. Therefore, always $C$.

(12)  a. $ALL[R \cap A][C]$.
b. $A$ (at speech time).
c. Therefore, $ALL[R][C]$.

(13)  a. If $A$, always $C$.
b. Not always $C$.
c. $\therefore$ Not $A$.

(14)  a. $ALL[R \cap A][C]$.
b. $\neg ALL[R][C]$.
c. $\therefore \neg A$ (at speech time).

In the MP counter-example (11), the item *always* is interpreted differently in the major premise and the conclusion—as the schematic interpretation in (12) makes clear. Specifically, the restriction of the $ALL$-quantifier that *always* denotes is different in the two sentences, due to the compositional interaction with the *if*-clause. In the MT counter-example (13), the instance of $A$ in the major premise again functions to restrict the $ALL$-quantifier over situations provided by *always*. This is a very different job, semantically, from the use of $A$ in the conclusion (13c) to predicate something of the current speech situation. A careful semantic analysis of these arguments reveals clearly why they are not intuitively compelling: both trade in a fallacy of equivocation. (As an exercise, the reader may wish to construct a similar analysis of the CD counter-example in (9).)

We are left with several possible interpretations of the situation so far. One option is to stick to our guns. The counter-examples were derived by uniform substitution of variables in the schematic argument forms by grammatical strings

---

[2] The notation $Q[R][N]$ is borrowed from generalized quantifier theory [1,31]. It is interpreted as follows: a quantifier $Q$ takes two arguments, a restriction $R$ and a nuclear scope $N$. For example, *All white dogs bark* would be translated as $ALL[\mathbf{white} \cap \mathbf{dog}][\mathbf{bark}]$, a claim that is true just in case the things that are both white and dogs are a subset of the things that bark.

of English. Since the result of such substitution sometimes results in clearly specious arguments, we have shown that these argument forms are invalid. Call this a **raw string-based** picture of validity. If this position is right, then MP, MT, and CD are all invalid. (Worryingly, so are even more basic principles like Reflexivity, $A$ implies $A$: see below.)

At the opposite extreme is a semantic or **interpretation-based** picture of validity, where the objects that we reason over are full semantic interpretations, with no information about strings. The basic idea is that validity is a relation among model-theoretic interpretations, involving truth-preservation; on this picture, formal relations among strings are neither necessary nor sufficient for argument validity.

Even if we can stomach the rejection of MP, MT, and CD, the raw string-based picture has some glaring problems. We did not quite use uniform substitution above, but rather replaced names by pronouns in some cases, assuming the reader would be able to fill in the appropriate interpretation (!) of the pronoun in each case. And we assumed that each token of ambiguous or polysemous words would be resolved in the same way. In addition, we relied on general principles of context-sensitive interpretation to determine, for example, when and where it is said to be raining/sunny/etc. [22]. When the strings we substitute have context-sensitive interpretations, even the simplest instances of MP, MT, and CD would be invalid if the context were not resolved in a uniform manner across sentences. But there is nothing in the strings themselves that tells us when we should hold context fixed across sentences, and when we should not.

Still, there seems to be a sense among theorists that ambiguity, anaphora resolution, and temporal/situational indexicality are fairly superficial aspects of interpretation. Perhaps the string-based picture can be maintained once these issues are patched up somehow—say, using the notion of "logical form" that I'll describe below.

Several recent papers seem to assume something like this in arguing against MT and CD. The counter-examples are strikingly similar to the ones given above, but they rely on a generalization of Lewis' observation by Kratzer [16,17]. Kratzer pointed out that *if*-clauses also restrict the domains of epistemic and deontic modal operators.

(15)   If he is rude, Mike ought to apologize.

    a.   **Wrong**: If he is in fact rude in the actual situation, the best situations (in general) are ones in which Mike apologizes.

    b.   **Right**: *Among situations in which Mike is rude*, the best are those in which he apologizes.

The incorrect interpretation in (15a) would imply that—if Mike is in fact rude— he ought to apologize, even in possible situations that are better than actuality- ones in which he has not been rude! The correct interpretation (15b) allows for the plausible option that Mike ought not to apologize in situations in which he hasn't done anything wrong.

Epistemic operators behave similarly: for instance, the intuitive interpretation of (16) makes reference to a conditional probability measure restricted by the antecedent *the die is above 3*, rather than an unconditional probability.

(16)  If the die is above 3, it's probably even.

    a.  **Wrong**: If above 3, $P(\textbf{even}) > .5$.

    b.  **Right**: $P(\textbf{even} \mid \textbf{above 3}) > .5$.

(16a) would have it that if the die is *in fact* above 3, then our *current* information implies that it's probably even—even if we have no idea what the outcome was. Arguably this is a *possible* interpretation, but if so it is certainly much less prominent than the one in (16b).

Yalcin [32] uses the interaction of epistemic operators and *if*-clauses to argue that MT is invalid, pointing to specious arguments along the lines of (17).

(17)  a.  If the die is above 3, it's probably even.

    b.  It's not the case that the die is probably even.

    c.  Therefore, the die is not above 3.

If a fair die has been rolled and we know nothing about how it came up, (17a) and (17b) are both true; and yet we clearly cannot conclude (17c). The semantic diagnosis is the same as it was in our AQ-based arguments. The instance of *probably* in (17a) picks out a certain condition on a probability measure whose domain has been restricted by the antecedent—that is, the conditional probability measure $P(\cdot \mid \textbf{above 3})$. In contrast, the instance of *probably* in (17b) picks out a condition on the corresponding unconditional probability measure, with an unrestricted domain.

Along similar lines, Kolodny & MacFarlane [15] give a counter-example to CD involving the deontic modal *ought*. Without going into the full detail of their now-famous "Miners' Puzzle", the argument has a schematic form that should look familiar from (9):

(18)  a.  If $A$, ought $C$.

    b.  If $B$, ought $D$.

    c.  Either $A$ or $B$.

    d.  Therefore, either ought $C$ or ought $D$.

Kolodny & MacFarlane observe that this argument fails intuitively in a scenario that they describe, and conclude that both CD and MP are invalid. Here again, the puzzle is generated by the fact that the instances of *ought* in (18d) talk about what is best globally, while those in (18a) and (18) talk about what is best in different restricted domains.

In both cases, the problematic examples involve a compositional semantic interaction *between* the two clauses that have been substituted in the antecedent and consequent of a conditional. A compositional interaction *within* one of these clauses would not have created a problem, since the interaction would be maintained in each instance of a string when we substitute it into the variables of an argument template.

# 3   Logical Form to the Rescue?

The examples adduced by Yalcin [32] and Kolodny & MacFarlane [15] would certainly undermine MT and CD if we thought of these principles as applying to arguments derived by uniform substitution of grammatical strings of English for sentence variables. But no one advocates a theory like this: it would be hopeless, rendering even the most basic principles of reasoning obviously invalid. For instance, Reflexivity—"$A$ implies $A$"—fails on raw strings because of the potential for context shifts.

(19)   a. He is thoughtful. (pointing to Barack Obama)

      b. Therefore, he is thoughtful. (pointing to Jair Bolsonaro)

The problem here is about what counts as "uniform substitution of sentences" in an argument template like Reflexivity or MP. Using string identity as the criterion, MP would of course be invalid—and every other logical principle you can think of.

(20)   a. If he is thoughtful, I'll vote for him. (pointing to Jair Bolsonaro)

      b. He is thoughtful. (pointing to Barack Obama)

      c. Therefore, I'll vote for him. (pointing to Jair Bolsonaro)

While reasoning over raw strings of English is a non-starter, the opposite extreme—reasoning over full semantic interpretations—is not much better for our purposes. According to the classic semantic definition of validity, an argument with premises $\{\gamma_1, \gamma_2, ...\}$ and conclusion $\delta$ is valid if and only if, for all situations $s$: if each of $\gamma_1, \gamma_2, ...$ is true in $s$, $\delta$ is true in $s$ as well. The semantic perspective is attractive because it automatically enforces some basic requirements for assessing validity of arguments. Since it makes crucial reference to *truth*, anaphora, ellipsis, indexicals, ambiguity, polysemy, quantifier scope, etc. must be resolved before the definition can even be applied. Sentences cannot be assessed for truth-in-a-situation without first performing these tasks.

With that said, the purely semantic approach to validity is also unsatisfying for many logical, psychological, and computational purposes. The basic question that we started with—"What is the status of principles like MP, MT, and CD?"—can't even be formulated within a strictly semantic approach. These are questions about the inferential relations among sentences with a particular *form*. The semantic picture of validity can see what sort of semantic object each premise denotes (e.g., a set of situations or "possible worlds" making the sentence true). It does not have access to information about whether a particular premise was formulated as a conditional, and this information is needed before we can pose the question of whether an example counts as an instance of MP. In other words, a theory at the pure semantic extreme would be theoretically handicapped because it would lack the resources to distinguish genuine instances of MP like (3) from specious instances like (20). In order to do this, a theory needs to pay attention to relevant aspects of the syntactic form of sentences (but not too much) in addition to their interpretations.

Note that this criticism applies to purely semantic theories of inferencing like the one that Johnson-Laird [14] seems to be arguing for, *even if* the theory is correct about the basic psychological mechanisms of reasoning. Simultaneous attention to the syntactic form of an argument and its semantics is needed in order to determine which apparent instances of an argument are genuinely relevant to our theoretical questions, and which are pseudo-arguments.[3]

To make their arguments against MT and CD work, then, Yalcin and Kolodny & MacFarlane must be assuming that validity is a property of syntactically structured representations. But these must be representations that crucially import *some* aspects of the semantic interpretation, to get the right criteria of identity for uniform substitution of sentences in argument templates. In logic and linguistics this task is often pursued by annotating pronouns and lexically ambiguous expressions with unpronounced numerical indices. Crucially, these elements are part of the syntactic form, but the way they are distributed has a purely semantic motivation—we assign indices depending on whether the pronouns differ in reference, or two instances of an ambiguous word have the same meaning, when the sentence is semantically interpreted. Helping ourselves to this additional information allows us to separate the specious argument in (19) from unproblematic instances of Reflexivity.

(21)   a. $He_1$ is thoughtful.
       b. **Good conclusion:** Therefore, $he_1$ is thoughtful.
       c. **Bad conclusion:** Therefore, $he_2$ is thoughtful.

Linguists and philosophers call these semantically enriched syntactic representations "logical form". On some theories, each sentence (as tokened in some context) has a grammatically privileged logical form. Logical forms vary from theory to theory, but they are usually assumed to contain unpronounced, semantically relevant information about how to resolve anaphors and ambiguities, fill in ellipsis sites, and disambiguate quantifier scope alternations, among other tasks. All of this information—and probably quite a lot more—is needed if we hope to formulate a reasoning theory that operates over syntactically structured representations, while avoiding classifying arguments like (19)–(20) as valid.

Here's the punch line. In light of the purported counter-examples described above, the question of whether MP, MT, and CD are valid depends essentially on a theoretical choice point: whether our logical forms include information about the quantificational domains of modals and AQs. If they do, then the problematic examples given in Sect. 1 are no longer problematic. The enriched representations with domain annotations would look something like this.

(22)   a. If it's raining, Mike always$^{(R \cap \textbf{rain})}$ carries an umbrella.

---

[3] With that said, the theory proposed by Johnson-Laird [13,14] is not actually a representative of the purely semantic extreme that the author appears to advocate in the surrounding prose. The theory's "models" contain syntactic annotations such as negation and ellipses, making them rather close in spirit to the "logical forms" that are widely employed in linguistic and philosophical work.

   b.  It's raining.

   c.  Therefore, Mike always$^R$ carries an umbrella.

This is no more problematic than the Obama/Bolsonaro argument in (19): the temptation to treat it as an instance of MP stems from a failure to attend to the silent domain index. Once we do, we see that this argument is not derived from the MP template by uniform substitution by logical forms of the appropriate type. The same diagnosis accounts for the AQ- and modal-based counter-examples to MT and CD. On the other hand, if we assume that logical forms do not contain information about quantificational domains, then all of these arguments are legitimate counter-examples, derived from the relevant argument templates by uniform substitution.

The fate of our bedrock reasoning principles hangs in the balance. How can we decide? If there is a linguistically and/or psychologically privileged level of logical form, the issue is an empirical one. We just need to do the empirical and theoretical work to find out whether domain variables do indeed exist at the level of logical form. A template for this kind of work is provided by Stanley and Szabó [29,30], who argue that the domain arguments of nominal quantifiers like *every* are explicitly present at logical form. These arguments likely extend to AQs, with the implication that all of the counter-examples we saw above are spurious if Stanley and Szabó are correct.

Many theorists have reasons to be skeptical about the linguistic or psychological reality of a privileged, syntactic level of logical form. If this skepticism is warranted, then the choice of whether or not to include domain variables as in (22) comes down to a modeling choice, made for theoretical or practical convenience. As an analogy, consider an important strand of computational work geared toward the Recognizing Textual Entailment task. One of the driving considerations of this literature is that reasoning over full semantic interpretations is computationally prohibitive, while string-based reasoning is relatively cheap (e.g., [25]). However, raw strings are not especially useful for this purpose for a variety of reasons, including those considered above. So, researchers in this tradition have suggested a variety of ways to implement reasoning systems that work with something close to the syntactic form of sentences of natural language, while adding annotations indicating semantic properties that are relevant to the task at hand.

One influential proposal along these lines, stemming from logic and later adopted into computational work, is the "monotonicity calculus" [3,12,24,25]. The basic idea is that we annotate strings of a natural language (say, English) with information about syntactic constituency and the monotonicity properties of the phrases. Here, the "monotonicity" of an environment means, essentially, whether it licenses valid inferences from a category to a subordinate category, a superordinate, or neither. For example, from the premise *A car went by* we can validly infer *A vehicle went by*, but not *A Honda went by*. So, the position of *car* is annotated with a +, indicating that it allows substitution by superordinates, but not subordinates. Phrases marked with − do the opposite, allowing inference to subordinates but not superordinates. So, *No car went by* implies *No Honda went by*, but does not imply *No vehicle went by*.

(23)  a. (A (car)$^+$ (went by)$^+$)$^+$
      b. (No (car)$^-$ (went by)$^-$)$^+$
      c. (Every (boy)$^-$ (laughed)$^+$)$^+$

Negation flips $+/-$ annotations, so that the opposite pattern of substitutions yields valid inferences:

(24)  a. (It's not true that (a (car)$^-$ (went by)$^-$ )$^-$ )$^+$
      b. **Good**: Therefore, it's not true that a Honda went by.
      c. **Bad**: Therefore, it's not true that a vehicle went by.

(25)  a. (It's not true that (no (car)$^+$ (went by)$^+$)$^-$ )$^+$
      b. **Good**: Therefore, it's not true that no vehicle went by.
      c. **Bad**: Therefore, it's not true that no Honda went by.

This approach has been very useful in computational tasks involving textual entailment. The necessary annotations are fairly straightforward to generate, and they allow us to construct large numbers of candidate valid inferences—and rule out many invalid inferences—without engaging in the laborious task of constructing a full semantic interpretation [24, 25].

Crucially, the monotonicity profiles of quantifiers are determined by their semantic properties, not their syntax. But importing this semantic information into an enriched level of syntax as in (23)–(25) is practically useful for certain purposes. Similarly, we might construe the choice of whether to include domain variables at logical form as a theoretical or practical convenience. If so, a disquieting conclusion would seem to follow: there is no fact of the matter about whether MP, MT, and CD are "really" valid. Validity *of an argument form* is a concept that only makes sense relative to a method of assigning logical forms to sentences. If our theoretical proclivities make us inclined to want these principles to come out valid—or, if we have some other motivation for annotating logical forms with domain variables—we can adopt a picture of logical form that does the task. On the other hand, if we choose logical forms that do not represent domain variables, then MP, MT, and CD are invalid, in line with the arguments of Yalcin and Kolodny/MacFarlane.

The reader can decide which way to go here: embrace a privileged level of logical form, or relativize the question of whether certain key reasoning principles are valid to an unforced theoretical choice. Either way, the startling conclusion is that the status of certain very basic reasoning principles—even modus ponens!—depends on fairly abstruse theoretical issues in natural language syntax and semantics.

## 4   Generics: Silent but Dangerous

Generic and habitual language provides a second advertisement for the importance of attending to compositional semantics. Certain sentences of English are
ambiguous between a one-off interpretation and an interpretation that expresses
a generalization about how certain kinds of events tend to go. For instance, *Bill
takes a walk after lunch* can be understood in either way, depending on context.

(26)   a.  (Why isn't Bill here now?) Bill takes a walk after lunch.

  b.  (What happens in the play's 4th act?) Bill takes a walk after lunch.

(In most languages, simple present sentences can also be interpreted as describing
an ongoing action. English, unusually, requires the present progressive for this:
*What is Bill doing? —He is taking a walk.*)

In formal semantics, the difference between generic and non-generic readings
is usually traced to a silent adverb of quantification *GEN* which is present at
logical form [18].

(27)   Birds have wings ⤳ *GEN* [$x$ is a bird] [$x$ has wings]
  "Generally, if something is a bird it has wings"

I use the ⤳ symbol to indicate a semantic interpretation (a fairly rough one,
chosen for expository purposes).

  Like overt AQs, the domain argument of *GEN* is often contextually provided.
(Here, $s$ is a variable over situations.)

(28)   I cycle to work ⤳ *GEN* [I go to work in $s$] [I cycle to work in $s$]
  "Generally, when it's time to go to work, I cycle"

In this example, *GEN* doesn't quantify over all situations: if it did, the sentence
would imply that I spend most of my time cycling to work. The intuitive interpretation involves a restriction to the (thankfully rare) moments when I need to
commute and have a choice of methods.

  If *GEN* is a covert AQ, we would expect that it interacts in the compositional
semantics with *if*-clauses in the same way that AQs do—and it does. Notice that
the content of the *if*-clauses in these examples ends up in the restriction of the
*GEN*-quantifier [7].

(29)   a.  Bartenders are happy if they get big tips.
    *GEN* [bartender $x$, $x$ gets a big tip in $s$] [$x$ is happy in $s$]

  b.  If it's raining, I take the bus to work.
    *GEN* [it's time to go to work in $s$, it's raining in $s$] [I take the bus
    in $s$]

There has been a lot of work in recent years about the social and cognitive
dangers of generic language (e.g., [4,9,21]). To these admittedly more pressing
issues, we can add several further dangers specifically for researchers interested
in reasoning. Because of their linguistic properties generics can mislead us in at
least three ways:

1. *GEN* is invisible but has major semantic effects. We have to examine each sentence carefully in order to discern whether it is present.
2. *GEN*'s contextual restriction is not fully predictable from the linguistic form.
3. *GEN* interacts compositionally with other expressions, including *if*-clauses.

Dangers 1 and 3 are particularly relevant to research on conditional reasoning. Any given instance of a conditional might, if it is generic, involve complex semantic interactions that are absent in overly similar sentences that are not generic. To illustrate why this is a problem, consider Douven's [5] arguments against the three-valued semantics for conditionals proposed by de Finetti [8]. In this theory, a conditional is true if it has a true antecedent and true consequent; false if it has a true antecedent and false consequent; and otherwise undefined. Douven points out that this semantics has the consequence that certain right- and left-nested conditionals should have the same interpretation as simple conditionals with a conjunctive antecedent.

(30)    a. **Right-nested**: If $A$, then (if $B$ then $C$)
        b. **Left-nested**: If ($B$ if $A$), then $C$
        c. **Conjunctive**: If ($A$ and $B$), then $C$

The equivalence between (30a) and (30c)—the so-called "Import-Export" property—is widely thought to be an empirically correct prediction. However, Douven points to a number of examples with in which left-nested conditionals are clearly not identical to the apparently matched conjunctive conditionals. For instance,

(31)    a. If this material becomes soft if it gets hot, it is not suited for our purposes.
        b. If this material gets hot and becomes soft, it is not suited for our purposes.

Similarly, the three-valued de Finetti semantics predicts that a left-and-right-nested conditional of the form in (32a) should be equivalent to a three-conjunct simple conditional with the form in (32b).

(32)    a. If ($B$ if $A$), then (if $C$ then $D$)
        b. If ($A$ and $B$ and $C$) then $D$

Douven gives the following counter-example, where the two sentences clearly have different truth-conditions.

(33)    a. If your mother gets angry if you come home with a B, then she'll get furious if you come home with a C.
        b. If you come home with a B and your mother gets angry and you come home with a C, then your mother will get furious.

In light of these and similar examples, Douven concludes that de Finetti's theory is "materially inadequate because it gets the truth conditions and probabilities of nested conditionals badly wrong". However, Lassiter & Baratgin [19] point out that all of Douven's counter-examples involve generic sentences. So, a possible way out for an advocate of the de Finetti theory is to claim that non-generic sentences behave as predicted, but the generic examples differ for principled semantic reasons. Specifically, on their most natural interpretations the sentences in (31) have two key differences: how many instances of *GEN* there are, and where; and how the *if*-clause(s) interact with *GEN*.

(34)     a.  (31a) ⤳ *GEN* (if (*GEN B* if *A*) then *C*)
         b.  (31b) ⤳ *GEN* (if (*A* and *B*) then *C*)

If these are the logical forms that we construct in interpreting these sentences, they are not substitution instances of the templates in (30b) and (30c). As a result, de Finetti's theory does not predict that they should be equivalent. Indeed, when de Finetti's theory is combined with off-the-shelf theories of genericity [18] and domain restriction [11], the predicted truth-conditions are sensible renditions of the intuitive interpretations of these sentences in (32) and (33) (see [19] for details).

     Lassiter & Baratgin note in addition that de Finetti's prediction of an equivalence between (30b) and (30c) is rather more plausible if we use sentences that are incompatible with a generic interpretation.

(35)     a.  If this material became soft at 3:05PM if it got hot at 3:04PM, our workers were not able to use it at 3:10PM.
         b.  If this material got hot at 3:04PM and became soft at 3:05PM, our workers were not able to use it at 3:10PM.

The use of specific time adverbials forces an episodic, non-generic interpretation, and the logical forms of these sentences are genuine instances of the templates in (30). Similarly, de Finetti's prediction that episodic left-and-right-nested conditionals are equivalent to thrice-conjunctive conditionals turns out to be very plausible when we enforce an episodic interpretation, ensuring that the example is a genuine substitution instance of the template in (32).

(36)     a.  If your mother got angry if you came home with a B last Thursday, then she'll get furious if you come home with a C tomorrow.
         b.  If you came home with a B last Thursday and your mother got angry and you come home with a C tomorrow, then your mother will get furious.

Attention to the non-obvious logical form of generic sentences, and to the compositional interaction of generic interpretation and *if*-clauses, is crucial to recognizing the difference between a spurious refutation like (31) and the genuine instance of the pattern like (35). The key remaining question is whether the sentences in (35) and (36) are indeed equivalent. In unpublished work with Maxime Bourlier, Baptiste Jacquet, and Jean Baratgin [20], we address this question in two experiments by asking participants to read pairs of sentences like (31) and (35). In the first experiment, participants made a binary judgment about whether the meanings were the same or different. In the second experiment, a different set of participants rated the same pairs on a continuous slider from "Absolutely the same" meaning to "Absolutely different". Each pair of sentences was either designed to receive a generic interpretation, or decorated with temporal adverbials in order to force a generic interpretation. We also manipulated whether the first sentence was left-nested, right-nested, or left-and-right nested.

Combining de Finetti's semantics with linguistic theories of genericity and domain restriction, these are the key predictions:

1. Left-nested conditionals like (30b), when episodic, should be rated as better paraphrases of their purported conjunctive equivalents like (30c) than the matched generic pairs.
2. Left-and-right-nested conditionals, when episodic, should be rated as better paraphrases of their purported conjunctive equivalents than the matched generic pairs.
3. The generic/episodic manipulation should make no difference when right-nested conditionals are compared to their conjunctive paraphrases.

Both experiments bore out the three predictions. When the first sentence was right-nested, participants largely rated the pairs as having the same meaning regardless of genericity. In contrast, participants gave significantly more "same" judgments for episodic right-nested conditionals than for their generic counterparts in experiment 1, and ratings significantly higher on the sameness scale in experiment 2. The same patterns held for episodic vs. generic left-and-right nested conditionals.

The data are consistent with the predictions described above. Far from refuting the de Finetti theory, the interpretation of nested conditionals seems to be precisely what that theory would predict when combined with insights from modern linguistics. But the road to this conclusion was fraught with danger. Due to *GEN*'s invisibility, and because of its compositional interaction with *if*-clauses, we would have ended up with an erroneous picture of the logic of conditionals if we had not paid close attention to the insights of linguistic research. In the voluminous logical and psychological literature on conditional reasoning, it remains to be seen where else incorrect conclusions may have been reached due to a failure to attend to the complexities of generic interpretation. Psychologists beware!

## 5   Conclusion

Linguistic semantics and psychology of reasoning have a lot to learn from each other. Focusing on conditionals, this paper has discussed two examples where careful attention to the details of compositional semantics is crucial in making sense of confusing patterns, and in separating genuine problems from pseudo-problems. Of course, there are many areas for fruitful interaction beyond the study of conditionals (e.g., [10, 26]). While I have focused on ways that reasoning can benefit from the insights of linguists, the flow of insights in the opposite direction—both methodological and theoretical—is no less crucial.

# References

1. Barwise, J., Cooper, R.: Generalized quantifiers and natural language. Linguist. Philos. **4**(2), 159–219 (1981)
2. Bennett, J.F.: A Philosophical Guide to Conditionals. Oxford University Press, Oxford (2003)
3. van Benthem, J.: Essays in Logical Semantics. Springer, Dordrecht (1986). https://doi.org/10.1007/978-94-009-4540-1
4. Berio, L., Musholt, K.: How language shapes our minds: on the relationship between generics, stereotypes and social norms. Mind Lang. **38**(4), 944–961 (2022)
5. Douven, I.: On de Finetti on iterated conditionals. In: Computational Models of Rationality: Essays Dedicated to Gabriele Kern-Isberner on the Occasion of her 60th Birthday, pp. 265–279. College Publications (2016)
6. Evans, J.S.B.T., Over, D.E.: If. Oxford University Press, Oxford (2004)
7. Farkas, D.F., Sugioka, Y.: Restrictive if/when clauses. Linguist. Philos. 225–258 (1983)
8. de Finetti, B.: La logique de la probabilité. In: Actes du congrès international de philosophie scientifique, vol. 4, pp. 1–9. Hermann Editeurs Paris (1936)
9. Gelman, S.A.: Generics in society. Lang. Soc. **50**(4), 517–532 (2021)
10. Geurts, B.: Reasoning with quantifiers. Cognition **86**(3), 223–251 (2003)
11. Huitink, J.: Modals, conditionals and compositionality. Ph.D. thesis, Radboud University Nijmegen (2008)
12. Icard, T.F., III., Moss, L.S.: Recent progress on monotonicity. Linguist. Issues Lang. Technol. **9**, 167–194 (2014)
13. Johnson-Laird, P.N.: Mental Models. Cambridge University Press, Cambridge (1983)
14. Johnson-Laird, P.N.: Against logical form. Psychologica Belgica **50**(3), 193–221 (2010)
15. Kolodny, N., MacFarlane, J.: Ifs and oughts. J. Philos. **107**(3), 115–143 (2010)
16. Kratzer, A.: Conditionals. In: von Stechow, A., Wunderlich, D. (eds.) Semantik: Ein internationales Handbuch der zeitgenössischen Forschung, pp. 651–656. Walter de Gruyter (1991)
17. Kratzer, A.: Modality. In: von Stechow, A., Wunderlich, D. (eds.) Semantik: Ein internationales Handbuch der zeitgenössischen Forschung, pp. 639–650. Walter de Gruyter (1991)
18. Krifka, M., Pelletier, F., Carlson, G., ter Meulen, A., Chierchia, G., Link, G.: Genericity: an introduction, pp. 1–124 (1995)
19. Lassiter, D., Baratgin, J.: Nested conditionals and genericity in the de Finetti semantics. Thought J. Philos. **10**(1), 42–52 (2021)
20. Lassiter, D., Bourlier, M., Jacquet, B., Baratgin, J.: How generics obscure the logic of conditionals, ms, University of Edinburgh and University of Paris (2023)
21. Leslie, S.J.: The original sin of cognition. J. Philos. **114**(8), 395–421 (2017)
22. Lewis, D.: Index, context and content. In: Philosophical Papers, volume 1: Papers in Philosophical Logic. Cambridge University Press, Cambridge (1997)
23. Lewis, D.: Adverbs of quantification. In: Keenan, E.L. (ed.) Formal Semantics of Natural Language, pp. 178–188. Cambridge University Press, Cambridge (1975)
24. MacCartney, B.: Natural language inference. Ph.D. thesis, Stanford University (2009)
25. MacCartney, B., Manning, C.D.: Natural logic for textual inference. In: Proceedings of the ACL-PASCAL Workshop on Textual Entailment and Paraphrasing, pp. 193–200 (2007)

26. Mascarenhas, S., Koralus, P.: Illusory inferences with quantifiers. Think. Reason. **23**(1), 33–48 (2017)
27. McGee, V.: A counterexample to modus ponens. J. Philos. **82**(9), 462–471 (1985)
28. Sinnott-Armstrong, W., Moor, J., Fogelin, R.: A defense of modus ponens. J. Philos. **83**(5), 296–300 (1986)
29. Stanley, J.: Context and logical form. Linguist. Philos. **23**(4), 391–434 (2000)
30. Stanley, J., Gendler Szabó, Z.: On quantifier domain restriction. Mind Lang. **15**(2–3), 219–261 (2000)
31. Szabolcsi, A.: Quantification. Cambridge University Press, Cambridge (2010)
32. Yalcin, S.: A counterexample to modus tollens. J. Philos. Log. **41**(6), 1001–1024 (2012)

# Technologies to Support Self-determination for People with Intellectual Disability and ASD

Florian Laronze(✉) 🆔, Audrey Landuran, and Bernard N'Kaoua

INSERM, Bordeaux Population Health Research Center, UMR 1219, University of Bordeaux, Bordeaux, France

{florian.laronze,bernard.nkaoua}@u-bordeaux.fr

**Abstract.** This article focuses on the concept of self-determination and the design and validation of digital tools intended to promote the self-determination of vulnerable people. Self-determination is an essential skill for carrying out daily activities. But in certain situations, and for certain populations, self-determination is lacking, which leads to the inability to live an independent life and in favorable conditions of well-being and health. In recent years, self-determination enhancing technologies have been developed and used to promote independent living among people with self-determination disorders. We will illustrate the main digital tools to support self-determination developed for two populations of people suffering from self-determination disorders: people with an intellectual disability and people with an autism spectrum disorder. The ability of these digital assistants to improve the comfort of life of these people will also be presented and discussed.

**Keywords:** Self-determination · digital assistance · intellectual disability · ASD

## 1 Self-determination

Wehmeyer (1992) developed a functional model of self-determination which is currently the most used (Landuran & N'Kaoua, 2018), particularly in the field of disability.

In this model, Wehmeyer (1999) proposes a definition of self-determination (completed a few years later) as "the set of skills and attitudes required in a person, allowing him to act directly on his life by making choices free, not influenced by undue external agents, with a view to maintaining or improving one's quality of life" (Wehmeyer, 2005). The principle of causal agent is a central element in this model. It implies that the person acts intentionally in order to bring about an effect to accomplish a specific objective or to bring about or create a desired change (Wehmeyer et al., 2011a, 2011b). Self-determination does not reflect a total absence of influence and interference, but rather consists of making choices and taking decisions without excessive and undue interference (Lachapelle et al., 2005).

According to this model, self-determined behavior refers to actions that are identified by four essential characteristics: (a) autonomy: the person is able to indicate preferences, make choices and initiate actions without external influence, (b) self-regulation: ability to regulate their behavior according to the characteristics of the environment and its

J. Baratgin et al. (Eds.): HAR 2023, LNCS 14522, pp. 19–35, 2024.
https://doi.org/10.1007/978-3-031-55245-8_2

behavioral repertoire in order to respond to a task, (c) psychological empowerment: the person has a feeling of control over their actions and the consequences of their actions on the environment, (d) self-realization: the person has a knowledge of herself (strengths, weaknesses, etc.) which allows her to adjust her choices and decisions according to her characteristics (Wehmeyer, 1999).

## 2 Intellectual Disability and Self-determination

Intellectual disability (ID) affects about 1% of the population, and involves problems that affect functioning in two areas: intellectual functioning (such as learning, problem solving, judgement) and adaptive functioning (activities of daily life such independent living). Additionally, these difficulties must have appeared early in the developmental period. The ID is extremely heterogeneous on the clinical and etiological levels and is characterized, on the cognitive level, by sensory and motor deficiencies (Bruni, 2006) in short-term, working and episodic memories (Jarrold & Baddeley 2001), language (Rondal, 1995), and executive functioning (Lanfranchi et al., 2010). Related to these cognitive difficulties, many studies have shown that people with ID encounter difficulties in their daily life (Van Gameren-Oosterom et al., 2013), and are less self-determined than their non-disabled peers (Wehmeyer 2013).

However, numerous studies have shown that being self-determined improves quality of life (Lachapelle et al., 2005), independent living or even academic and professional success (Wehmeyer & Schwartz, 1997). Supporting the self-determination of people with ID is therefore a major issue. It is a question of allowing these people to live as much as possible according to their choices, their wishes, their desires, their preferences and their aspirations, without the handicap being a factor of exclusion (Nirje, 1972).

### 2.1 Technologies to Support Self-determination and Intellectual Disability

The rise of digital technology has made it possible to open up new and extremely promising lines of investigation. Indeed, in recent years, digital technologies, and in particular self-determination support technologies, have shown extremely positive results in people with ID, particularly with regard to social inclusion or community participation (Lachapelle et al., 2013).

These technologies now cover many areas of daily life. For example, communication support technologies offer expression aids to translate non-verbal communication behaviors (pressing an image, symbol, etc.) into synthesized or digitized verbal messages (Soto et al., 1993), computer-assisted reading devices (Sorrell et al., 2007) or even Enhanced and Alternative Communication devices that complement or replace the production language, for example, through the use of pictograms (Kagohara et al., 2013). Assistive technologies for social interactions can assist people in recognizing simple emotions (fear, joy, sadness, anger), solving interpersonal problems or even conversation skills (Wert & Neisworth, 2003). Learning support technologies enable the acquisition of skills such as mathematics (Bouck et al., 2009), reading (Haro et al., 2012), cognitive skills (Brandão et al., 2010), motor skills or even the capacity for sensory integration

(Wuang et al., 2011) which corresponds to the ability to interpret and organize effectively the information captured by the senses.

Other technologies aim to support daily activities, such as: managing one's budget (Mechling, 2008), using an automated banking machine (Alberto et al., 2005), paying for purchases (Ayres, et al., 2006), running errands (Bramlett et al., 2011), doing laundry, washing dishes (Cannella-Malone et al., 2011), setting the table (Lancioni et al., 2000; Ayres et al., 2010), cleaning up (Wu et al., 2016), putting away groceries (Cannella-Malone et al., 2006), using the bus (Davies et al., 2010), learning new routes (Brown et al., 2011), making navigation decisions autonomously in order to reach unfamiliar places (McMahon et al., 2015) or even time management (Ruiz et al., 2009).

We should also note the support technologies for professional activities allowing the acquisition of professional skills (Allen et al., 2012), work-related social skills (Gilson, Carter, 2016), or making appropriate decisions when performing a professional task (Davies et al., 2003).

Finally, support technologies for leisure activities allow, for example, the learning of complex game sequences (D'Ateno et al., 2003), the use of and access to entertainment videos (Kagohara, 2011), transferring music to an MP3 (Lachapelle et al., 2013), community inclusion in the library (Taber-Doughty et al., 2008), access to digital documents and the Internet (Stock et al., 2006).

Positive repercussions have been noted in people using these technologies, such as an increase in self-confidence, sense of self-efficacy, motivation, self-esteem, identity development, self-determination or quality of life (Näslund & Gardelli, 2013).

While assistive technologies for self-determination have been proposed to help carry out many activities of daily living, no digital assistance has been proposed to help people with disabilities project themselves into the future, set goals and to develop life projects. It is in this context that we designed the digital assistant "It's my life! I choose it" for help with decision-making and the development of a life plan, for people with disabilities (Landuran & N'Kaoua, 2021).

## 2.2  Design and Validation of a Digital Assistant for Decision-Making and Development of a Life Project

If the notion of self-determination emphasizes the importance of freely consented choices, the notion of life project brings a dimension of projection into the future, of capacities to imagine one's own life according to one's desires and expectations. In the field of disability, the life project occupies a central place in supporting people (Nair, 2003). Many studies have shown that the development of a life plan, life goals or even personal goals has positive consequences, in particular on health, well-being, personal development or even quality of life (Cross & Markus, 2010).

The digital assistant "It's my life! I choose it" has 4 main sections. The first "Choosing, what is it?" consists in defining and proposing exercises on the notion of choice (what is a choice, what can we choose, what can't we choose, etc.). In the second part "My life", the user is invited to answer questions about his past and his present, in order to help him project himself into the future. The third part "What is important to me" makes it possible to reflect on the notion of value (what are the important values, what are the values that we do not share, etc.). Finally, in the fourth part "My project", the participants are encouraged to reflect on what they wish to do later in many areas such as family life, leisure, education, professional life, etc. The person is invited to consider the projects they wish to carry out in different areas (housing, work, affective life and leisure), to classify them according to their interests, and to try to define the needs necessary to achieve them (Fig. 1).

**Fig. 1.** Examples of digital assistant pages. On the top left-hand side: Home screen containing the 4 sections of the digital assistant. On the top right-hand side: An example of a question from the "Choosing, what is it?" section (*Do you choose your clothes: "Yes", "No", "Not always", "I don't know"*). On the bottom left-hand side: An example of a question from the "What is important to me" section *(Choose an image that represents "being with others")*. On the bottom right-hand side: An example of a question from the "My project" section (Choose the project you want: "housing", "hobbies", "work", "emotional life").

The repercussions of the use of the digital assistant on self-esteem, the level of worry, self-determination, psychological well-being and the formulation of the life project were evaluated with adults presenting a Down's syndrome.

The descriptive analysis of the results shows that the assistant makes it possible to: 1) improve the richness of the life project of adults with Down's syndrome in the different areas of the life project (Housing, Work, Emotional Life, Leisure); 2) increase the feeling of well-being in the dimensions of autonomy, meaning in life and self-acceptance; 3) increase self-esteem; 4) reduce feeling of worry (Fig. 2.). A study currently underway should provide additional data to enable statistical analyses to be carried out.

Another disabling activity limitation linked to intellectual disability is the inability, for some young adults, to access residential autonomy, which forces them to live with their parents or in specialized accommodation. In this context, the use of smart homes has been proposed as a means of improving access to home autonomy for people with ID by promoting their autonomy and control over their environment and by adapting to their lifestyle and their abilities (Lussier-Desrochers et al., 2008).

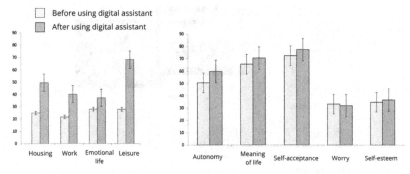

**Fig. 2.** Before-and-after evolution of the different factors assessed with the implementation of the digital assistant. The y-axis represents participants' responses to the different measurement scales used for each item (homogenised out of 100).

### 2.3 Adaptation and Validation of a Digital Home Assistance Platform (Landuran et al., 2023)

Access to autonomous housing for people with disabilities is one of the issues related to inclusion and social participation (French law of 2005; Convention on the rights of people with disabilities, 2006) and constitutes a major step in personal development. Creating a "home" that meets people's needs has important implications for psychological well-being, community participation, daily autonomy and the emergence of self-determination (Fänge & Iwarsson, 2003).

However, studies show that people with ID do not fully participate in their lives and communities (Foley et al., 2013). Most adults live with their parents and not in self-contained accommodation (Foley et al., 2013). The lack of suitable housing (Bigby, 2010) and the cognitive and adaptation difficulties of this population can constitute an obstacle important in access to this residential autonomy.

As part of the work we present here (Landuran et al., 2023) we adapted the assisted living platform called DomAssist to the specificities of people with ID (Dupuy et al., 2016). It consists of sensors installed in the house (for example, motion sensors, contact sensors or electricity consumption sensors) and digital tablets allowing interactions between the user and the platform (Fig. 3.).

The support services offered by this platform cover three areas (Fig. 4.): security and safety (monitoring the front door and electrical appliances); assistance in carrying out daily activities (reminder of activities or appointments); assistance with social participation (simplified email, information on social events, etc.).

We assessed the impact of prolonged (six-month period) use of the platform on home living skills, self-determination, quality of life, self-esteem, level of worry, and well-being. Participants with ID were divided into two groups, the first equipped with the home assistance platform, the second (control group) playing games on a digital tablet. The evaluations (t-test analyses) showed (Fig. 5.) a significative improvement on average for the experimental group compared to the control group for: home skills (domiciliary hability), self-determination, autonomy, competence and meaning of life. A (non-significant) trend towards increased of self-esteem and quality of life as well

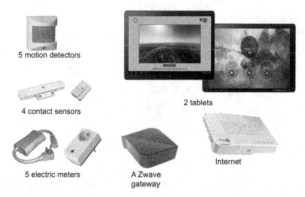

5 motion detectors

4 contact sensors

2 tablets

5 electric meters

A Zwave gateway

Internet

**Fig. 3.** The different elements of the home platform.

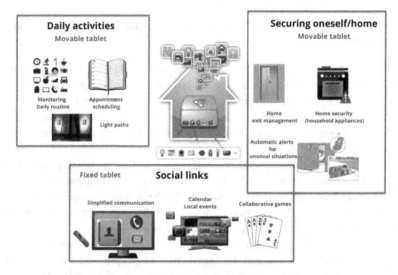

**Daily activities**
Movable tablet

Monitoring Daily routine    Appointment scheduling

Light paths

**Securing oneself/home**
Movable tablet

Home exit management    Home security (household appliances)

Automatic alerts for unusual situations

Fixed tablet    **Social links**

Simplified communication    Calendar Local events    Collaborative games

**Fig. 4.** Services offered by the DomAssist home assistance platform.

as a reduced worry were also observed in the experimental group, and will need to be specified in future investigations.

## 3  Autism Spectrum Disorder and Self-determination

Autism Spectrum Disorder (ASD) is defined as a lifelong neurodevelopmental disorder characterized by two main symptoms: persistent deficits in social communication/interaction and restricted, repetitive behaviors (Bougeard et al., 2021). People with ASD may also have deficits intelligence and motor functions, as well as high levels of anxiety, stress, depression and isolation (Hudson et al., 2019; Salari et al., 2022). In

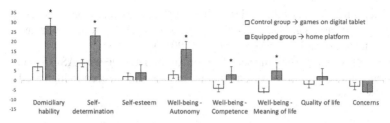

**Fig. 5.** Evolution of perceptions, in the different areas studied, after using the home assistance platform, for the two groups. The y-axis represents participants' responses to the different measurement scales used for each item (homogenised out of 100). * = $p < 0.05$ (t-test analyses).

addition to these various issues, much research has also focused on the level of self-determination and its consequences for people with ASD at different stages of their lives.

Some research has notified than children and youth with ASD have limited opportunities and supports to engage in self-determined actions in their environments (Morán et al., 2021) as well as lower levels of self-determination in comparison with their peers without disabilities (Shogren et al., 2018) and with other disabilities (Chou et al., 2017; Shogren et al., 2018). For example, Chou et al. (2017) showed that high school students with ASD had lower levels of autonomy and psychological empowerment than students with intellectual or learning disabilities. This lower level of self-determination is also linked to a lower level of quality of life and lower satisfaction with life (Shogren et al., 2008; White et al., 2018). Certain factors also appear to further reduce the level of self-determination, such as the severity of the disorder (White et al., 2022), young age (Wehmeyer et al., 2011a, 2011b) and male gender (Carter et al., 2013).

In adulthood, self-determination is also essential for people with ASD. Kim (2019) identified in a qualitative review that self-determined behaviors of adults with autism as goal-setting, decision making, problem solving, and self-management had positive influences on their employment status, social participation, positive identity, and stress management. Young adults with Autism Spectrum Disorder (ASD) also frequently encounter a reduced quality of life compared to their peers of similar age and abilities (Bishop-Fitzpatrick et al., 2018). Additionally, they often face challenges in attaining independence in their living arrangements (Steinhausen et al., 2016), and struggle to reach the typical developmental milestones associated with adulthood (Picci & Scherf, 2015). A large number of studies carried out with people with disabilities, including ASD, have also shown that the consequences of a low level of self-determination and independence can lead to poorer in-school and post-school outcomes such as advancing to higher education and gaining employment (Morán et al., 2021).

At university, students with ASD also face many challenges including difficulties with group work, peer relationships, ambiguous instructions, time management, navigating a lack of structure at university, etc. (McPeake et al., 2023). These various academic difficulties can lead ASD students to experience fatigue, anxiety, depression, burn-out, dropping out of classes, suicidal ideation and attempting suicide, at higher levels than neurotypical students (Jackson et al., 2018; McPeake et al., 2023). They are also led to graduate at significantly lower rates when compared to their typically developing peers and students with other disabilities (Davis et al., 2021) And even if they manage to graduate, among all disabled graduates (at all qualification levels), graduates with autism are most likely to be unemployed (Allen & Coney, 2019).

In this context, supporting the self-determination of people with ASD is therefore a central issue, in particular by enabling them to access and succeed in higher education and the jobs of their choice.

### 3.1  Assistive Technologies and Self-Determination in College Students with ASD

Given the difficulties encountered by ASD students at university, various technologies are being developed to assist and help them in different areas of their life as university students.

Some technologies (e.g.: mobile app, software, learning management systems, etc.) are developed to assist learning for students with ASD: 1) to reinforce the acquisition of certain concepts (McMahon et al., 2016; Begel et al., 2021; Dahlstrom-Hakki & Wallace, 2022); 2) to facilitate student engagement during the course session (Francis et al., 2018; Huffman et al., 2019) as well as outside by improving their revisions (Francis et al., 2018; O'Neill & Smyth, 2023); 3) to follow distance education (Satterfield et al., 2015; Richardson, 2017; Adams et al., 2019; Madaus et al., 2022). For example, Huffman et al, (2019) showed the efficacy of a self-monitoring application in university lectures to assist a student with ASD by allowing him to be more engaged and concentrated during the lesson.

Over technologies and digital tools are developed to assist or improve communication and interaction skills of students with ASD at university, mainly involving training in social interaction and communication via online course modules (PPT, videos, etc.), coaching (telecoaching, audioching) and video-modeling (Mason et al., 2012; Mason et al., 2020; Gregori et al., 2022), often combined together.

Other technologies are designed to enhance different cognitive and executive abilities, mainly on spatial navigation through mobile applications that can help students with ASD to get around more easily and to carry out activities, particularly on the university campus (McMahon et al., 2015; Kearney et al., 2021; Wright et al., 2022). For example, McMahon et al. (2015) showed the effectiveness of a location-based augmented reality application to teach one college student with ASD to navigate a city independently to find local employment opportunities.

Finally, Hrabal et al. (2023) showed that certain technologies (e.g.: video modeling), which are not specifically intended for college students, can help them improve daily living skills such as meal preparation and housekeeping tasks.

In general, these technology and digital tools appear to benefit university students with ASD, both in terms of effectiveness and satisfaction with their use, particularly for abilities essential to self-determination (learning, self-management, communication, spatial navigation, etc.). However, it is important to exercise caution when interpreting these findings for two principal reasons: 1) most of the studies involve very few participants (1 to 4) and are case studies; 2) there is considerable variability in the technologies and abilities targeted.

### 3.2 Atypie-Friendly and Development of a Virtual Visit System for the University of Bordeaux

In France, Atypie-Friendly is a national program involving 25 universities, which aims to promote the inclusion and support of people with neurodevelopmental disorders, particularly ASD, at university. The Atypie-Friendly program addresses many aspects of students' lives, including the development of digital tools and projects to support the transition from high school to university.

Indeed, arriving at university is a crucial time for students. It involves radical changes compared with life at high school, with new teaching styles, new people, new facilities and so on. This transition can lead to difficulties in establishing new routines and can cause anxiety, particularly for students with ASD (Jill Boucher, 1977; van Steensel et al., 2011). These students may also experience difficulties with spatial navigation (Ring et al., 2018). A number of schemes have been put in place to overcome these difficulties. Universities generally organize open days to allow college students to discover the main services in their future environment. Individual detailed visits to their study site, to familiarize them with their future environment and their main contacts, can also be offered if required. However, it is difficult to envisage the possibility of carrying out visits for all new students. In addition, a number of factors (availability, geographical distance, etc.) mean that not all students who so wish can take part in these open days or individual visits.

Based on this premise, we are currently developing a virtual visit of key services at the University of Bordeaux (Health Service, Disability Service, Housing Service, University Restaurant). These visits will focus on 4 key themes necessary for the development of ASD student's self-determination (health and well-being, studies and disability, accommodation and catering).

Each visit will comprise (Fig. 6.): a "passive" guided digital visit in first person view of the key service (the person will be able to watch the journey scroll from a given point to the destination); an active visit (the person will be able to complete the journey themselves using mouse and keyboard); once at the destination, a presentation of the main tasks of the service by a resource person (passive mode: video; active mode: interactive video with clickable questions). The aim of setting up an active visit is to improve information retention, particularly spatial navigation (Cogné et al., 2017).

In order to test the effectiveness of these virtual visits, a protocol will be put in place to compare these three systems: 1) the current University system (information on the University website and use of Google Maps); 2) the passive virtual visit system; 3) the active virtual visit system. The effectiveness of these systems will be compared in terms of information retention (remembering the route, information about the service) and usability.

One of the project's future enhancements will be to integrate a conversational user interface including a conversational agent based on artificial intelligence into the virtual visit system. Conversational user interfaces, whether through text-based chat or voice recognition with synthesized responses, offer flexibility, personalization, and alternative communication methods to perform tasks for users (Iniesto et al., 2023). This conversational agent could complement the virtual visit system by accompanying students on their administrative processes (e.g.: assistance in disability disclosure and support) and in presenting in more details university services, resources and the administrative procedures necessary to access it. This point is extremely important to the extent that the complexity of administrative procedures constitutes a significant obstacle to the entry into university of students with ASD, and more generally of students with disabilities (Iniesto et al., 2023).

**Fig. 6.** Virtual visit to the Student Health Centre at the University of Bordeaux. Left: illustration of the passive virtual visit with 3 intersections. Right: illustration of the active virtual visit.

## 4   Conclusion

Autonomy and support for the self-determination of vulnerable people is one of the major societal challenges of the years to come.

Nirje (1972) was one of the first to use the term self-determination for people with disabilities in his chapter "The right to self-determination" which deals with the principle of normalization. This author clearly expressed the importance of personal self-determination for all, without excluding people with mental retardation or other disabilities. For him, a major challenge of the principle of normalization consists of creating conditions by which a person with disabilities experiences the respect and dignity to which every human being has the right. Thus, any action that concerns a person must take into account their choices, wishes, desires, preferences and aspirations.

In France, this idea is expressed through the law on "equal rights and opportunities, participation and citizenship of disabled people", adopted in 2005. In this law, the ability to express a life project is the starting point for procedures allowing disabled people to obtain compensation (financial, human, etc.), in order to help them realize their projects. But despite this impulse, too few people with disabilities live independently, are employed full time, hold paid employment or pursue postsecondary education (Steinhausen et al., 2016; Morán et al., 2021). Furthermore, large numbers of people with disabilities remain dependent on caregivers, service providers, and overburdened social systems (Wehmeyer, 1992), and have few opportunities to make choices based on their interests and abilities and find their lives controlled by other people who too often make decisions for them (Kozleski and Sands, 1992; Stancliffe, 1995).

In recent years, technologies supporting self-determination have proven to be extremely promising solutions. In this article we have seen that many assistive technologies are being tested in the scientific literature to assist people with ID and ASD in many areas of life (Näslund and Gardelli, 2013; McMahon et al., 2016). However, several limitations in access and use of digital technologies by these populations have been identified in the literature. They concern, in part, the lack of accessibility and the lack of technologies adapted to people's needs (Caton and Chapman, 2016). The authors also underline the importance of taking precautions given the great variability of the technologies tested and the small number of participants often included in the protocols (McMahon et al., 2015). To address these limitations, authors have proposed the use of universal design methodology (Wehmeyer et al., 2012), as well as different types of methodologies that largely include the user in the design cycle (Greenbaum and Kyng, 2020).

In our digital tool design activities, we combine different methods such as a user-centered design approach including participatory design, principles of universal design, recommendations from the literature and results from our empirical studies. Future users, families and professionals are placed at the center of the design process, through an iterative evaluation of the proposed solutions, at each stage of the process. This approach ensures the best possible usability of the tools developed.

But in general, these digital tools are far from being adapted to all possible forms of user disability. In this area, technological progress and prospects for development are considerable in all areas of digital personal assistance (home automation, robotics, social communication, digital tablets, etc.). The need to adapt human-machine interfaces in order to promote the autonomy of different populations of people is an economic, societal and public health issue. Recent progress in Artificial Intelligence and the perspectives offered by intelligent systems or conversational agents, or even robotics in cooperation with humans, open extremely promising issues in helping autonomy and self-determination for vulnerable people but also for everyone.

# References

Adams, D., Simpson, K., Davies, L., Campbell, C., Macdonald, L.: Online learning for university students on the autism spectrum: a systematic review and questionnaire study. Australas. J. Educ. Technol. 35(6), 111–131 (2019). https://doi.org/10.14742/ajet.5483

Alberto, P.A., Cihak, D.F., Gama, R.I.: Use of static picture prompts versus video modeling during simulation instruction. Res. Dev. Disabil. 26(4), 327–339 (2005). https://doi.org/10.1016/j.ridd.2004.11.002

Allen, K.D., Burke, R.V., Howard, M.R., Wallace, D.P., Bowen, S.L.: Use of audio cuing to expand employment opportunities for adolescents with autism spectrum disorders and intellectual disabilities. J. Autism Dev. Disord. 42(11), 2410–2419 (2012). https://doi.org/10.1007/s10803-012-1519-7

Allen, M., Coney, K.: What Happens Next? A Report on the First Destinations of 2017 Disabled Graduates (2019). https://www.agcas.org.uk/write/MediaUploads/Resources/Disability%20TG/What_Happens_Next_report_2019.pdf

Ayres, K., Cihak, D.: Computer- and video-based instruction of food-preparation skills: acquisition, generalization, and maintenance. Intellect. Dev. Disabil. 48(3), 195–208 (2010). https://doi.org/10.1352/1944-7558-48.3.195

Ayres, K.M., Langone, J., Boon, R.T., Norman, A.: Computer-based instruction for purchasing skills. Educ. Train. Dev. Disabil. 41(3), 253–263 (2006)

Begel, A., Dominic, J., Phillis, C., Beeson, T., Rodeghero, P.: How a remote video game coding camp improved autistic college students' self-efficacy in communication. In: Proceedings of the 52nd ACM Technical Symposium on Computer Science Education, pp. 142–148 (2021). https://doi.org/10.1145/3408877.3432516

Bigby, C.: A five-country comparative review of accommodation support policies for older people with intellectual disability. J. Policy Pract. Intellect. Disabil. **7**(1), 3–15 (2010). https://doi.org/10.1111/j.1741-1130.2010.00242.x

Bishop-Fitzpatrick, L., Mazefsky, C.A., Eack, S.M.: The combined impact of social support and perceived stress on quality of life in adults with autism spectrum disorder and without intellectual disability. Autism **22**(6), 703–711 (2018). https://doi.org/10.1177/1362361317703090

Boucher, J.: Alternation and sequencing behaviour, and response to novelty in autistic children. J. Child Psychol. Psychiatry **18**(1), 67–72 (1977). https://doi.org/10.1111/j.1469-7610.1977.tb00417.x

Bouck, E.C., Bassette, L., Taber-Doughty, T., Flanagan, S.M., Szwed, K.: Pentop computers as tools for teaching multiplication to students with mild intellectual disabilities. Educ. Train. Dev. Disabil. **44**(3), 367–380 (2009)

Bougeard, C., Picarel-Blanchot, F., Schmid, R., Campbell, R., Buitelaar, J.: Prevalence of autism spectrum disorder and co-morbidities in children and adolescents: a systematic literature review. Front. Psychiatry **12**, 744709 (2021). https://doi.org/10.3389/fpsyt.2021.744709

Bramlett, V., Ayres, K.M., Cihak, D.F., Douglas, K.H.: Effects of computer and classroom simulations to teach students with various exceptionalities to locate apparel sizes. Educ. Training Autism Dev. Disabil. **46**(3), 454–469 (2011). http://www.jstor.org/stable/23880598

Brandão, A., et al.: Semiotic inspection of a game for children with down syndrome. In: 2010 Brazilian Symposium on Games and Digital Entertainment, pp. 199–210 (2010). https://doi.org/10.1109/SBGAMES.2010.24

Brown, D.J., McHugh, D., Standen, P., Evett, L., Shopland, N., Battersby, S.: Designing location-based learning experiences for people with intellectual disabilities and additional sensory impairments. Comput. Educ. **56**(1), 11–20 (2011). https://doi.org/10.1016/j.compedu.2010.04.014

Bruni, M.: Fine Motor Skills for Children with Down Syndrome, 2nd edn. Woodbine House, Bethesda, MD (2006)

Cannella-Malone, H.I., Fleming, C., Chung, Y.-C., Wheeler, G.M., Basbagill, A.R., Singh, A.H.: Teaching daily living skills to seven individuals with severe intellectual disabilities: a comparison of video prompting to video modeling. J. Posit. Behav. Interv. **13**(3), 144–153 (2011). https://doi.org/10.1177/1098300710366593

Cannella-Malone, H., Sigafoos, J., O'Reilly, M., de la Cruz, B., Edrisinha, C., Lancioni, G.E.: Comparing video prompting to video modeling for teaching daily living skills to six adults with developmental disabilities. Educ. Train. Dev. Disabil. **41**(4), 344–356 (2006)

Carter, E.W., Lane, K.L., Cooney, M., Weir, K., Moss, C.K., Machalicek, W.: Parent assessments of self-determination importance and performance for students with autism or intellectual disability. Am. J. Intellect. Dev. Disabil. **118**(1), 16–31 (2013). https://doi.org/10.1352/1944-7558-118.1.16

Caton, S., Chapman, M.: The use of social media and people with intellectual disability: a systematic review and thematic analysis. J. Intellect. Dev. Disabil. **41**(2), 125–139 (2016). https://doi.org/10.3109/13668250.2016.1153052

Chou, Y.-C., Wehmeyer, M.L., Palmer, S.B., Lee, J.: Comparisons of self-determination among students with autism, intellectual disability, and learning disabilities: a multivariate analysis. Focus Autism Other Dev. Disabil. **32**(2), 124–132 (2017). https://doi.org/10.1177/1088357615625059

Cogné, M., et al.: The contribution of virtual reality to the diagnosis of spatial navigation disorders and to the study of the role of navigational aids: a systematic literature review. Ann. Phys. Rehabil. Med. **60**(3), 164–176 (2017). https://doi.org/10.1016/j.rehab.2015.12.004

Convention on the Rights of Persons with Disabilities (CRPD) | Division for Inclusive Social Development (DISD), https://social.desa.un.org/issues/disability/crpd/convention-on-the-rights-of-persons-with-disabilities-crpd (2006). Retrieved 25 Sep 2023

Cross, S., Markus, H.: Possible selves across the life span. Hum. Dev. **34**(4), 230–255 (2010). https://doi.org/10.1159/000277058

Dahlstrom-Hakki, I., Wallace, M.L.: Teaching statistics to struggling students: lessons learned from students with LD, ADHD, and autism. J. Stat. Data Sci. Educ. **30**(2), 127–137 (2022). https://doi.org/10.1080/26939169.2022.2082601

D'Ateno, P., Mangiapanello, K., Taylor, B.A.: Using video modeling to teach complex play sequences to a preschooler with Autism. J. Posit. Behav. Interv. **5**(1), 5–11 (2003). https://doi.org/10.1177/10983007030050010801

Davies, D.K., Stock, S.E., Holloway, S., Wehmeyer, M.L.: Evaluating a GPS-based transportation device to support independent bus travel by people with intellectual disability. Intellect. Dev. Disabil. **48**(6), 454–463 (2010). https://doi.org/10.1352/1934-9556-48.6.454

Davies, D.K., Stock, S.E., Wehmeyer, M.L.: A palmtop computer-based intelligent aid for individuals with intellectual disabilities to increase independent decision making. Res. Pract. Persons Severe Disabil. **28**(4), 182–193 (2003). https://doi.org/10.2511/rpsd.28.4.182

Davis, M.T., Watts, G.W., López, E.J.: A systematic review of firsthand experiences and supports for students with autism spectrum disorder in higher education. Res. Autism Spectr. Disord. **84**, 101769 (2021). https://doi.org/10.1016/j.rasd.2021.101769

Dupuy, L., Consel, C., Sauzéon, H.: Une assistance numérique pour les personnes âgées: Le projet DomAssist (2016). https://inria.hal.science/hal-01278203

Fänge, A., Iwarsson, S.: Accessibility and usability in housing: construct validity and implications for research and practice. Disabil. Rehabil. **25**(23), 1316–1325 (2003). https://doi.org/10.1080/09638280310001616286

Foley, K.-R., et al.: Functioning and post-school transition outcomes for young people with Down syndrome. Child: Care, Health Dev. **39**(6), 789–800 (2013). https://doi.org/10.1111/cch.12019

Francis, G.L., Duke, J.M., Kliethermes, A., Demetro, K., Graff, H.: Apps to support a successful transition to college for students with ASD. Teach. Except. Child. **51**(2), 111–124 (2018). https://doi.org/10.1177/0040059918802768

Gilson, C.B., Carter, E.W.: Promoting social interactions and job independence for college students with autism or intellectual disability: a pilot study. J. Autism Dev. Disord. **46**(11), 3583–3596 (2016). https://doi.org/10.1007/s10803-016-2894-2

Greenbaum, J., Kyng, M. (eds.): Design at Work: Cooperative Design of Computer Systems. CRC Press (2020)

Gregori, E., Mason, R., Wang, D., Griffin, Z., Iriarte, A.: Effects of telecoaching on conversation skills for high school and college students with autism spectrum disorder. J. Spec. Educ. Technol. **37**(2), 241–252 (2022). https://doi.org/10.1177/01626434211002151

Haro, B.P.M., Santana, P.C., Magaña, M.A.: Developing reading skills in children with Down syndrome through tangible interfaces. In: Proceedings of the 4th Mexican Conference on Human-Computer Interaction, pp. 28–34 (2012). https://doi.org/10.1145/2382176.2382183

Hrabal, J.M., Davis, T.N., Wicker, M.R.: The use of technology to teach daily living skills for adults with autism: a systematic review. Adv. Neurodev. Disord. **7**(3), 443–458 (2023). https://doi.org/10.1007/s41252-022-00255-9

Hudson, C.C., Hall, L., Harkness, K.L.: Prevalence of depressive disorders in individuals with autism spectrum disorder: a meta-analysis. J. Abnorm. Child Psychol. **47**(1), 165–175 (2019). https://doi.org/10.1007/s10802-018-0402-1

Huffman, J.M., Bross, L.A., Watson, E.K., Wills, H.P., Mason, R.A.: Preliminary investigation of a self-monitoring application for a postsecondary student with autism. Adv. Neurodev. Disord. **3**(4), 423–433 (2019). https://doi.org/10.1007/s41252-019-00124-y

Iniesto, F., et al.: Creating 'a simple conversation': designing a conversational user interface to improve the experience of accessing support for study. ACM Trans. Access. Comput. **16**(1), 1–29 (2023). https://doi.org/10.1145/3568166

Jackson, S.L.J., Hart, L., Brown, J.T., Volkmar, F.R.: Brief report: self-reported academic, social, and mental health experiences of post-secondary students with autism spectrum disorder. J. Autism Dev. Disord. **48**(3), 643–650 (2018). https://doi.org/10.1007/s10803-017-3315-x

Jarrold, C., Baddeley, A.: Short-term memory in Down syndrome: applying the working memory model. Down Syndr. Res. Pract. **7**(1), 17–23 (2001). https://doi.org/10.3104/reviews.110

Kagohara, D.M.: Three students with developmental disabilities learn to operate an ipod to access age-appropriate entertainment videos. J. Behav. Educ. **20**(1), 33–43 (2011). https://doi.org/10.1007/s10864-010-9115-4

Kagohara, D.M., et al.: Using iPods® and iPads® in teaching programs for individuals with developmental disabilities: a systematic review. Res. Dev. Disabil. **34**(1), 147–156 (2013). https://doi.org/10.1016/j.ridd.2012.07.027

Kearney, K.B., Joseph, B., Finnegan, L., Wood, J.: Using a peer-mediated instructional package to teach college students with intellectual and developmental disabilities to navigate an inclusive university campus. The J. Spec. Educ. **55**(1), 45–54 (2021). https://doi.org/10.1177/0022466920937469

Kim, S.Y.: The experiences of adults with autism spectrum disorder: Self-determination and quality of life. Res. Autism Spectr. Disord. **60**, 1–15 (2019). https://doi.org/10.1016/j.rasd.2018.12.002

Kozleski, E.B., Sands, D.J.: The yardstick of social validity: evaluating quality of life as perceived by adults without disabilities. Educ. Train. Ment. Retard. **27**(2), 119–131 (1992)

Lachapelle, Y., Lussier-Desrochers, D., Caouette, M., Therrien-Bélec, M.: Expérimentation d'une technologie mobile d'assistance à la réalisation de tâches pour soutenir l'autodétermination de personnes présentant une déficience intellectuelle. Revue francophone de la déficience intellectuelle **24**, 96–107 (2013). https://doi.org/10.7202/1021267ar

Lachapelle, Y., et al.: The relationship between quality of life and self-determination: an international study. J. Intellect. Disabil. Res. **49**(10), 740–744 (2005). https://doi.org/10.1111/j.1365-2788.2005.00743.x

Lancioni, G.E., O'Reilly, M.F., Dijkstra, A.W., Groeneweg, J., Van den Hof, E.: Frequent versus nonfrequent verbal prompts delivered unobtrusively: their impact on the task performance of adults with intellectual disability. Educ. Train. Ment. Retard. Dev. Disabil. **35**(4), 428–433 (2000)

Landuran, A., N'Kaoua, B.: Projet de vie: Regard des adultes avec une trisomie 21 et des aidants (familles et professionnels). Revue francophone de la déficience intellectuelle **28**, 61–69 (2018). https://doi.org/10.7202/1051099ar

Landuran, A., N'Kaoua, B.: Designing a digital assistant for developing a life plan. Int. J. Human-Comput. Interact. **37**(18), 1749–1759 (2021). https://doi.org/10.1080/10447318.2021.1908669

Landuran, A., Sauzéon, H., Consel, C., N'Kaoua, B.: Evaluation of a smart home platform for adults with Down syndrome. Assist. Technol. **35**(4), 347–357 (2023). https://doi.org/10.1080/10400435.2022.2075487

Lanfranchi, S., Jerman, O., Dal Pont, E., Alberti, A., Vianello, R.: Executive function in adolescents with Down Syndrome. J. Intellect. Disabil. Res. **54**(4), 308–319 (2010). https://doi.org/10.1111/j.1365-2788.2010.01262.x

Lussier-Desrochers, D., Lachapelle, Y., Pigot, H., Beauchet, J.: Des habitats intelligents pour promouvoir l'autodétermination et l'inclusion sociale. Des Habitats Intelligents Pour Promouvoir l'autodétermination et l'inclusion Sociale **18**, 53–64 (2008)

LOI n° 2005–102 du 11 février 2005 pour l'égalité des droits et des chances, la participation et la citoyenneté des personnes handicapées et liens vers les décrets d'application—Dossiers législatifs—Légifrance, https://www.legifrance.gouv.fr/dossierlegislatif/JORFDOLE0000177 59074/ (2005). Retrieved 25 September 2023

Madaus, J., Cascio, A., Gelbar, N.: Perceptions of college students with autism spectrum disorder on the transition to remote learning during the COVID-19 pandemic. Dev. Disabil. Netw. J. **2**(2), 5 (2022). https://digitalcommons.usu.edu/ddnj/vol2/iss2/5

Mason, R.A., Gregori, E., Wills, H.P., Kamps, D., Huffman, J.: Covert audio coaching to increase question asking by female college students with autism: proof of concept. J. Dev. Phys. Disabil. **32**(1), 75–91 (2020). https://doi.org/10.1007/s10882-019-09684-2

Mason, R.A., Rispoli, M., Ganz, J.B., Boles, M.B., Orr, K.: Effects of video modeling on communicative social skills of college students with asperger syndrome. Dev. Neurorehabil. **15**(6), 425–434 (2012). https://doi.org/10.3109/17518423.2012.704530

McMahon, D., Cihak, D.F., Wright, R.: Augmented reality as a navigation tool to employment opportunities for postsecondary education students with intellectual disabilities and autism. J. Res. Technol. Educ. **47**(3), 157–172 (2015). https://doi.org/10.1080/15391523.2015.1047698

McMahon, D.D., Cihak, D.F., Wright, R.E., Bell, S.M.: Augmented reality for teaching science vocabulary to postsecondary education students with intellectual disabilities and autism. J. Res. Technol. Educ. **48**(1), 38–56 (2016). https://doi.org/10.1080/15391523.2015.1103149

McPeake, E., et al.: "I just need a little more support": a thematic analysis of autistic students' experience of university in France. Res. Autism Spectr. Disord. **105**, 102172 (2023). https://doi.org/10.1016/j.rasd.2023.102172

Mechling, L.C.: Thirty year review of safety skill instruction for persons with intellectual disabilities. Educ. Train. Dev. Disabil. **43**(3), 311–323 (2008)

Morán, M.L., et al.: Self-determination of students with autism spectrum disorder: a systematic review. J. Dev. Phys. Disabil. **33**(6), 887–908 (2021). https://doi.org/10.1007/s10882-020-097 79-1

Nair, K.P.S.: Life goals: the concept and its relevance to rehabilitation. Clin. Rehabil. **17**(2), 192–202 (2003). https://doi.org/10.1191/0269215503cr599oa

Näslund, R., Gardelli, Å.: 'I know, I can, I will try': youths and adults with intellectual disabilities in Sweden using information and communication technology in their everyday life. Disabil. Soc. **28**(1), 28–40 (2013). https://doi.org/10.1080/09687599.2012.695528

Nirje, B.: The right to serf-determination. In: Wolfensberger, W. (ed.) Normalization: The principle of normalization, pp. 176–200. National Institute on Mental Retardation, Toronto (1972)

O'Neill, S.J., Smyth, S.: Using off-the-shelf solutions as assistive technology to support the self-management of academic tasks for autistic university students. Assist. Technol. (2023). https://doi.org/10.1080/10400435.2023.2230480

Picci, G., Scherf, K.S.: A Two-hit model of autism: adolescence as the second hit. Clin. Psychol. Sci. **3**(3), 349–371 (2015). https://doi.org/10.1177/2167702614540646

Richardson, J.T.E.: Academic attainment in students with autism spectrum disorders in distance education. Open Learn.: The J. Open, Distance e-Learn. **32**(1), 81–91 (2017). https://doi.org/10.1080/02680513.2016.1272446

Ring, M., Gaigg, S.B., de Condappa, O., Wiener, J.M., Bowler, D.M.: Spatial navigation from same and different directions: the role of executive functions, memory and attention in adults with autism spectrum disorder. Autism Res. **11**(5), 798–810 (2018). https://doi.org/10.1002/aur.1924

Rondal, J.A.: Exceptional Language Development in Down Syndrome: Implications for the Cognition-Language Relationship. Cambridge University Press (1995). https://doi.org/10.1017/CBO9780511582189

Ruiz, I., García, B., Méndez, A.: Technological solution for independent living of intellectual disabled people. In: Omatu, S. et al. (eds.) IWANN 2009. LNCS, vol. 5518, pp. 859–862. Springer, Heidelberg (1999). https://doi.org/10.1007/978-3-642-02481-8_130

Salari, N., et al.: The global prevalence of autism spectrum disorder: a comprehensive systematic review and meta-analysis. Ital. J. Pediatr. **48**(1), 112 (2022). https://doi.org/10.1186/s13052-022-01310-w

Satterfield, D., Lepage, C., Ladjahasan, N.: Preferences for online course delivery methods in higher education for students with autism spectrum disorders. Procedia Manufact. **3**, 3651–3656 (2015). https://doi.org/10.1016/j.promfg.2015.07.758

Shogren, K.A., Shaw, L.A., Raley, S.K., Wehmeyer, M.L.: Exploring the effect of disability, race-ethnicity, and socioeconomic status on scores on the self-determination inventory: student report. Except. Child. **85**(1), 10–27 (2018). https://doi.org/10.1177/0014402918782150

Shogren, K.A., et al.: Understanding the construct of self-determination: examining the relationship between the arc's self-determination scale and the American institutes for research self-determination scale. Assess. Eff. Interv. **33**(2), 94–107 (2008). https://doi.org/10.1177/1534508407311395

Sorrell, C.A., Bell, S.M., McCallum, R.S.: Reading rate and comprehension as a function of computerized versus traditional presentation mode: a preliminary study. J. Spec. Educ. Technol. **22**(1), 1–12 (2007). https://doi.org/10.1177/016264340702200101

Soto, G., Belfiore, P.J., Schlosser, R.W., Haynes, C.: Teaching specific requests: a comparative analysis on skill acquisition and preference using two augmentative and alternative communication aids. Educ. Train. Ment. Retard. **28**(2), 169–178 (1993)

Stancliffe, R.J.: Assessing opportunities for choice-making: a comparison of self- and staff reports. Am. J. Ment. Retard. **99**(4), 418–429 (1995)

Steinhausen, H.-C., Mohr Jensen, C., Lauritsen, M.B.: A systematic review and meta-analysis of the long-term overall outcome of autism spectrum disorders in adolescence and adulthood. Acta Psychiatr. Scand. **133**(6), 445–452 (2016). https://doi.org/10.1111/acps.12559

Stock, S.E., Davies, D.K., Davies, K.R., Wehmeyer, M.L.: Evaluation of an application for making palmtop computers accessible to individuals with intellectual disabilities. J. Intellect. Dev. Disabil. **31**(1), 39–46 (2006). https://doi.org/10.1080/13668250500488645

Taber-Doughty, T., Patton, S.E., Brennan, S.: Simultaneous and delayed video modeling: an examination of system effectiveness and student preferences. J. Spec. Educ. Technol. **23**(1), 1–18 (2008). https://doi.org/10.1177/016264340802300101

Van Gameren-Oosterom, H.B.M., et al.: Practical and social skills of 16–19-year-olds with Down syndrome: independence still far away. Res. Dev. Disabil. **34**(12), 4599–4607 (2013). https://doi.org/10.1016/j.ridd.2013.09.041

van Steensel, F.J.A., Bögels, S.M., Perrin, S.: Anxiety disorders in children and adolescents with autistic spectrum disorders: a meta-analysis. Clin. Child. Fam. Psychol. Rev. **14**(3), 302–317 (2011). https://doi.org/10.1007/s10567-011-0097-0

Wehmeyer, M.L.: Self-determination and the education of students with mental retardation. Educ. Train. Ment. Retard. **27**(4), 302–314 (1992)

Wehmeyer, M.L.: A functional model of self-determination: describing development and implementing instruction. Focus Autism Other Dev. Disabil. **14**(1), 53–61 (1999). https://doi.org/10.1177/108835769901400107

Wehmeyer, M.L.: Self-determination and individuals with severe disabilities: re-examining meanings and misinterpretations. Res. Pract. Persons Severe Disabil. **30**(3), 113–120 (2005). https://doi.org/10.2511/rpsd.30.3.113

Wehmeyer, M.L. (ed.): The Oxford Handbook of Positive Psychology and Disability. Oxford University Press, USA (2013). https://doi.org/10.1093/oxfordhb/9780195398786.001.0001

Wehmeyer, M.L., et al.: Personal self-determination and moderating variables that impact efforts to promote self-determination. Exceptionality **19**(1), 19–30 (2011). https://doi.org/10.1080/09362835.2011.537225

Wehmeyer, M.L., Tassé, M.J., Davies, D.K., Stock, S.: Support needs of adults with intellectual disability across domains: the role of technology. J. Spec. Educ. Technol. **27**(2), 11–21 (2012)

Wehmeyer, M.L., Palmer, S.B., Lee, Y., Williams-Diehm, K., Shogren, K.: A randomized-trial evaluation of the effect of whose future is it anyway? On self-determination. Career Dev. Except. Individuals **34**(1), 45–56 (2011). https://doi.org/10.1177/0885728810383559

Wehmeyer, M., Schwartz, M.: Self-determination and positive adult outcomes: a follow-up study of youth with mental retardation or learning disabilities. Except. Child. **63**(2), 245–255 (1997). https://doi.org/10.1177/001440299706300207

Wert, B.Y., Neisworth, J.T.: Effects of video self-modeling on spontaneous requesting in children with autism. J. Posit. Behav. Interv. **5**(1), 30–34 (2003). https://doi.org/10.1177/10983007030050010501

White, K., Flanagan, T.D., Nadig, A.: Examining the relationship between self-determination and quality of life in young adults with autism spectrum disorder. J. Dev. Phys. Disabil. **30**(6), 735–754 (2018). https://doi.org/10.1007/s10882-018-9616-y

White, S.W., Smith, I., Brewe, A.M.: Brief report: the influence of autism severity and depression on self-determination among young adults with autism spectrum disorder. J. Autism Dev. Disord. **52**(6), 2825–2830 (2022). https://doi.org/10.1007/s10803-021-05145-y

Wright, R.E., McMahon, D.D., Cihak, D.F., Hirschfelder, K.: Smartwatch executive function supports for students with ID and ASD. J. Spec. Educ. Technol. **37**(1), 63–73 (2022). https://doi.org/10.1177/0162643420950027

Wu, P.-F., Cannella-Malone, H.I., Wheaton, J.E., Tullis, C.A.: Using video prompting with different fading procedures to teach daily living skills: a preliminary examination. Focus Autism Other Dev. Disabl. **31**(2), 129–139 (2016). https://doi.org/10.1177/1088357614533594

Wuang, Y.-P., Chiang, C.-S., Su, C.-Y., Wang, C.-C.: Effectiveness of virtual reality using Wii gaming technology in children with Down syndrome. Res. Dev. Disabil. **32**(1), 312–321 (2011). https://doi.org/10.1016/j.ridd.2010.10.002

# Human and Artificial Thinking

# Can Machines and Humans Use Negation When Describing Images?

Yuri Sato[1]([✉]) [iD] and Koji Mineshima[2] [iD]

[1] Ochanomizu University, Tokyo, Japan
`sato.yuri@ocha.ac.jp`
[2] Keio University, Tokyo, Japan
`minesima@abelard.flet.keio.ac.jp`

**Abstract.** Can negation be depicted? It has been claimed in various areas, including philosophy, cognitive science, and AI, that depicting negation through visual expressions such images and pictures is challenging. Recent empirical findings have shown that humans can indeed understand certain images as expressing negation, whereas this ability is not exhibited by machine learning models trained on image data. To elucidate the computational ability underlying the understanding of negation in images, this study first focuses on the image captioning task, specifically the performance of models pre-trained on large linguistic and image datasets for generating text from images. Our experiment demonstrates that a state-of-the-art model achieves some success in generating consistent captions from images, particularly in photographs rather than illustrations. However, when it comes to generating captions containing negation from images, the model is not as proficient as humans. To further investigate the performance of machine learning models in a more controlled setting, we conducted an additional analysis using a Visual Question Answering (VQA) task. This task enables us to specify where in the image the model should focus its attention when answering a question. As a result of this setting, the model's performance was improved. These results will shed light on the disparities in the attentional focus between humans and machine learning models.

**Keywords:** negation · grounding · image captioning · visual question answering · machine learning · cognitive science

## 1 Introduction

Human-human communication involves the expression of negation, e.g., *This bottle is empty* or *Nothing happened*, which is the competence base for higher cognition in humans [2,8]. However, it is not yet clear what, if anything, can be said to constitute understanding of the meaning of a negation. Let us suppose that we follow the standard referential semantics according to which understanding meaning consists in making a correspondence between words and things in the world, or grounding them [18]. For example, if you say, "There is a cup,"

© The Author(s), under exclusive license to Springer Nature Switzerland AG 2024
J. Baratgin et al. (Eds.): HAR 2023, LNCS 14522, pp. 39–47, 2024.
https://doi.org/10.1007/978-3-031-55245-8_3

you understand the sentence if you can pick out the cup from among several objects [4,22]. However, there is literally no corresponding object in the world for the expression "no" when one says "there is no cup" or for the expression "empty" when one says "the cup is empty". Accordingly it is not clear how to test the understanding of negation by picking up the object depicted. It is well known that Wittgenstein [24] made this point with the phrase "a picture cannot be negated".

Despite these difficulties, it seems that we humans actually understand the meaning of negation correctly and use it in communication without problems. What is it about our cognitive abilities that makes this possible? To address this question, this study takes the approach of comparing the performance of humans and machine learning-based AI models on two major tasks related to understanding negation in images, i.e., the task of generating an appropriate caption from an image and the task of answering a question based on the content of an image. By conducting this comparison, the study aims to gain insights into the differences between human and AI comprehension of negation in visual contexts.

The technology of describing images in natural languages is one of the most intensively developed areas of AI research in recent years, and offers interesting directions in relation to human-machine interaction research. For example, research on a task called "vision and language navigation" in which humans give linguistic instructions and robot agents act accordingly in a real-world or visual virtual environment [1,3]. In the context of such interactions, the theme of this research, "negation," is also attracting attention. It is research related to the task of comparing scene images at one point in time with scene images at another point in time and detecting what has changed [15,16]. This includes the task associated with negation, which is to guess what is missing. In contrast, our study focuses on a simpler setting: being able to understand negation from a single image rather than multiple images (cf. [14,20]).

Sato, Mineshima, and Ueda [21] conducted a study in which they collected two sets of images: negation images, which were understood as expressing negation, and negation-free images, which were not associated with negation. For comic illustration images, the determination of negation images or negation-free images was based on the descriptions provided by crowd workers (approximately 24 individuals per image). As for photograph images, the labels provided by annotators (with a minimum agreement of 2/3) were used to determine whether an image belonged to the negation or negation-free category. These collected images were then employed as training data for human and machine learning models, using state-of-the-art models such as convolutional neural networks (CNN) and vision transformers (ViT). The task involved classifying new images as either negation or negation-free. The results showed that the performance of the machine learning models was not as good as that of humans and merely reached chance-level accuracy. The findings suggest that models trained solely on image features encounter difficulties in accurately determining whether an image contains negation or not. Indeed, it remains an open question whether a machine learning model trained on both linguistic and visual data can comprehend that a specific image conveys negation.

In Sect. 2, we investigate this question by focusing on the image captioning task, which involves generating text descriptions from images. Unlike the image classification task, where the training data consists of image-label pairs, the image captioning task combines image data with descriptive text during training. In particular, we focus on the CLIP Interrogator (CLIP-I) model.[1,2] This model generates the input text for prompts in text-to-image generation AI models such as Stable Diffusion, which we use as an image captioning model. This model is mainly composed of CLIP (Contrastive Language-Image Pre-training) [17] and BLIP (Bootstrapping Language-Image Pre-training) [10], building a large dataset of image-text pairs from the Internet and incorporates them as training data. In this respect, the model has acquired relevance among linguistic expressions in some depth, which is expected to play an important role in understanding negation. The model is also envisioned to be used as a zero-shot transfer for a wide variety of tasks. Therefore, our study directly transposed the CLIP-I model trained on the pre-training data to the image captioning task, without using a specific training dataset for negation understanding.

If the performance of captioning in the AI model such as CLIP-I is comparable to that of humans, it suggests that background knowledge or commonsense provided by linguistic information plays an important role in understanding negation in images. In addition, in Sect. 3, we conducted an analysis using a visual question answering task (VQA), which allows the model to specify where in the image to focus its attention in answering a question, since the VQA task allows the model to freely set up a set of questions. Therefore, if the AI model's performance improves in this form of task, it suggests that information about which parts of the image to pay attention to, in addition to linguistic information, plays an important role in understanding negation in the image.

## 2    Experiment 1: Image Captioning Task

### 2.1    Experimental Setting

For comic illustration images, we targeted 18 negation images identified in [21]. These were collected from their created masterpiece list and the existing Manga 109 dataset [13]. For photograph images, 7 images were targeted. These were the negation images used in [21] that furthermore contained more than a certain number of negation expressions in the caption text. This criterion was defined as more than half of the total 10 English and Japanese captions of MS-COCO images (both of which were given 5 caption data per image) [11,23]. The training model for CLIP-I was ViT-L (best for Stable Diffusion 1).

---

[1] https://github.com/pharmapsychotic/clip-interrogator.
[2] https://huggingface.co/spaces/pharma/CLIP-Interrogator.

42      Y. Sato and K. Mineshima

photo 7

Human-EN #5: An empty room with a hard wood floor next to a kitchen.
Human-JA #5: A new room with nothing but gentle light.
CLIP-I ⟨5⟩: an empty living room with hard wood floors

photo 2

Human-EN #1: A chair and umbrella next to a empty pool.
Human-JA #1: A ladder attached to a pool that is not filled with water.
CLIP-I ⟨4⟩: a white chair sitting next to a swimming pool

photo 3

Human-EN #3: An empty suitcase sitting on a wood floor
Human-JA #2: Nothing in the suitcase.
CLIP-I ⟨5⟩: a piece of luggage sitting on the floor next to a radiator

photo 1

Human-EN #5: A restroom toilet missing a lid next to a trash can.
Human-JA #3: Flowing water in a western style toilet without lid and toilet seat.
CLIP-I ⟨4⟩: a white toilet sitting in a bathroom next to a trash can

illust 1

Human-JA #4: Human: He came home and found no one there
CLIP-I ⟨2⟩: a black and white drawing of a person in a room

illust 2

Human-JA #14: There is no toilet paper
CLIP-I ⟨5⟩: a cartoon of a man sitting on a toilet

**Fig. 1.** Image captions generated by the English human annotator (first column), the Japanese human annotator (second column), and the CLIP-I model (third column in blue; ⟨number⟩ represents the median of the evaluated values; prompt words listed after the first comma were omitted). Human caption data for the illustration images were collected from [21]. Human caption data for the Photo images were collected from [11, 23]. (Color figure online)

## 2.2 Supplementary Evaluation Analysis

The output of the CLIP-I model was assessed by five evaluators, who rated the consistency between the generated captions and the images on a 5-point scale.

**Method.** Three of the five evaluators were undergraduate students in philosophy courses and in their 20s, while the other two were workers in their 40s. They had no experience as annotators in image captioning experiments, nor did they have any expertise in AI research. They were not instructed that the given caption sentences were generated by AI models, but were simply asked to evaluate the caption sentences. A rating of 1 indicated strong inconsistency, while a rating of 5 indicated strong consistency.

**Results.** For comic illustration images, 13 of the 18 images had a median rating of 1 or 2. For photograph images, the median rating was 4 or 5 for all 7 images, indicating a high level of consistency between the generated captions and the images. See Fig. 1 and the Appendix for the median rating of each image.

## 2.3 Results and Discussion

Of all 25 images, only in one case, Photo 7, did the CLIP-I model generate a caption sentence that included a negation, just as humans do. "Empty room" was a negation expression used by all five of the English annotators (three of the five Japanese annotators used the negation expression "nothing" in this image). Similarly, the CLIP-I model also generated the negation expression "empty room". This case is a negation in the sense that the object occupying a particular space is not present, which is a common idea with the negation expression by shading regions in the Venn diagrams [19]. Negation in this sense appeared in human-generated captions in other images (e.g., photo 2, photo 3, and illustration 1), but not in captions generated by the CLIP-I model. This suggests that the model's ability to understand this type of negation in the image is not a robust result.

The occurrences of negation clauses in human-generated image captions were highest in the case of illustration 2. There, all 22 participants generated caption sentences that contained negation, with 19 participants using expressions such as "there is no toilet paper" and the remaining 3 participants using "the toilet paper was out". However, the CLIP-I model generated no caption sentences containing negations about toilet paper, only the description of a person sitting on the toilet. Negation here is used in the sense that a particular object is missing. Negation in this sense is also seen in the caption sentence generated by the human in the case of the photo image 1 (no toilet lid). See the Appendix for human and model caption text for the other 19 images.

The consistency rating of the model-generated caption text for the illustration images was low, but this could be due to the model's inability to understand Japanese comic conventions [6] as in the illustrations 5, 6, 10, 14, 15 and Japanese lifestyle as in the illustrations 1, 5, 7, 12, 17, 18. On the other hand, half of the eight illustration images (e.g., illustration 2) and all of the photograph images, which were less affected by these local/cultural factors, had sufficiently high consistency ratings. In this respect, the model achieves some success in generating consistent captions from images. However, when it comes to generating captions containing negation from images, the model is not as proficient as humans. This is not to say that the model-generated captions are describing the incorrect things, but rather that the targets described by models and humans are different.

# 3    Experiment 2: Visual Question Answering Task

Solving questions such as why this dissociation between model and human areas of attention occurs, and how models can be built that pay attention to human-like areas, is the next step that emerges to elucidate the computational process of negation comprehension in images. One approach we can take is to specify, as a task setting, what to focus on in the image and then describe it in words. The task format of visual question answering (VQA) fits this purpose [5,12]. Here, rather than describing the entire image as in image captioning, we can prepare a question about what is in a particular area of the image (kitchen, pool, suitcase, etc.) by operating the prompt. We can determine if the model was able to generate a negation by checking if the model's answer contains a negation word (e.g., "nothing").

## 3.1    Experimental Setting

We used the ViLT model [9] in a visual question answering task.[3,4] Since the pre-training of this model included MS-COCO image and caption data, we did not use the photograph images, but only targeted the comic illustration images in our experiments. As in Experiment 1,we used the 18 comic illustration images in [21].

The basic format of the questions was "what is in $\varphi$?", where $\varphi$ was the name of the relevant domain based on the human-generated captions. Each question text was manually created. For example, the questions were prepared as follows: "What is in the room in front of the boy?" for the illustration 1 in Fig. 1, What is beyond the boy's left hand? for the illustration 2 in Fig. 1, What is in the seat next to the woman? for the illustration 3 in Fig. 2. A related previous study of VQA is [7], where the question "Is the man not wearing shoes?" was answered in a "Yes/No" format. However, since the purpose of this study was to see if it is possible to generate negation from images, we did not use this question-answer format. Rather, we used the above question-answer format in order to examine whether negation appears in the answer part rather than the question part.

## 3.2    Results

As shown in Fig. 2, the model outputs multiple answers along with scores. If we only count the highest scoring answer as the model's output, the results showed

---

[3] https://github.com/dandelin/vilt.
[4] https://huggingface.co/dandelin/vilt-b32-finetuned-vqa.

Question: What is in the seat next to the woman?
Answer (Score): nothing (0.1402), ball (0.0541), pillow (0.0517),
stuffed animal (0.0393), teddy bear (0.0383)

Question: What is beyond the boy's left hand?
Answer (Score): basket (0.138), paper (0.097), trash can (0.096),
bowl (0.073), container (0.064)

**Fig. 2.** Illustration images with given questions and output answers in the VQA task. The illustration 3 is counted as a correct answer case. The illustration 2 is counted as an incorrect answer case.

that 3 out of the total 18 images answered correctly as expressing negation. The illustration 3 in Fig. 2 shows an example where the models gives a correct answer ("nothing"), while the illustration 2 in Fig. 2 shows an example where the models gives a wrong answer. We also loosen the criterion for correct answer in line with the 3/5 criterion in the image captioning task: the model's output is considered correct if the negation word appears within 40% of the total score of the top answers. Then the result showed that 5 out of the total 18 images surpassed this criterion. See the Appendix for the other 4 images.

## 4   General Discussions

Let us summarize the results we obtained:

(1) Humans generated caption sentences containing negation words to a certain degree of standard with 18 illustration images and 7 photo images [21].
(2) Out of the above 25 images used as analysis targets, the CLIP Interrogator model generated caption sentences containing negation words for only one photo image (Experiment 1).
(3) As a result of performing the VQA task in the ViLT model, using the above 18 illustration images as the analysis target and answering the question that indicated the image part to be paid attention to, responses including negation words were obtained in 5 images with a certain level of criteria (Experiment 2).

As discussed in detail in [21], background knowledge plays an essential role in the human understanding of negation from images. For instance, the information "there is nothing on the plate" can be inferred from an image because of the common sense that something is typically placed on a plate in dinner, which in turn creates an expectation that the plate should have something on it. When this common sense comes into play, people naturally and effortlessly focus on the plate within the entire image. The results of Experiment 1, in which sentence generation was performed from the entire image, indicate a disparity in performance between humans and machine learning models in this respect. In Experiment 2, we tested the machine learning model's negation recognition ability in a more controlled setting by presenting a question specifying which part of the image to focus on. The results showed improved performance when the model's attention was guided by the question text.

These findings suggest that supplementary linguistic information can assist AI models in better focusing on the image and extracting relevant information, including negation. Humans draw such "natural questions" from images alone and can extract complex logical information. The challenge of improving AI models by emulating these human abilities to understand language and images holds significant potential for future research.

**Acknowledgements.** All comic images in this paper are from the Manga-109 dataset and are licensed for use. "HighschoolKimengumi vol. 20" p. 139 © Motoei Niizawa/Shueisha for illust 1 of Fig. 1, "MoeruOnisan vol19" p. 58 © Tadashi Sato/Shueisha for illust 2 of Fig. 1, "OL Lunch" p. 9 © Yoko Sanri/Shogakukan. Regarding photographs, they are retrieved from MS-COCO. The COCO image id is #449681 for Photo 7, #163084 for Photo 2, #51587 for Photo 3, #65737 for Photo 1.

This study was supported by Grant-in-Aid for JSPS KAKENHI Grant Number JP20K12782 and JP21K00016 as well as JST CREST Grant Number JPMJCR2114.

## A    Appendix

https://www.dropbox.com/scl/fi/pv5kyujtsz1a3l76kkn9t/app_HAR23.pdf?rlkey=tpgclqok7w49pqmg7cf1g9423&dl=0.

## References

1. Ahn, M., et al.: Do as I can, not as I say: grounding language in robotic affordances. In: CoRL 2023, PMLR, vol. 205, pp. 287–318 (2023)
2. Altman, S.: The structure of primate social communication. In: Altman, S. (ed.) Social Communication Among Primates, pp. 325–362. University of Chicago Press, Chicago (1967)
3. Anderson, P., et al.: Vision-and-language navigation: interpreting visually-grounded navigation instructions in real environments. In: CVPR 2018, pp. 3674–3683. IEEE (2018)
4. Bender, E. M., Koller, A.: Climbing towards NLU: on meaning, form, and understanding in the age of data. In: ACL 2020, pp. 5185–5198 (2020)
5. Bernardi, R., Pezzelle, S.: Linguistic issues behind visual question answering. Lang. Linguist. Compass **15**(6), elnc3.12417 (2021). https://doi.org/10.1111/lnc3.12417
6. Cohn, N.: The Visual Language of Comics: Introduction to the Structure and Cognition of Sequential Images. Bloomsbury Academic, London (2013)
7. Gokhale, T., Banerjee, P., Baral, C., Yang, Y.: VQA-LOL: visual question answering under the lens of logic. In: Vedaldi, A., Bischof, H., Brox, T., Frahm, J.-M. (eds.) ECCV 2020. LNCS, vol. 12366, pp. 379–396. Springer, Cham (2020). https://doi.org/10.1007/978-3-030-58589-1_23

8. Horn, L.R.: A Natural History of Negation. University of Chicago Press, Chicago (1989)
9. Kim, W., Son, B., Kim, I.: ViLT: vision-and-language transformer without convolution or region supervision. In: ICML 2021. PMLR, vol. 139, pp. 5583–5594 (2021)
10. Li, J., Li, D., Xiong, C., Hoi, S.: BLIP: bootstrapping language-image pre-training for unified vision-language understanding and generation. In: ICML 2022. PMLR vol. 162, pp. 12888–12900 (2022)
11. Lin, T.-Y., et al.: Microsoft COCO: common objects in context. In: Fleet, D., Pajdla, T., Schiele, B., Tuytelaars, T. (eds.) ECCV 2014. LNCS, vol. 8693, pp. 740–755. Springer, Cham (2014). https://doi.org/10.1007/978-3-319-10602-1_48
12. Manmadhan, S., Kovoor, B.C.: Visual question answering: a state-of-the-art review. Artif. Intell. Rev. **53**, 5705–5745 (2020). https://doi.org/10.1007/s10462-020-09832-7
13. Matsui, Y., et al.: Sketch-based manga retrieval using manga109 dataset. Multimed. Tools Appl. **76**(20), 21811–21838 (2017). https://doi.org/10.1007/s11042-016-4020-z
14. van Miltenburg, E., Morante, R., Elliott, D.: Pragmatic factors in image description: the case of negations. In: VL 2016, pp. 54–59. ACL (2016)
15. Park, D. H., Darrell, T., Rohrbach, A.: Robust change captioning. In: ICCV 2019, pp. 4624–4633, IEEE (2019)
16. Qiu, Y., Satoh, Y., Suzuki, R., Iwata, K., Kataoka, H.: Indoor scene change captioning based on multimodality data. Sensors **20**(17), 4761 (2020). https://doi.org/10.3390/s20174761
17. Radford, A., et al.: Learning transferable visual models from natural language supervision. In: ICML 2021. PMLR vol. 139, pp. 8748–8763 (2021)
18. Russell, B.: The Problems of Philosophy. Oxford University Press, Oxford (1912)
19. Sato, Y., Mineshima, K.: How diagrams can support syllogistic reasoning: an experimental study. J. Log. Lang. Inf. **24**, 409–455 (2015). https://doi.org/10.1007/s10849-015-9225-4
20. Sato, Y., Mineshima, K.: Visually analyzing universal quantifiers in photograph captions. In: Giardino, V., Linker, S., Burns, R., Bellucci, F., Boucheix, J.M., Viana, P. (eds.) Diagrams 2022. LNCS, vol. 13462, pp. 373–377. Springer, Cham (2022). https://doi.org/10.1007/978-3-031-15146-0_34
21. Sato, Y., Mineshima, K., Ueda, K.: Can negation be depicted? Comparing human and machine understanding of visual representations. Cogn. Sci. **47**(3), e13258 (2023). https://doi.org/10.1111/cogs.13258
22. Søgaard, A.: Grounding the vector space of an octopus: word meaning from raw text. Minds Mach. **33**(1), 33–54 (2023). https://doi.org/10.1007/s11023-023-09622-4
23. Yoshikawa, Y., Shigeto, Y., Takeuchi, A.: STAIR captions: constructing a large-scale Japanese image caption dataset. In: ACL 2017, pp. 417–421 (2017)
24. Wittgenstein, L.: Notebooks 1914-1916. In: Anscombe, G.E.M., von Wright, G.H. (eds.) University of Chicago Press, Chicago (1984). (Original Work Published 1914)

# An Incremental Diagnosis Algorithm of Human Erroneous Decision Making

Valentin Fouillard[1,2(✉)], Nicolas Sabouret[1], Safouan Taha[2], and Frédéric Boulanger[2]

[1] Université Paris-Saclay, CNRS, Laboratoire Interdisciplinaire des Sciences du Numérique, 91405 Orsay, France
{valentin.fouillard,nicolas.sabouret}@universite-paris-saclay.fr
[2] Université Paris-Saclay, CNRS, ENS Paris-Saclay, CentraleSupélec, Laboratoire Méthodes Formelles, 91190 Gif-sur-Yvette, France
{safouan.taha,frederic.boulanger}@universite-paris-saclay.fr

**Abstract.** This paper presents an incremental consistency-based diagnosis (CBD) algorithm that studies and provides explanations for erroneous human decision-making. Our approach relies on minimal correction sets to compute belief states that are consistent with the recorded human actions and observations. We demonstrate that our incremental algorithm is correct and complete *wrt* classical CBD. Moreover, it is capable of distinguishing between different types of human errors that cannot be captured by classical CBD.

**Keywords:** Diagnosis · Human errors · Belief revision

## 1 Introduction

Erroneous decision making can cover many situations, from information misinterpretation to distraction or even rule-breaking [10]. Such errors led to well-documented critical accidents in nuclear safety, air transport, medical care, or aerospace engineering. Understanding what happened in such accidents is essential to preventing it from happening again and designing new systems that consider human fallibility [19]. This process is called Human Error Analysis [31].

Human error analysts try to precisely understand the situation and guess the operators' belief states that can explain their errors. For example, in the context of air transports, they use flight recorders (that record information from the flight instruments and all conversations in the cockpit) to understand what the pilots thought and why they adopted an erroneous course of action.

This reconstruction of human beliefs is done manually. However, we claim that logic-based modeling could help analysts identify possible mental states that can explain the accident. Indeed, logic-based diagnosis has been a very active field of AI since the 1980s with many famous frameworks [17,21,23]. However, applying logic-based models to human error analysis is difficult because these frameworks implement purely rational reasoning. In contrast, human factors that

J. Baratgin et al. (Eds.): HAR 2023, LNCS 14522, pp. 48–63, 2024.
https://doi.org/10.1007/978-3-031-55245-8_4

come into play in erroneous decision-making tend to break the basic rationality principles. Logic-based diagnosis of human error thus requires taking into account possible deviations from rationality. This is the goal of our research.

This paper is organized as follows: Sects. 2 and 3 present different types of human errors that we consider in this paper, and classical logic-based models that serve as a basis for our framework. Section 4 and 5 present our model and incremental diagnosis algorithm. Sections 6 and 7 demonstrate the correctness and completeness of our algorithm. Section 8 discusses the related work, and the last section presents the perspectives of our approach.

## 2   Different Types of Human Errors

While human error covers a wide range of situations [10], our research focuses on four human cognitive mechanisms which we find present in our context of application (airplane accidents): selection, memory, attention and reasoning. All these mechanisms can produce incorrect beliefs and lead to critical erroneous decisions. Two accidents will serve as examples in this paper: the Air Inter Flight 148 in 1992 (Mont Saint Odile)[1] and the Air France Flight 447 in 2009 (Rio-Paris)[2]

**Information Selection and Preference.** When facing contradictory information, human beings tend to prefer the one that confirms their own beliefs or preferences, possibly falling into a confirmation bias [20]. This is one possible explanation to what happened in AF447 when the pilots ignored information that was in contradiction with their initial interpretation of the situation.

**Forgetting and False Memories.** Human beings are not omniscient and can misremember some information, forget it entirely or even build false memories [11]. This happened in AI148 when the pilots forgot that they had previously configured the system in Vertical Speed mode instead of Flight Path Angle.

**Attention Error.** Human beings have limited attention capacities. Consequently, they might miss some critical information and make erroneous decisions based on incomplete information [5]. For example in AF447, even though the stall alarm rang more than 75 times, the pilots concentrated all their attention on the overspeed information.

**Reasoning Error.** The principle of bounded rationality [25] tells us that we cannot always process information with complete and perfect reasoning, which can lead us to draw incorrect conclusions from accurate information. For example in AF447, the pilots observed a vibration on the control stick and should have

---

[1] https://bea.aero/uploads/tx_elydbrapports/F-GGED.pdf.
[2] https://bea.aero/docspa/2009/f-cp090601/pdf/f-cp090601.pdf.

concluded to a stall. However, the investigators showed that the pilots probably concluded to an overspeed situation due to erroneous reasoning.

Our research consists in taking into account such human errors in logic-based diagnosis. The following section shows how the four errors presented above relate to some classical theories and models in the field of Knowledge Representation and Reasoning.

# 3   Logic-Based Modeling of Human Reasoning

## 3.1   Information Selection and Preferences

Computing changes in an agent's belief base when facing new information has been studied in the 1980s as the problem of the *belief revision* [1]. This problem consists in restoring the consistency of the agent's belief base when confronted with new (and conflicting) information.

Belief revision models seem like a promising direction to capture the information preference errors presented in Sect. 2. Indeed, [7] showed empirically that the *screened revision* operator can capture the *belief bias*, namely the tendency to judge arguments based on the plausibility of the conclusion instead of how well they support the conclusion. This revision operator can take into account a set of *screened* propositions, which are immune to revision (i.e. they cannot be among the eliminated propositions). This makes it possible to consider that, from the agent's point of view, certain beliefs cannot be abandoned because of their importance.

## 3.2   Forgetting and False Memories

The *Frame Problem* is a seminal to the field of Reasoning about Actions and Changes. Introduced by McCarthy and Hayes [16], it can be summarized as the challenge of representing the effects of an action without explicitly representing a large number of intuitively obvious non-effects. Three different aspects appear when studying the Frame Problem: *inertia, update* and *extrapolation*.

*Inertia.* McCarthy and Hayes showed that logic modeling requires to describe explicitly the inertia of beliefs. More formally, if $\varphi$ holds in state $s$, then $\varphi$ must still hold in the state $do(a, s)$ resulting from the execution of action $a$, unless $a$ explicitly modifies $\varphi$. Several solutions were proposed to address the Frame Problem [18]. The general idea is to model the inertia of beliefs using inertia clauses of the form $\varphi_{t+1} = \varphi_t$ (depending on the representation of time and states). The difficulty is now to correctly update a belief when the agent acts upon or perceives its environment.

*Update.* Assume that action $a$ changes the value of $\varphi$. Applying the effects of $a$ in the presence of a general inertia clause results in an inconsistent belief base in which both $\varphi_{t+1}$ and $\neg\varphi_{t+1}$ hold. The KM theory [12] describes a set of axioms that a logical operator must verify to restore consistency when performing a belief update.

*Extrapolation.* Agents can also observe changes in the world's state caused by other agents. The integration of the new information that conflicts with the inertia clauses is known as the *extrapolation* process. [6] showed that this extrapolation process is an instance of a belief revision on the clauses of inertia of a temporally indexed logic. A belief revision operator can then be used to perform belief extrapolation.

**Relation Between Inertia and Human Errors Modeling.** We claim that inertia and its operations (*update, extrapolation*) can be a way of capturing forgetting and false memories. More precisely, an error in the *update* or *extrapolation* operation can result in this type of memory error. For example in AI148, the pilots didn't perform a correct *belief update* about the autopilot configuration, which led them to believe that they were still in Vertical Speed mode, through inertia. Therefore, we define both an *extrapolation* operator and an *update* operator that support such distortions (we call them *"frame distortions"*). To implement this, we propose to rely on *circumscription*.

**Circumscription.** Proposed by [15] as a first attempt to deal with the Frame Problem, *circumscription* consists in selecting models that minimize the number of changes in the world. While this method does not correctly solve the Frame Problem because it computes some changes or non-changes in the inertia that are not expected in a rational reasoning [9], it can be useful to compute our *frame distortions* that represent human omissions or false memories.

Moreover, [13] showed that *circumscription* is equivalent to a belief revision operator. We might thus use the same revision operator to capture information preferences, memory errors, and information preference errors!

### 3.3    Attention and Reasoning Errors

Logic-based diagnosis can refer to different approaches. Deduction [4] and abduction [22] consist in computing possible explanations given some knowledge about errors and symptoms of these errors. On the contrary, consistency-based diagnosis [23] considers only knowledge about "how the system usually works" (without any information about the possible errors). The goal is to identify deviations from the system's expected behavior.

This approach (consistency-based diagnosis or CBD) is well suited for Human Error Analysis since analysts usually don't know all possible errors and symptoms that can explain an operator's erroneous decision: all they know is which decision was expected. This is why we will use CBD in our work.

The logic-based framework proposed by [23] consists in restoring consistency (hence the name) between the description of the system's expected behavior and the observations of the system's behavior. It is composed of three elements. First, a set of logic formulas $SD$ that describe the system. Second, a set of predicates $ASS$ that describe the *assumables* of the form $\neg ab(c)$ which represent the fact that component $c$ is supposed to behave normally. Third, a conjunction of predicates $OBS$ that describe an observation of the system.

When $SD \cup ASS \cup OBS$ is inconsistent, a diagnosis $\Delta$ is a minimal set of *assumables* such that $SD \cup (ASS \backslash \Delta) \cup OBS$ is consistent. In other words, a diagnosis is a minimal set of elements that must be assumed "abnormal" to be consistent with the observations.

Our proposal is to capture attention and reasoning error by applying such a CBD. Indeed, attention errors can be seen as ignoring some information and reasoning errors as inference rules that weren't applied. Let us assume that:

- $OBS$ contains a representation of the observed action (*i.e.* the result of the human operator's erroneous decision).
- $SD$ contains the operator's inference rules and information available to him (*i.e* observations) that allow him to make a decision.
- $ASS = SD$ which means that we assume all available information and all inference rules of the human operator to behave normally, *i.e.* they are used and don't infer anything beyond the scope of their behavior.

The diagnosis $\Delta$ will thus contain the information and rules that the human operator ignored to make his decision and perform his action.

**Belief Revision and Consistency-Based Diagnosis.** Several research emphasize the strong connection between a belief revision operator and a CBD [3,30]. They show that a belief revision operator can be used to compute a CBD and conversely, a CBD can be used as a belief revision operator. The benefit of CBD is that there are algorithms for calculating the diagnosis, in particular the Liffiton algorithm that we will be using, which is based on MCS (see Sect. 4.3).

This means that we can use a CBD algorithm to capture all four kinds of human errors: not only attention errors and reasoning errors, but also information preferences (which correspond to belief revisions) and memory errors (which correspond to frame distortions, captured by circumscription, also equivalent to a belief revision operator). The following sections present our framework.

## 4   Diagnosis Framework

Our approach relies on restoring consistency in a set of logical formulas that represent the beliefs of the agent at a given time, the information it received, the inference rules he could use and the action he selected eventually.

### 4.1   Logical Modeling

Our model is based on the continuity of Reiter's situation calculus [24]: we give ourselves a starting situation $S_0$ (which is a set of fluents) and a set of actions (which are exactly the trace of the operator's actions that led to the accident) with $S_t$ the situation "at time $t$" resulting from performing action $a_t$ in situation

$S_{t-1}$. We extend this model by adding information communicated to the operator (that we call *observations*). Moreover, we allow each fluent to be time-indexed to represent beliefs of the form "in situation $S_t$, the agent believes that at time $t'$, fluent $\varphi$ was true".

This paper focuses on the CBD algorithm for one single time step. We thus only consider one situation and one action performed by the agent. This leads us to propose the following model.

**Model.** We consider a set of propositions $\mathcal{P}$, all indexed temporally. $\varphi_t \in \mathcal{P}$ represents the concept "$\varphi$ holds at time $t$". Based on these propositions, we define the language $\mathcal{L}_0$ with the following grammar:

$$\alpha ::= \varphi_t \mid \bot \mid \top \mid \neg\alpha \mid \alpha \wedge \alpha \mid \alpha \vee \alpha$$

where $\alpha$ is a valid formula from $\mathcal{L}_0$, $\varphi_t \in \mathcal{P}$, $\bot$ is always false and $\top$ always true.

We also consider a set $\mathcal{A}ct$ of atomic propositions representing the actions. We define the language $\mathcal{L}$ as an extension of $\mathcal{L}_0$ by adding the three following operators:

$$\phi ::= \alpha_1 \rightarrow \alpha_2 \mid \{\alpha\}act \mid act \wedge \alpha::\varphi_{t+n}$$

with $\alpha \in \mathcal{L}_0$, $\alpha_1 \in \mathcal{L}_0$, $\alpha_2 \in \mathcal{L}_0$, $act \in \mathcal{A}ct$ and $n \in \mathbb{N}^*$.

- $\alpha_1 \rightarrow \alpha_2$ means that the agent can infer $\alpha_2$ from $\alpha_1$.
- $\{\alpha\}act$ mean that $\alpha$ is the precondition of action $act$. In other words, $\alpha$ must be true for the action to be done.
- $act \wedge \alpha::\varphi_{t+n}$ mean that $\varphi_{t+n}$ is the effect of the action $act$ when $\alpha$ is true. In other words, $\varphi_{t+n}$ is true if $act$ is done and $\alpha$ is true.

### 4.2   Definition of the Diagnosis Problem

Let us consider the following elements:

- The belief state $B_{t-1} \in 2^{\mathcal{P}}$ which correspond to the current situation.
- A set of rules $\mathcal{R} \in 2^{\mathcal{L}}$ that the agent can use. For example, $alarm_t \rightarrow stall_t$ says that if the agent believes that there is an alarm, it should believes that he is in a stall situation.
- A set of possible observations in the environment $Obs \in 2^{\mathcal{P}}$. For example, $alarm_t \in Obs$ mean that the agent could observe an alarm.
- $a \in \mathcal{A}ct$ the action selected by the agent. For example, $a = Push$ means that the agent decided, based on his beliefs, observations and rules, to push the plane's control stick.
- In addition, we introduce the set of inertia clauses:

$$\mathfrak{K} = \{\varphi_t = \varphi_{t-1}\}_{\forall \varphi \in \mathcal{P}}$$

Given these elements, our problem is to compute a new belief state $B_t$ that is consistent and that integrates all these elements. Formally, we apply a consistency-based diagnosis, *i.e.* we compute the minimum $\Delta$ such that:

$$B_t = ((B_{t-1} \cup \mathcal{R} \cup Obs \cup \mathfrak{K}) \backslash \Delta) \cup \{a\} \text{ is consistent}$$

In other word, we want to compute the set $\Delta$ of previous propositions, inference rules or possible observations that should be ignored by the agent to perform the action $a$.

The connection with human error diagnosis is the following: experts know the possible observations ($Obs$) and the action performed by the operator ($a$), as well as the domain rules ($\mathcal{R}$), and must find out which rules, observations, previous beliefs or inertia clauses were ignored by the operator. Note that the action is not an *assumable*: it was actually done and cannot be ignored.

## 4.3 Diagnosis Computation

To compute $\Delta$ using CBD, we use the notion of *Minimal Correction Set* (MCS). For a given system $\Phi = \{\phi_1, \phi_2 \ldots \phi_n\}$, $M \subseteq \Phi$ is a MCS of $\Phi$ if and only if $\Phi \backslash M$ is consistent and $\forall \phi_i \in M, (\Phi \backslash M) \cup \{\phi_i\}$ is inconsistent.

We use the algorithm proposed by [14] which supports *screened revision*. We have implemented this algorithm with the help of the SMT-solver Z3. The implementation of our algorithm in C# is available on a git repository[3]. 

We note $\mathfrak{M}(\Phi, screened)$ the set of MCSes, with $\Phi$ the system to be corrected, and $screened \subset \Phi$ the set of propositions and rules that cannot be removed by the MCS algorithm (i.e. $\mathfrak{M}(\Phi, screened) \cap screened = \emptyset$). We have our diagnostic reference Algorithm 1 defined as:

$$B_t = (B_{t-1} \cup \mathcal{R} \cup Obs \cup \mathfrak{K} \cup \{a\}) \backslash \Delta$$

where $\Delta \in \mathfrak{M}(\Phi, screened)$, $\Phi = \{B_{t-1} \cup \mathcal{R} \cup Obs \cup \mathfrak{K} \cup \{a\}\}$ and $screened = \{a\}$.

## 4.4 Considering Different Types of Error

As presented in Sect. 3, many errors can explain an erroneous decision and can be captured by a CBD operator. However it is difficult to know which element in $\Delta$ corresponds to which error (information preferences, memory, attention or reasoning). For example, if $\Delta$ contains some element from $Obs$, we can't know if this is due to a preference error (selection between two contradictory pieces of information) or to an attention error. Yet this a vital information for human error analysis.

To ease the understanding of the belief states and errors of the agent, we propose to use an incremental diagnosis algorithm. The idea of this algorithm is to perform a computation of the MCSes by increment, where each of the increment is focused on a specific error. Hence we can easily find the increment at the origin of a proposal in an MCS. The following section presents this algorithm.

---

[3] https://gitlab.dsi.universite-paris-saclay.fr/valentin.fouillard/humandiagnosis.

# 5    Incremental Diagnosis Algorithm

Our algorithm works in four steps: 1) Detection of MCSes that correspond to belief revisions, 2) MCSes related to an erroneous decision, 3) MCSes corresponding to extrapolation 4) MCSes corresponding to an update or a frame distortion. To illustrate our algorithm, we shall use the following example, which is a very simplified representation of the AF447 (Rio-Paris flight) situation:

$$B_{t-1} = \{\neg \text{acceleration}_{t-1}, \neg \text{alarm}_{t-1}, \text{buffet}_{t-1}\}$$
$$Obs = \{\text{acceleration}_t, \text{alarm}_t\}$$

$$\mathcal{R} = \left\{ \begin{array}{l} R^1 \equiv \text{alarm}_t \rightarrow \text{stall}_t \\ R^2 \equiv \text{acceleration}_t \rightarrow \text{overspeed}_t \\ R^3 \equiv \text{buffet}_t \rightarrow \text{stall}_t \\ R^4 \equiv \{\neg \text{overspeed}_t\} \text{ Push} \\ R^5 \equiv \{\neg \text{stall}_t\} \text{ Pull} \\ R^6 \equiv \text{stall}_t \wedge \text{overspeed}_t \rightarrow \bot \end{array} \right\}$$

$$a = \quad \text{Pull}$$

In this situation the pilot believes that there is no sign of acceleration and no stall alarm, the control stick is vibrating (*a.k.a.* buffet). This is given in $B_{t-1}$. The pilot can observe both an acceleration and a stall alarm (*Obs*) and decides to pull the stick ($a = $ Pull). The set of rules $\mathcal{R}$ represents classical pilot knowledge about stall and overspeed situations. In particular, it is expected to push the stick in case of stall, and to pull it in case of overspeed, not the contrary.

## 5.1    Information Preference

Starting from $B_{t-1}$, we add only the observations *Obs* and the rules $\mathcal{R}$ in the belief base of the agent. This ensures that the MCS captures only inconsistencies due to the observations (and possible reasoning about these observations), not the action or the inertia clauses. If we are in a situation where the agent observes two contradictory information or an information inconsistent with his beliefs, we compute a belief revision to build all possible revisions:

$$B_t^{rev} = \Phi \backslash M_{rev} \tag{1}$$
$$\text{with } M_{rev} \in \mathfrak{M}(\Phi, screened), \Phi = \{B_{t-1} \cup \mathcal{R} \cup Obs\}, screened = \{\emptyset\}$$

In the example introduced in the beginning of this section, the observations of the alarm, the buffet and the acceleration are inconsistent with each other. Therefore a possible MCS $M_{rev}$ computed at this step is $\{\text{acceleration}_t\}$: the pilot prefers to keep the stall alarm rather the acceleration information.

Note that there are several possible MCSes and thus several $B_t^{rev}$. For instance, the alternative correction (ignoring the alarm instead of the acceleration) is also a possible belief state. This is true for all steps of our algorithm.

## 5.2   Attention and Reasoning Error

For each possible belief base $B_t^{rev}$, we introduce the action in it. Since observations inconsistencies were already corrected, we ensure that the MCSes detected at this step are related to the action (*i.e.* reasoning error). We compute the belief base $B_t^{diag}$ resulting from this erroneous decision by:

$$B_t^{diag} = \Phi \backslash M_{diag}$$

with $M_{diag} \in \mathfrak{M}(\Phi, screened), \Phi = \{B_t^{rev} \cup \{a\}\}, screened = \{a\}$

(2)

To illustrate this step, let's consider that the previous step computes $M_{rev} = \{acceleration_t\}$. Adding the action in the system creates an inconsistency: from $R^1$ and $R^5$, we can infer that we should not perform the *Pull* action. One of the possible MCS computed at this step is $M_{diag} = \{R^1\}$: the pilot draws an incorrect conclusion about the alarm.

## 5.3   Extrapolation

From each possible belief base $B_t^{diag}$, we introduce the inertia clauses $\mathfrak{K}$. For each proposition $\varphi_{t-1}$ in $B_{t-1}$, $\mathfrak{K}$ contains the clause $\varphi_t = \varphi_{t-1}$. We also remove the action done by the agent from the belief base. This ensures that the inconsistencies will be related to the observations and the inertia clauses, and thus to the changes in the world that the agent has to consider. In other words, we compute an extrapolation to build:

$$B_t^{ext} = \Phi \backslash M_{ext}$$

with $M_{ext} \in \mathfrak{M}(\Phi, screened), \Phi = \{(B_t^{diag} \backslash \{a\}) \cup \mathfrak{K}\}, screened = \{B_t^{diag}\}$

(3)

To illustrate this step, let's start from the previous corrections on our example (*i.e.* ignoring $acceleration_t$ and $R^1$). By adding the inertia clauses in the logic system, we create an inconsistency on the alarm belief: we can deduce $\neg alarm_t$ from $B_{t-1}$ but we have $alarm_t$ in *Obs*. One possible MCS computed at this step is $M_{ext} = \{alarm_t = alarm_{t-1}\}$: the pilot simply updates their belief base with the new information about the alarm (and this is not an error).

## 5.4   Update and Distortion

For each possible belief base $B_t^{ext}$, we re-introduce the previously removed action. Since inconsistencies related to observations and the inertia were already solved, we ensure that the MCSes detected in this step correspond either to an update (the action needs to change the inertia) or a frame distortion (the action should not impact the inertia but it is inconsistent with it), which captures a possible memory error. Indeed as we performed a circumscription (equivalent to CBD, see Sect. 3), both are captured. We compute the final belief base of the agent:

$$B_t = \Phi \backslash M_{dist}$$

with $M_{dist} \in \mathfrak{M}(\Phi, screened), \Phi = \{B_t^{ext} \cup \{a\}\}, screened = \{a\}$

(4)

To illustrate this step, let's consider the previous corrections. When adding the action in the system, we have an inconsistency between the buffet and the pull action, via rules $R^5$ and $R^3$. One possible MCS for this step is $M_{dist} = \{\text{buffet}_{t-1} = \text{buffet}_t\}$: the pilot believes that the truth value of the buffet has changed between the two time steps, without any reason (which might be explained by forgetting the previous information).

### 5.5   Resulting Explanation

Each successive correction to reach $B_t$ from $B_{t-1}$ leads to several possible solutions. Therefore, we can say that a possible belief state $B_t$ is computed through a sequence $x$ of MCS choices. We note $\Delta^x$ the union of all MCSes in the sequence $x$: $\Delta^x = \{M^x_{rev} \cup M^x_{diag} \cup M^x_{ext} \cup M^x_{dist}\}$ where $x$ is a sequence computed by our incremental algorithm. There are as many possible $\Delta^x$ as correction choices at each step in the algorithm.

From this algorithm, we can easily, determine which type of error corresponds to a proposition $\varphi \in \Delta^x$, by finding the corresponding subset. In the next section, we prove that our algorithm correctly captures a consistency-based diagnosis.

## 6   Correctness and Completeness

In this section we consider the results of our incremental algorithm (Algorithm 2) compared to the results of the reference Algorithm 1, defined Sect. 4.3. Our algorithm is correct if its solutions are effective corrections of $\Phi$ and it is complete if its solutions include all the solutions provided by Algorithm 1. However even if our algorithm is correct, it can compute non-minimal corrections, contrary to Algorithm 1.

### 6.1   Correctness

Since Algorithm 1 returns a minimal set of ignorance to make the beliefs consistent, we need to check that each $\Delta^x$ proposed by Algorithm 2 ignores at least the same thing as one of the possible $\Delta$ given by Algorithm 1 to make the beliefs consistent. It can ignore more (this would be a non-minimal solution) but not less (this would not be sound).

Let's consider that Algorithm 1 consists in correcting the set B to make it consistent. Then, Algorithm 2 corresponds to four increments for the form: (1) add some propositions from $B$ to a subset $A$ of $B$; (2) correct $A$. These increments are repeated until all the propositions missing from the subset $A$ of the first increment have been added to $B$. Thus, if we show the soundness and the completeness over two increments, by induction, our 4-step Algorithm 2 is also complete (we are only adding propositions to a subset to arrive at $B$: it is a recursive algorithm).

Let consider two increments: first in computing an MCS $M_1$ on a set $A$ then computing an MCS $M_2$ on the set $B \backslash M_1$ where $A \subseteq B$. The ignorance which results from these two time steps is then the union of the two MCses: $M_1 \cup M_2$. On the contrary, Algorithm 1 consists in computing an MCS $M$ on the set $B$.

We must prove that $M \subseteq M_1 \cup M_2$, i.e., Algorithm 2 ignores at least the same propositions as Algorithm 1. More formally, we must prove theorem (T1):

$$A \subseteq B \wedge M_1 \in \mathfrak{M}(A, \emptyset) \wedge M_2 \in \mathfrak{M}(B \backslash M_1, \emptyset) \\ \Rightarrow \exists M \in \mathfrak{M}(B, \emptyset), \ M \subseteq M_1 \cup M_2 \tag{T1}$$

Under the premises, we can say that:

(a) $B \backslash (M_1 \cup M_2) \subseteq B$ since removing sets from $B$ can only give a subset of $B$.
(b) $B \backslash (M_1 \cup M_2) \nvdash \bot$ because, by construction, $M_2$ makes $B \backslash M_1$ consistent (it is a MCS). Let alone $B \backslash (M_1 \cup M_2)$ can only be consistent.

We know that M is minimal, or that $B \backslash M$ is a maximal subset of $B$ which does not imply $\bot$. Moreover, we know from (a) and (b) that $B \backslash (M_1 \cup M_2)$ is a subset of $B$ which does not imply $\bot$. Therefore, $M_1 \cup M_2$ can only be a superset of an MCS $M$ of $B$. Indeed, because $M$ is minimal and makes $B$ consistent, as well as because $M_1 \cup M_2$ also makes $B$ consistent, $M_1 \cup M_2$ can only be a superset of $M$. Otherwise, $M$ would not be minimal, which is a contradiction.

We can thus conclude that $\exists M \in \mathfrak{M}(B, \emptyset), M \subseteq M_1 \cup M_2$, which proves theorem $(T1)$.

### 6.2    Completeness

Theorem $(T1)$ Subsect. 6.1 tells us that the solutions found by Algorithm 2 contain at least the propositions corrected by Algorithm 1. Moreover, it tells us that the solutions found by increments are not necessarily minimal, *i.e.* increments can lead us to find solutions where the agent ignores more than necessary. Determining whether Algorithm 2 is complete is then equivalent to determining whether it computes *all* minimal solutions given by Algorithm 1. More formally, theorem (T2) states that, for any minimal solution $M$, there exist $M_1$ and $M_2$ obtained by Algorithm 2 such that $M_1 \cup M_2 = M$:

$$A \subseteq B \wedge M \in \mathfrak{M}(B, \emptyset) \\ \Rightarrow \exists M_1, M_2, \ M_1 \in \mathfrak{M}(A, \emptyset) \wedge M_2 \in \mathfrak{M}(B \backslash M_1, \emptyset) \wedge M = M_1 \cup M_2 \tag{T2}$$

To begin with, let us note that if $B \backslash M$ is consistent with $A \subseteq B$, then not only is $A \backslash M$ consistent (there are fewer propositions) but also there is a $M' \backslash M$ such that $A \backslash M'$ is consistent (some of the propositions of $M$ are not present in $A$ so we can remove them). Thus, we can state that:

(a) $\forall A \subseteq B, \exists M_1 \in \mathfrak{M}(A, \emptyset)$ with $M_1 \subseteq M$.
(b) $\forall M_1, \exists M_2 \in \mathfrak{M}(B, \emptyset)$ with $M_2 \subseteq M$.

We then deduce:

(c) $M_1 \cup M_2 \subseteq M$ by (a) and (b).

(d) by (T1), $\exists M' \in \mathfrak{M}(B, \emptyset)$ such that $M_1 \cup M_2 = M'$.

(e) We thus have $M' = M_1 \cup M_2$ and $M' \subseteq M$. However since both $M$ and $M'$ are MCSes, by definition of minimality, $M' = M$.

Algorithm 2 is thus complete in the sense that it returns, like Algorithm 1, all possible minimal solutions to make the agent's beliefs consistent ($T2$). These proofs have been verified in Isabelle/HOL and are available on a git repository[4].

However, theorems ($T1$) and ($T2$) tell us that it also computes some non-minimal solutions which cannot be computed by Algorithm 1. The next section discusses these non-minimal solutions.

# 7 Discussion

From a purely logic-based diagnosis point of view in logic, the CBD defines the best solutions to explain a non-expected behavior as the minimal solutions that allow to recover the consistency.

However, the solutions obtained by our incremental algorithm are far from uninteresting. To illustrate this, let us consider another simplified representation of the AF447 situation:

$$Obs = \{\text{alarm}_t, \text{acceleration}_t\}$$

$$\mathcal{R} = \left\{ \begin{array}{l} R^a = \text{alarm}_t \rightarrow \text{stall}_t \\ R^b = \text{acceleration}_t \rightarrow \neg \text{stall}_t \\ R^c = \{\neg \text{stall}_t\} \text{Pull} \end{array} \right\}$$

$$a = \text{Pull}$$

In this example, the agent decides to pull the control stick when faced with two contradictory pieces of information. Algorithm 1 will compute the possible MCSes:

$\Delta^a = \{\text{alarm}_1\}$, $\Delta^b = \{R_1^a\}$, $\Delta^c = \{R_1^c, R_1^b\}$, $\Delta^d = \{R_1^c, \text{acceleration}_1\}$

With Algorithm 2, we obtain:

$\Delta^a = \{\text{alarm}_1\}$ $\Delta^b = \{R_1^a\}$ $\Delta^c = \{R_1^c, R_1^b\}$ $\Delta^d = \{R_1^c, \text{acceleration}_1\}$

$\Delta^e = \{\text{acceleration}_1, \text{alarm}_1\}$, $\Delta^f = \{\text{acceleration}_1, R_1^a\}$, $\Delta^g = \{R_1^b, R_1^a\}$,

$\Delta^h = \{R_1^b, \text{alarm}_1\}$

Algorithm 2 returns the same solutions as Algorithm 1 but also gives non minimal solutions $\Delta^e$ to $\Delta^h$. These solutions explore belief revisions that go against the decision made by the agent. For example, in $\Delta^e$ and $\Delta^f$, the agent does not take into account the acceleration information, even though it is consistent with their decision to pull the control stick.

In other words, while it is not necessary to ignore the acceleration to restore consistency, since it does not ultimately contradict the agent's decision, these corrections come from the fact that the agent had to manage an inconsistency during the revision phase. Algorithm 1 cannot explore such corrections: the revision of belief chosen by the agent is always consistent with their decision. On

---

[4] https://gitlab.dsi.universite-paris-saclay.fr/valentin.fouillard/incrementalcompleteness.

the contrary, Algorithm 2 explores more complex behaviors where for example the agent prefers one piece of information rather than another while ignoring reasoning rules allowing him to use this information correctly. For example, we can consider that $\Delta^f$ means that the agent pays attention to the alarm instead of the acceleration, but considers that the alarm is faulty and that it does not indicate a stall.

# 8    Related Work

Finding explanations for a situation through logic modeling, namely diagnosis, has thrived since the 1980s [23]. However to our knowledge, none of the proposed models are applied in the context of erroneous human decision-making. While several works proposed solutions for diagnosing dynamic systems (*i.e.* taking actions and changes into account) [17,27], all assume that the solution must comply with the frame problem's inertia. However, unlike logic-based models, human beings sometime forget information, which conflicts with the frame inertia, namely, a *frame distortion*. One of our contributions is to take into account such distortions in the logical model when diagnosing human errors.

Research in AI has attempted to model human reasoning errors or, more generally, human reasoning limitations, for predictive purposes in simulation. For instance, [29] uses a finite state automaton to simulate opinion dynamics regarding vaccination. Their model supports the decision of non-vaccination even when the rational information should lead the agent to accept it. In a different context, [2] uses the BDI paradigm to implement probabilistic functions that lead to erroneous beliefs in reaction to bushfires. All these models propose valuable solutions to simulate human decisions, but they cannot be used for diagnosis purposes in general cases.

Another approach for capturing false beliefs and human reasoning errors is to get rid of *logical omniscience*, *i.e.* the capacity to infer all the consequences of a belief $\varphi$. For example, [26] proposes a framework based on the *impossible world* (*i.e.* worlds that are not closed under logical consequences) to simulate reasoning errors. They associate *resources consumption* to each reasoning rule, which limits the applicability of lengthy inferences. However, the computation of all *impossible* worlds to select the most plausible one requires exponential computation power. Moreover, their model does not consider actions and changes.

All these approaches give interesting, yet partial, solutions to our problem: they neither handle the frame distortion problem, nor do they work for diagnosis purposes. Our framework combines these ideas to model erroneous decision making and to compute diagnosis that take human errors into account.

# 9    Conclusion and Perspectives

We proposed an *incremental consistency-based diagnosis* to compute belief states that could explain erroneous decision making of a human operator. This Algorithm is based on the computation of Minimal Correction Sets. The resulting belief states are consistent with the observations and actions performed by the operator and take into account four kinds of human errors: information preference, memory, attention and reasoning errors. While this paper presented the algorithm on one single time step, it has been implemented and works with several successive states, thus building a tree of all possible successive belief states of the agent.

In its current version, our model computes all possible scenarios, but it does not identify the most "plausible" ones. For example, the complete model of the Rio-Paris crash returns over 9000 scenarios, which is overwhelming for a human expert. To address this limitation, we propose filtering this set to extract classical human errors, identified in the literature as cognitive biases [28]. To this goal, [8] proposed some logic-based patterns to identify cognitive biases in accident scenarios. We propose to include such patterns in our model and extend them to capture other cognitive biases, so as to reduce the set of possible scenarios for the experts.

> **Abbreviation definitions**
>
> - CBD: Consistency Based diagnosis, a logical framework that restores consistency (hence the name) between the description of the system's expected behavior and the observations of the system's behavior (see Sect. 3.3 for a complete definition).
> - MCS: A set of minimal corrections to be removed from a system to restore coherence (see Sect. 4.3 for a complete definition).

# References

1. Alchourrón, C.E., Gärdenfors, P., Makinson, D.: On the logic of theory change: partial meet contraction and revision functions. J. Symb. Log. **50**(2), 510–530 (1985)
2. Arnaud, M., Adam, C., Dugdale, J.: The role of cognitive biases in reactions to bushfires. In: ISCRAM, Albi, France (2017)
3. Boutilier, C., Beche, V.: Abduction as belief revision. Artif. Intell. **77**(1), 43–94 (1995)
4. Buchanan, B., Shortliffe, E.: Rule-based expert system - the MYCIN experiments of the stanford heuristic programming project (1984)
5. Cisler, J.M., Koster, E.H.: Mechanisms of attentional biases towards threat in anxiety disorders: an integrative review. Clin. Psychol. Rev. **30**(2), 203–216 (2010)
6. Dupin de Saint-Cyr, F., Lang, J.: Belief extrapolation (or how to reason about observations and unpredicted change). Artif. Intell. **175**(2), 760–790 (2011)
7. Dutilh Novaes, C., Veluwenkamp, H.: Reasoning biases, non-monotonic logics and belief revision. Theoria **83** (2016)
8. Fouillard, V., Sabouret, N., Taha, S., Boulanger, F.: Catching cognitive biases in an erroneous decision making process. In: IEEE International Conference on Systems, Man and Cybernetics (SMC) (2021)

9. Hanks, S., McDermott, D.: Nonmonotonic logic and temporal projection. Artif. Intell. **33**(3), 379–412 (1987)
10. Hollnagel, E.: Cognitive Reliability and Error Analysis Method (CREAM). Elsevier, Amsterdam (1998)
11. Kaplan, R.L., Van Damme, I., Levine, L.J., Loftus, E.F.: Emotion and false memory. Emot. Rev. **8**(1), 8–13 (2016)
12. Katsuno, H., Mendelzon, A.O.: On the difference between updating a knowledge base and revising it. In: Cambridge Tracts in Theoretical Computer Science, pp. 183–203. Cambridge University Press (1992)
13. Liberatore, P., Schaerf, M.: Reducing belief revision to circumscription (and vice versa). Artif. Intell. **93**(1–2), 261–296 (1997)
14. Liffiton, M.H., Sakallah, K.A.: Algorithms for computing minimal unsatisfiable subsets of constraints. J. Autom. Reason. **40**(1), 1–33 (2008)
15. McCarthy, J.: Applications of circumscription to formalizing common-sense knowledge. Artif. Intell. **28**(1), 89–116 (1986)
16. McCarthy, J., Hayes, P.J.: Some philosophical problems from the standpoint of artificial intelligence. Mach. Intell. 463–502 (1969)
17. McIlraith, S.A.: Explanatory diagnosis: conjecturing actions to explain observations. In: Levesque, H.J., Pirri, F. (eds.) Logical Foundations for Cognitive Agents. Artificial Intelligence, pp. 155–172. Springer, Berlin (1999). https://doi.org/10.1007/978-3-642-60211-5_13
18. Morgenstern, L.: The problem with solutions to the frame problem. In: The Robot's Dilemma Revisited: The Frame Problem in Artificial Intelligence, pp. pp. 99–133. Ablex Publishing Co., Norwood (1996)
19. Murata, A., Nakamura, T., Karwowski, W.: Influence of cognitive biases in distorting decision making and leading to critical unfavorable incidents. Safety **1**(1), 44–58 (2015)
20. Nickerson, R.S.: Confirmation bias: a ubiquitous phenomenon in many guises. Rev. Gen. Psychol. **2**(2), 175–220 (1998)
21. Paul, G.: Approaches to abductive reasoning: an overview. Artif. Intell. Rev. **7**(2), 109–152 (1993)
22. Poole, D.: Representing diagnosis knowledge. Ann. Math. Artif. Intell. **11**(1), 33–50 (1994)
23. Reiter, R.: A theory of diagnosis from first principles. Artif. Intell. **32**(1), 57–95 (1987)
24. Reiter, R.: The frame problem in the situation calculus: a simple solution (sometimes) and a completeness result for goal regression. In: Artificial and Mathematical Theory of Computation, pp. 359–380. Citeseer (1991)
25. Simon, H.A.: Bounded rationality. In: Eatwell, J., Milgate, M., Newman, P. (eds.) Utility and Probability, pp. 15–18. Springer, Heidelberg (1990). https://doi.org/10.1007/978-1-349-20568-4_5
26. Solaki, A., Berto, F., Smets, S.: The logic of fast and slow thinking. Erkenntnis **86**(3), 733–762 (2021)
27. Thielscher, M.: A theory of dynamic diagnosis. Electron. Trans. Artif. Intell. **1**(4), 73–104 (1997)
28. Tversky, A., Kahneman, D.: Judgment under uncertainty: heuristics and biases. Science **185**(4157), 1124–1131 (1974)
29. Voinson, M., Billiard, S., Alvergne, A.: Beyond rational decision-making: modelling the influence of cognitive biases on the dynamics of vaccination coverage. PloS One **10**(11) (2015)

30. Wassermann, R.: An algorithm for belief revision. In: Proceedings of the Seventh International Conference on Principles of Knowledge Representation and Reasoning, pp. 345–352 (2000)
31. Wiegmann, D.A., Shappell, S.A.: Human error analysis of commercial aviation accidents using the human factors analysis and classification system (HFACS). Technical report, United States. Office of Aviation Medicine (2001)

# Relationship Between Theory of Mind and Judgement Based on Intention in 4–7 Y.O. Children

Véronique Salvano-Pardieu[✉] and Valérie Pennequin

PAVEA Université de Tours, 3 rue des Tanneurs, 37000 Tours, France
{veronique.pardieu,valerie.pennequin}@univ-tours.fr

**Abstract.** In this research we compare the ability to take into account the intent in moral judgement with the level of Theory of Mind (ToM) in Typically Developing children (4 to 7 y.o.). In the first test we evaluated the moral algebra: combination of intent and consequence. Children had to judge social interactions between two characters in four situations: intentional harm, attempted harm, accidental harm and no harm and for two levels of aggressiveness: low and high. In the second test the ToM of the participants was measured with the scale of Wellman & Liu (2004). They had to answer 7 questions measuring their ability to take the perspective of a character. The results show an effect of ToM on the ability to take into account the intention. Children with the best level of ToM were the most able to judge social interaction according to the intent of the character. In addition, this effect increases with age. The oldest children, who usually have the best scores in ToM take more intent into account and discriminate more accurately levels of aggressiveness.

**Keywords:** Moral Judgement · Theory of Mind · Children's cognitive development

## 1 Introduction

### 1.1 Moral Judgement

Moral judgment is the ability to judge what is considered "good" or "bad" according to the values of a social group. Piaget (1932) and later Kohlberg (1964), Kohlberg & Kramer (1969) studied moral judgment using semi-structured interviews. This method was mainly based on verbal explanations following long fictional stories or dilemmas. A more recent approach: Anderson's Theory of Cognition (1981, 1996, 2014) measures judgment on an intensity scale by weighing factors such as intention and consequence. This method uses short stories describing everyday life situations and for this reason avoids shortcomings of the semi-structured interviews and fictive dilemmas used in moral judgment experiments (Hommers and Lee 2010, Hommers, Lewand and Ehrmann 2012). The advantage of Anderson's method also lies in its ability to explain the evolution in the consideration of intention and consequence in moral judgment. Indeed, Anderson (1981, 1996, 2014) defines moral judgment as a decision-making mechanism determined by two factors: intention of the actor and consequence of the action. In social interactions, and especially in judgment of blame, a person judges a situation by taking into account the presence or absence of harmful intention and the presence or absence of consequences to the action. The importance given to each factor leads to a combination called moral algebra. If we use two factors such as "bad intention" and "negative consequence" and two modalities such as "with" and "without", we obtain four possibilities: 1- with a bad intention – with a negative consequence ("intentional harm"), 2- with bad intention – without negative consequence ("attempted harm"), 3- without bad intention – with negative consequence ("accidental harm") and 4- without bad intention – without negative consequence ("absence of harm"). The weight given to the "Intent" and "Consequence" factors are assessed globally for each situation. In Anderson's functional theory of cognition, each participant must choose on an intensity scale, the level of blame they want to assign to each combination of intention and consequence (i.e. to each of the four situations presented above). When the intention and consequence factors are taken into account but evaluated independently, that is, when there is no interaction between these two factors, moral algebra is additive. This additive moral algebra is mainly observed in children (Przygotzki and Mullet, 1997, Salvano-Pardieu et al 2016). On the other hand, when an interaction is observed between the "intention" and "consequence" factors and therefore, when the weight of one of the factors is modulated by the other, moral algebra is multiplicative.

This multiplicative moral algebra is mostly seen in teenagers and adults. Przygotzki & Mullet (1997) showed that children, adolescents and adults use the same rule of moral judgment by weighing the intention and consequence factors, but that the weight given to each factor changes with age. Children generally place more importance on consequence than on intent, unlike adolescents and adults. Several studies on the moral judgment of children confirm this result (Killen Mulvey, Richardson, Jampol and Woodward, 2011; Rogé & Mullet, 2011; Cushman 2013; Salvano-Pardieu et al. 2016; Salvano-Pardieu, Olivrie, Pennequin & Pulford 2020). By referring to Anderson's theory and using his method to evaluate moral algebra, we can analyze precisely how children of different ages evaluate intention and consequence and how the weight given to each factor changes. Differences should therefore appear in the moral algebra of children aged 4 to 7, particularly in relation to the weighing of intention.

When one blames an action, one refers to known social rules by assessing whether these rules were followed or not, whether the action was deliberate or accidental and by evaluating the importance of its consequence. To decide whether the action is deliberate or not, one must be able to understand the intention of the actor, i.e., to take into account his point of view. This ability to understand the other's point of view, to "take their perspective" and to attribute thoughts, beliefs and emotions to them is based on the theory of mind, defined as the ability to infer to oneself and others mental states and therefore take into account at the cognitive level, the point of view of another. Several authors (Coricelli, 2005; Duval et al, 2011; Hynes, Baird and Grafton, 2006; Kalbe, Schlegel, Sack, Nowak, Dafotakis, Bangard, Brand, Shamay-Tsoory, Onur, and Kessler, 2010) distinguish two aspects of theory of mind: Cognitive theory of mind (ability to infer mental states: thoughts, beliefs, reasoning, to oneself and others) and affective theory of mind (ability to infer affective and emotional states in others and to understand their emotion). These two aspects of the theory of mind seem to have different functions corresponding to different neurological structures. Indeed, research in functional neuro-imaging (Hynes et al., 2006; Kalbe et al., 2010) distinguishes at the level of the orbito-frontal lobe an involvement of the dorso-lateral prefrontal cortex in cognitive theory of mind and an involvement of the ventromedial prefrontal cortex in affective theory of mind. Theory of mind, in general, would help understand our own mental states and explain our own behaviors. But it would also make it possible to infer mental states in others, which would make it possible to predict and explain their behavior (Melot, 1999). This cognitive skill would be at the origin of social adaptation (Deneault and Morin, 2007) and would be essential in the development of social skills (Tourette, Recordon, Barbe and Soares-Boucaud, 2000). Barisnikov, Van der Linder and Detraux (2002) establish a link between the development of theory of mind and social adaptation in individuals.

Several authors have been interested in the development of the theory of mind and many agree that this cognitive capacity emerges around the age of 4–5 y.o., sometimes earlier, but continues to develop thanks to social interactions during adolescence and early adulthood (Hughes & Leekam, 2004; Gweon et al, 2012; Miller & Marcovitch, (2012). Miller, 2012; Valle, Massaro, Castelli and Marchetti, 2015). Wellman (2002) proposes a hierarchy in the different stages of the theory of mind from the understanding, around 2–3 y.o., that people, have different desires and different beliefs to the understanding, around 6–7 y.o., that people can hide their emotion.

## 1.2 Theory of Mind

For Wellman (2002) this skill, which he calls the "first theory of the mind", would only make it possible to conceive mental states as representations of reality. He believes that for young children aged 2–3 years, perceived mental states are necessarily real and cannot be the result of false beliefs, their cognitive system not being developed enough to impute false beliefs to others. For Wimmer and Perner (1983), Astington, & Jenkins (1999); Astington (2003); Wellman (2002), it was necessary to wait until the child was 4–5 y.o. for him to understand that the beliefs, his own or those of others, are not necessarily true. For these authors, it is from this age that the child understands that mental states, beliefs or emotions can be built on false representations, false beliefs or wrong information. He also understands that a false belief can be at the origin of the behavior of others. Wellman and Liu (2004) propose a scale to assess the level of 1st order theory of mind according to child's development.

The first order theory of mind is the ability to infer another person's mental state, i.e., what a person thinks, believes or feels in a specific situation. For example: Mary thinks that Peter will choose the chocolate cake and not the lemon pie because she knows that Peter likes chocolate cake. First order ToM develops early in childhood and is acquired around 9 y.o. Later, from 8–9 y.o. to adulthood, the second order ToM develops (Perner and Wimmer, 1985; Duval et al., 2011; Fu et al., 2014). The second order ToM is the ability to infer a person's mental state with another person's mental state. For example: Mary believes that Peter believes that... This second order ToM is not studied in this research, we focus on the first order ToM, measured with the Wellman and Liu's scale.

In this scale, several skills of the first order ToM are evaluated from 3 y.o. to 9 y.o.

1) The diversity of desires: the child understands that two people can have different desires. 2) The diversity of beliefs: the child understands that two people can have different knowledge and beliefs on the same subject. 3) Access to knowledge, access to information: the child sees what is in a box and understands that another person who does not see what is in the box does not have the same knowledge as him. He understands that not everyone has access to the same information. 4) False beliefs: the child can predict what another person thinks even if it is a false belief, i.e., a belief that does not correspond to reality. 5) Explicit false beliefs: the child can predict the behavior of another person based on what they believe even if the child knows that this belief is false. 6) Belief emotion: The child is able to understand what a person feels according to his beliefs and what he imagines, even if his beliefs are false. 7) The hidden emotion: The child understands that a person can feel an emotion but apparently show a different emotion.

Several studies link the development of theory of mind to moral judgment. Indeed, the skills in theory of mind allowing to infer the point of view of the other, seem essential to the understanding and the judgment of the intention of the actor in the judgment of blame (Rogé and Mullet, 2011; Buon et al., 2016; Salvano-Pardieu et al., 2016; Valle et al., 2015). Buon et al., (2016) with reference to Kahneman's dual system of thought (2011), develop a model presenting a dual system of moral judgment. These authors hypothesize that the judgment of blame is governed by two subsystems, one evaluating the intention of the actor, the other the consequence of the action. This "ETIC" model would involve 3 components: Emotion, Theory of Mind and Inhibitory Control. Emotion would be involved in the action evaluation system and would increase according to the severity of the consequence, especially during intentional or accidental harm. The theory of mind would be involved in the system of evaluation of intention observed in situations of intentional harm and attempted harm. In the attempted harm situation, the negative consequence being absent, only theory of mind would be involved in the attribution of blame.

Finally, in the event of accidental harm, inhibitory control, if sufficiently developed, would make it possible to reduce the impact of the emotion linked to the severity of the consequence and to focus the judgment on the intention of the actor, involving the theory of the mind. These authors therefore establish very clearly, a link between the ability to take intention into account in moral judgment and theory of mind, just like other authors before them (Baird and Astington, 2004; Killen et al., 2011). In their study, Killen et al. (2011), evaluate through a task of false belief and a task of moral judgment presenting an action of transgression and accidental harm, the link between the theory of mind and the consideration of intention in the judgment of blame. Their results show that children with the lowest skills in the false belief task are those who blame accidental actions the most, unlike those with the highest scores in the false belief task. Indeed, children with the lowest theory of mind scores judge the action on the presence of a negative consequence and not on the neutral intention of the character and for this reason blame this action more. Conversely, children with higher scores in theory of mind take more account of neutral intention and blame the action less even if it leads to negative consequences. These results confirm the link between theory of mind and the ability to take into account the actor's intention in moral judgment. However, these studies often only use a false belief task and certain moral judgment situations (intentional harm or accidental harm) to assess the link between theory of mind and intention consideration.

The connection between the theory of mind and the 4 possibilities of moral judgment situations, as presented in Anderson's method, has not yet been established experimentally. This study aims to meet this objective and to highlight that the increase in the consideration of intention in the judgment of blame can be explained by an increase in skills in theory of mind. We hypothesize a positive correlation between taking intention into account in the judgment of blame and performance in theory of mind. The choice of Anderson's method is justified by the evaluation of four judgment situations allowing the comparison of deliberate negative actions with or without a consequence, with accidental actions with or without a consequence. This difference between deliberate and accidental actions allows us to assess, for each participant, the effect of intention. In addition, the analysis of the 4 judgment situations will allow us to observe whether the level of theory of mind has an effect on the algebraic structure of the judgment of blame. We hypothesize that moral algebra will evolve with the level of theory of mind, the higher the level of theory of mind, the more children will blame actions based on intention, even in the absence of negative consequences and the less they will blame accidental actions followed by negative consequences. Finally, in order to best measure the theory of mind, we chose not to rely solely on the notion of false beliefs as may have been the case in previous studies (Wimmer and Perner, 1983; Flavell et al., 1990; Flavell 1999; Killen et al., 2011), but to use the scale of Wellmann and Liu (2004). Indeed, this scale covers a wider range of concepts from the theory of mind (various desires, diverse beliefs, access to knowledge, etc.) and makes it possible to assess progress, the various questions asked being, for the authors, of increasing difficulty.

In addition, this scale is also one of the most suitable in terms of age since it covers the development of children from 3 to 8 y.o. and is administered fairly quickly in about fifteen minutes. The use of this scale also allowed us to consider a more precise analysis of the relationship between the theory of mind and the consideration of intention in the judgment of blame. Indeed, among the 7 questions of the Wellman and Liu scale, some evaluate the cognitive theory of mind (Diverse Desires, Diverse Beliefs; Access to Belief; Content False Belief and Explicit False Belief) while others assess affective theory of mind (Belief Emotion and Hidden Emotion).

## 2 Method

### 2.1 Participants

The children participating in our study were enrolled in preschool and primary schools in the Touraine conurbation. These schools were located either in urban areas or in rural areas, from a middle social background. 105 children (47 boys and 58 girls), aged 4 to 7 y. o., with typical development, participated in this experiment (M = 5.4 y.o.; SD = 10.4 m.o.). These children were divided into three groups according to their age and class level. Two groups were in preschool and one group in reception year in primary school.

The first group was a preschool group (M = 4.4 y.o.; SD = 3.2 m.o.). This group consisted of 35 children, 20 girls (M = 4.4 y.o.; SD = 3.0 m.o.) and 15 boys (M = 4.3 y.o.; SD = 3.5 m.o.).

The second group was a preschool group too (M = 5.3 y.o.; SD = 3.3 m.o.). This group consisted of 35 children: 18 girls (M = 5.2 y.o.; SD: 2.8 m.o.) and 17 boys (M = 5.4 y.o.; SD = 3.8 m.o.).

Finally, the last group was a reception year group in primary school, (M = 6.4 y.o.; SD = 4.2 m.o.). This group of 35 children consisted of 21 girls (M = 6.3 y.o.; SD = 4.3 m.o.) and 14 boys (M = 6.5 y.o.; SD = 4.0 m.o.).

All participants were French with French mother tongue and gave their consent with the written consent of their parents to participate in this experiment. This agreement was based on the confidentiality and anonymity of the data. All participants were volunteers receiving no payment. All the children in this study participated in this experiment at school with the agreement of their teachers and the school principal.

## 2.2  Material

### Material Used to Assess Moral Algebra in Blame Judgement

The material consisted of eight short stories describing in a few lines simple and concrete social interactions between two characters. Each story contained the following information: (a) the main character's negative intention (with or without); (b) the negative consequence of the action (with or without); (c) the level of aggressiveness of the action (low: pushing, high: punching).

These eight stories presented in appendix 1 were obtained by orthogonally crossing these three factors and their modalities. Four situations were presented for each level of aggressiveness: intentional harm (with negative intention and with negative consequence), attempted harm: (with negative intention but without consequence); accidental harm (without negative intention but with a negative consequence) and absence of harm: (without negative intention and without consequence). Intentional harm and attempted harm are deliberate actions with a negative intention, while accidental harm and no harm are accidental actions without negative intention.

In the stories, the protagonists are friends when the action happens accidentally but do not like each other, since there is an aggressor and a victim, when the action happens deliberately. The result of a pre-experiment with some children having shown that they had difficulty understanding situations of accidental harm and especially attempted harm, these two situations have been made explicit in the text to avoid any possible ambiguity on the intention of the main character. In addition, accidental and deliberate actions were clearly presented in the scenarios to help children better understand them.

The low level of aggression corresponded to the story "push". The push was accidental between two friends or deliberate between two students who did not like each other. The consequence was a fall, with or without knee pain.

Example of the "push" story in the "intentional harm" situation.

Marie and Fanny play together in the playground. Claire comes to join them but Fanny doesn't like Claire and isn't happy. Fanny pushes Claire. Claire falls to the ground and hurts her knee. How do you blame Fanny?

The high level of aggression matched the "punch" story. The punch was accidental between two friends playing punching bag or intentional between two boys not liking each other, one punching the other. The consequence present or not was a broken nose.

Example of the "punch" story for the "attempted harm" situation.

Philippe does not like Franck. During recess, Philippe in a bad mood argues with Franck. He tries to punch him in the face but Franck turns his head at the last moment. Franck is not injured. How do you blame Philippe?

For each story, children had to blame the main character by choosing the intensity of the blame on a scale ranging from "0" no blame to "16" highest blame. The children had to give their answer by putting a cross on a scale of 16 color levels ranging from green to yellow, orange and red. Green meant no blame or weak blame, yellow meant medium blame, orange meant strong blame, and red meant very strong blame.

The experimenter explained to the children how to use this scale. It also explained to children the intensity of blame based on their experiences at school or at home when they were scolded or punished for doing something wrong. It was specified to each child that he had to answer only on the intensity of the blame he wanted to give. The experimenter ensured that the child understood how to use the scale by asking him to use this scale with two stories presented during the familiarization phase and asking him to justify his answer, before starting the experiment.

**Material Used to Assess the Level of Theory of Mind**

To evaluate the Theory of Mind, we used Wellman and Liu's scale of theory of mind (2004) which we translated into French. This scale presents a series of 7 questions, each of which calls upon a capacity of the theory of mind. To illustrate each question, we have used pictures or toys figures.

Question 1: "Diverse Desires" (DD) assesses the ability to understand that we have different desires ones from another. Question 2: "Divers Beliefs" (DB), assesses the ability to understand that we can have different beliefs. Question 3: "Knowledge Access" (KA) assesses the ability to understand that we do not all have the same level of knowledge. Question 4: "Content False Beliefs" (CFB) and question 5 "Explicit False Belief" (EFB) evaluate false beliefs, i.e., the ability to understand that one can have false beliefs either in the contents of a box or container (CFB), or on the place where a person is looking for something (EFB). Question 6: " Belief Emotion" (BE) assesses the ability to understand a person's emotion according to her mistaken belief. Finally, question 7: "Real-Apparent Emotion" (RAE) assesses the ability to understand that we can hide our emotions and display an emotion that does not correspond to our real emotion, we can lie about our emotions.

The use of this scale had the advantage of presenting concrete material (boxes, figurines, images or drawings, etc.) which made it possible to eliminate the biases due to difficulties of abstraction that 4-year-old children may encounter.

## 2.3  Procedure

The 2 tests presented to the children: judgment of blame and theory of mind were counterbalanced among the participants to avoid a test order effect. The test was carried out individually with the two experimenters in parallel and took place over several days for all the participants.

**Test of Judgement of Blame**

In this test, each participant completed the questionnaire individually with an experimenter who explained the instructions, starting with the familiarization phase during which he read two short stories, then asked the child to blame the character of the story by ticking an intensity level on the scale. During the experimental phase, the 8 stories were presented in a booklet, in a random order for each participant in order to avoid any order effect. The experimenter read each story aloud to the child before the child assigned blame to the character in the story by placing a cross on the blame intensity scale. During the experimental phase, participants were not allowed to compare their answers or to go back or change previous assessments. Each participant's results were converted into a numerical value expressing the distance (measured in cm) between their mark on the response scale and the left anchor "0". No participant consistently answered "0: no blame" or "16: very severe blame". They used the full scale from minimum "0" to maximum "16" and no floor or ceiling effect was observed. Finally, the children had no time limit to answer, but this test lasted on average about ten minutes.

**Test of ToM**

The Theory of Mind test contains 7 questions, all rated 0 or 1. Children got 1 point if they answered from the point of view of the character in the story and if they were able to answer according to the characters' desires, beliefs or emotions and not from their own desires, beliefs, feelings or emotions. The maximum score was 7 points. This test was carried out individually and, lasted about fifteen minutes.

# 3   Results

The analysis of these results will be presented in three parts. In a first part we will evaluate, by an analysis of variance, the effect of age on the judgment of blame and on the evolution of the moral algebra of children then we will analyze the effect of age on the scores in theory of mind. In a second part, the regression analysis will allow us to establish a link between the children's scores of ToM and their ability to judge social interaction according to the intention of the actor. Finally, in a third part, we will compare, with an analysis of variance, moral algebra of children according to their level of ToM (low or high).

## 3.1   First Part

**Effect of Age on Judgment of Blame**

A repeated measures ANOVA with 3 age groups (4 y.o., vs 5 y.o., vs 6 y.o.) × 2 Levels of Aggressiveness (story): (push vs punch) × 2 Intentions: (with vs without) × 2 Consequences: (with vs without) was performed on the entire sample of participants. In this ANOVA, the factor: Age Group was a between-subjects factor, while the factors: Level of Aggressiveness (story), Intention and Consequence were within-subjects factors. The size of the effect of each of these factors was estimated with a partial $\eta^2$.

## Main Effects of Factors: Age, Level of Aggressiveness (Stories), Intention and Consequence

### Effect of Age

On average, children aged 4 y.o., blame scenarios more severely (M = 11.6) than children aged 5 y.o., (M = 9.4) or 6 y.o., (M = 7.3). This difference is statistically significant [F (2, 102) = 26.71; p < 0.0001] ($\eta_p^2 = 0.34$).

### Effect of Level of Aggressiveness, Intention and Consequence

On average, the children blame the punch story (M = 10.08) more severely than the push story (M = 8.8), this difference is statistically significant [F (1, 102) = 22.81; P < 0.0001], ($\eta_p^2 = 0.18$). Unlike the children of 5 and 6 y.o., children of 4 y.o., do not discriminate between the two levels of aggressiveness and blame at the same intensity (11.6) both stories "push" and "punch". Finally, children blame deliberate negative actions (M = 11.2) more than accidental ones (M = 7.7), [F (1, 102) = 98.61; p < 0.0001] ($\eta_p^2 = 0.49$); they also blame actions with a negative consequence (M = 11.4) more than actions without a negative consequence (M = 7.4), [F(1, 102) = 163.06; p < 0.0001]; ($\eta_p^2 = 0.62$). Even if the size of the effects shows that children attach more importance to the consequence ($\eta_p^2 = 0.62$) than to the intention ($\eta_p^2 = 0.49$), these first results confirm that children, even younger, blame social interaction situations based on the intention of the actor and the consequence of the action.

### Interaction Effect: "Age group × Intention × Consequence"

As shown in Fig. 1, moral algebra, that is, how children combine intention and consequence factors, varies with age. For children of 4 y.o., there is very little difference between the four situations: "intentional harm", "attempted harm", accidental harm" and "no harm". On the other hand, for children aged 5 y.o. and 6 y.o., noticeable differences are observed for these different situations. This difference in moral algebra observed between the different age groups is statistically significant [F (2, 102) = 8.47; p < 0.001]. Although all children of all ages blame the "intentional harm" situation most severely and "no harm" the least, this effect increases with age.

A post hoc analysis carried out with a Fisher LSD test shows that unlike the youngest children who do not make any difference between the situations of "intentional harm", "attempted harm" and "accidental harm", the children of both other groups make a clear difference between these three situations. The oldest children (6 y.o.) blame "intentional harm" more than "attempted harm", p < 0.0001, but blame "attempted harm" and "accidental harm" at the same level. Unlike 6 y.o. children, 5 y.o. children blame "accidental harm" more than "attempted harm", p < 0.05. This result shows that they favor the consequence of the action rather than the intention of the actor.

### Theory of Mind Test Results

A factorial ANOVA with 3 age groups (4–5 y.o. vs 5–6 y.o. vs 6–7 y.o.) × Total ToM score, was performed on the entire sample of participants. In this ANOVA, Factor: Age

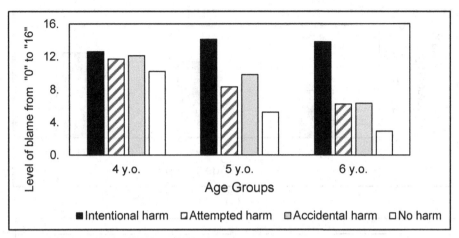

**Fig. 1.** Moral algebra for each age group according to the 4 situations: Intentional harm, Attempted harm, Accidental harm and No harm.

Group was a between-subjects factor, while Factor: Total ToM Score was a within-subjects factor. An ANOVA was also performed with the 3 age groups for each question of the scale.

**Effect of Age Group on Theory of Mind Score**
As we can see in Fig. 2, the overall ToM test score (Fig. 2A) as well as the score per question (Fig. 2B) is assessed according to the age of the children. The youngest children, 4 y.o., have the lowest ToM score while the oldest children have the highest score, with the highest score observed for the children's group of 6 y.o. This difference in performance in the ToM, according to the age of the children, is statistically significant [F (2, 102) = 20.39; p < 0.0001]. In addition, performance also evolves according to the questions in the scale. The two first questions: Divers Desires (DD) and Divers Beliefs (DB), present the best results, they are successful by more than 70% of children aged 4 and more than 85% of children aged 5 and 6. No significant difference in performance is observed for these questions between the 3 age groups. Although the performances are lower, around 60%, for the Explicit False Beliefs (EFB) question, we also did not observe significant differences in performance between the 3 age groups.

Conversely, for the other 4 questions on the scale: Knowledge Access (KA), Content False Belief (CFB), Belief Emotion (BE) and Real-Apparent Emotion (RAE), performance increases with age. The youngest do not exceed 30% of success in these questions while the oldest have performances between 50 and 90% depending on the questions. This difference in performance according to age is statistically significant for KA: [F (2; 102) = 20.11; p < 0.0001]; CFB [F(2;102) = 11.01; p < 0.0001]; BE [F (2; 102) = 15.07; p < 0.0001] and RAE [F (2; 102) = 4.39; p < 0.05].

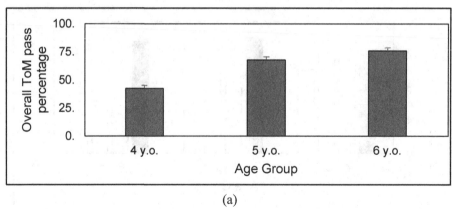

(a)

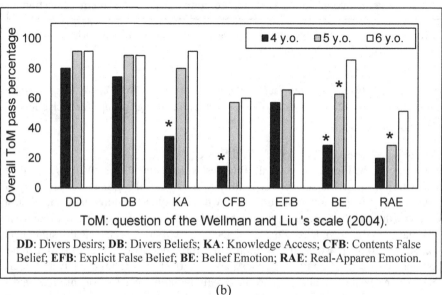

ToM: question of the Wellman and Liu 's scale (2004).

**DD**: Divers Desirs; **DB**: Divers Beliefs; **KA**: Knowledge Access; **CFB**: Contents False Belief; **EFB**: Explicit False Belief; **BE**: Belief Emotion; **RAE**: Real-Apparen Emotion.

(b)

**Fig. 2.** Overall theory of mind pass rate for each age group (A), and for each age group based on each question (B). Significant difference (* $p < .05$) is observed between the group of children of 4 y.o., and the two others age groups for the questions Knowledge Access (KA), Contents False Belief (CFB), and Belief Emotion (BE). A significant difference is observed between the group of children of 5 y.o., and the group of children of 6 y.o., for the two last questions: Belief Emotion (BE) and Real-Apparent Emotion (RAE), evaluating the emotional component of ToM.

## 3.2   Second Part

### Link Between Level of ToM and Consideration of the Actor's Intention in Social Interactions

To answer our hypothesis linking the child's skills in theory of mind and his consideration of the actor's intention in the judgment of blame, a regression analysis was carried out between the total score of theory of mind and the difference between the average score of blame for stories with bad intent (intentional harm and attempted harm) and the average score of blame for stories without bad intent (accidental harm and no harm). This difference of score between the two situations with bad intent and the two situations without bad intent gives the "intention value" which was calculated for each child. The correlation between the overall theory of mind score and the "intent value" is $r = 0.37$, $p < 0.0001$. This result shows that the higher the children's theory of mind scores, the more they blame based on intent. To refine this analysis, we carried out correlations between the "intent value" and each question of the scale of the theory of mind. We wanted to know on the one hand, if specific questions were correlated with the consideration of intention and on the other hand, if the two components of the scale (the questions relating to the cognitive component of ToM and the questions relating to the emotional component of ToM) were correlated with consideration of intention. Among the 7 questions of the scale, 3 present a significant correlation with the "intent value": question Knowledge Access (KA): $r = 0.27$; $p < 0.01$; question Contents false Belief (CFB): $r = 0.23$; $p < 0.05$ and question Belief Emotion (BE): $r = 0.32$; $p < 0.001$. These results show that taking intention into account is correlated with the cognitive and affective component of the theory of mind.

### 3.3  Third Part

**Effect of Theory of Mind Level on Children's Moral Algebra**
To be able to evaluate the effect of the theory of mind on the moral algebra of children aged 4 to 7, we no longer divided the children according to their age but according to their overall score in the theory of mind. We divided them into two groups: group 1 (low level), made up of 51 children with a score of ToM less than or equal to 4, and group 2 (high level) made up of 54 children with a score of ToM greater than 4. A repeated measures ANOVA with 2 groups of TOM score (G1: low level, G2: high level) × 2 levels of Aggressiveness of the action (story): push vs punch) × 2 Intentions: (with vs without) × 2 Consequences: (with vs without) was performed on the entire sample of participants. In this ANOVA, the factor: Group was a between-subjects factor, while the factors: Level of Aggressiveness (story), Intention and Consequence were within-subjects factors. The size of the effect of each of these factors was estimated with a partial $\eta^2$.

**Global Analysis**
*ToM Score Effect*
On average, children with the lowest score on ToM blame more severely ($M = 10.6$) than children with the highest score ($M = 8.4$). This difference is statistically significant [$F (1,87) = 13.27$; $p < .0001$].

In addition, the statistically significant interaction: *"ToM score group × level of aggressiveness of the action"* shows that children with the lowest ToM score distinguish less the severity of the consequence between the "push" story ($M = 10.0$) and the "punch" story ($M = 10.7$) compared to children who have a high theory of mind score and who themselves differentiate more between the seriousness of a push ($M = 7, 4$) and a punch ($M = 9.2$).

### Interaction Effect: Group (ToM Score) × Intention

The statistically significant interaction between "Group" and "Intention" shows that children who have a high level of ToM take intention more into account in the judgment of blame than children who have a low level of ToM. Indeed, as we observe in Fig. 3, although children with a low level of theory of mind blame deliberate actions (M = 11.4) more than accidental actions (M = 9.2), this difference of 2.2 which refers to "intent value" remains low. By contrast, children with a high level of theory of mind blame actions with negative intention much more (M = 10.6) than accidental actions (M = 6.0) and this difference (4.6) or "intent value" is higher.

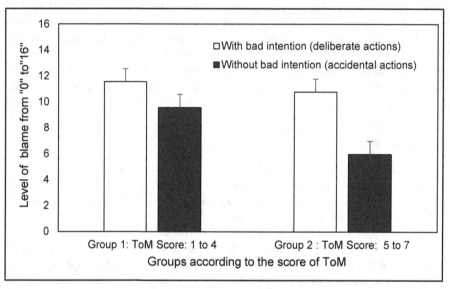

**Fig. 3.** Level of blame and standard error for the low and high score groups of ToM according to the level of intention of the actor (with or without bad intention).

### Interaction: Groups × Intention × Consequences

As we can see in Fig. 4, moral algebra varies across groups. While children in group 1 blame accidental harm more than attempted harm, children in group 2 blame both situations equally. Moreover, they blame the absence of harm less severely than the children in group 1. This difference is statistically significant [F (1, 103) = 6.39; p < 0.05]. A post-hoc analysis: Fisher's LSD, revealed that the difference between the two situations "attempted harm" and "accidental harm" was statistically significant for group 1 (low level of ToM); p < 0.01, but no significant difference was observed between these two situations for group 2 (high level of ToM). In addition, the moral algebra of the children with a low level of ToM is additive [F (1, 50) = 0.18; NS], while the moral algebra of those with a high level of ToM is multiplicative with a significant interaction between Intent and Consequences factors: [F (1, 53) = 11.55, p < 0.01].

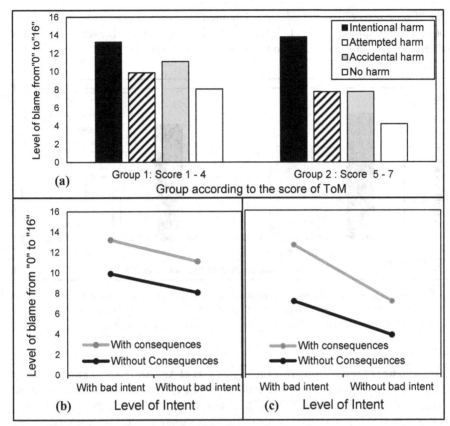

**Fig. 4.** Moral algebra of both groups of children according to their ToM score. In (A) the moral algebra is displayed according to the four situations of judgement. (B) presents the additive moral algebra for the group with a low-level score of ToM (no interaction between the two factors). (C) presents a multiplicative moral algebra for the group with a high-level score of ToM with an interaction between intent and consequences factors.

## Specific Analysis

A repeated measure ANOVA with 2 groups (G1: score 1 to 4 vs G2: score 5 to 7) × 2 Intentions: (with vs without) × 2 Consequences: (with vs without) was conducted on the entire sample of participants for each level of aggressivity (stories). In this ANOVA group was a between-subject factor, but Intent and Consequence were within-subjects factors. The effect size of each of these factors was estimated with $\eta^2$ partiel.

## Moral Algebra of Both Groups for the Low Level of Aggressiveness: Push Story

As shown in Fig. 5, different moral algebra is observed between both groups. While moral algebra of the group with a low score of ToM is additive, [$F(1, 50) = 3.05$; NS], moral algebra of the group with a high-score of ToM is multiplicative, [$F(1, 53) = 5.21$, $p < 0.05$]. There is also a variation in the weight given to the "Intent" and "Consequence" factors depending on the level of theory of mind. Children in group 1 give less weight to the "Consequence" ($\eta^2 = 0.32$) and "Intention" ($\eta^2 = 0.31$) factors than children in group 2 ("Consequence" ($\eta^2 = 0.55$) and "Intention" ($\eta^2 = 0.52$) However, children in both groups give as much weight to intention as to consequence.

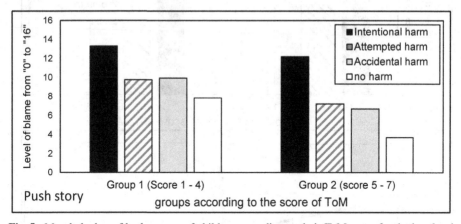

**Fig. 5.** Moral algebra of both groups of children according to their ToM score for the low level of aggressiveness: "Push story". The moral algebra is displayed according to the four situations of judgement: Intentional harm, attempted harm, accidental harm and no harm.

**Moral Algebra of Both Groups for the High Level of Aggressiveness: Punch History**
As shown in Fig. 6, moral algebra differs between the two groups. While group 1 moral algebra is additive, [$F(1, 50) = 0.46$; NS], the moral algebra of group 2 is multiplicative, [$F (1, 53) = 10.57$, $p < 0.01$]. We also observe a variation in the weight given to the "Intent" and "Consequence" factors, depending on the level of theory of mind. The children of group 2 (high score of ToM), give more weight to the "intention" ($\eta^2 = 0.62$) and "consequences" ($\eta^2 = 0.57$) factors than the children of group 1 (Low score of ToM). The latter give almost no weight to the intention ($\eta^2 = 0.06$) but give a much greater weight to the consequence ($\eta^2 = 0.34$) and blame more severely accidental harm than attempted harm, unlike the children of group 2 who have a higher level of ToM.

## 4  Discussion

In this experiment, we wanted to study the effect of the development of theory of mind on the ability to take intention into account in the judgment of blame. We hypothesized that skills in theory of mind would favor in moral judgment the choice to blame the intention of the actor rather than the consequence of the action. This relationship would also develop with age, the oldest children would be the most competent in theory of mind and those who would give the strongest weight to intention in the judgment of blame.

### 4.1  Effect of Age

On average, children give more weight to the consequence of the action than to the intention of the actor, which confirms the results already observed in the literature (Killen

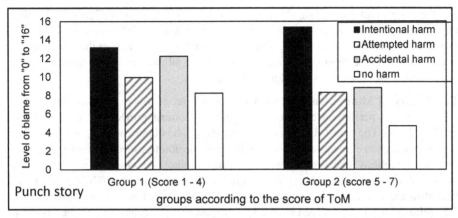

**Fig. 6.** Moral algebra of both group of children according to their ToM score for the high level of aggressiveness: "Punch story". The moral algebra is displayed according to the four situations of judgement: Intentional harm, attempted harm, accidental harm and no harm.

Mulvey, Richardson, Jampol and Woodward, 2011; Rogé & Mullet, 2011; Cushman, 2013; Salvano-Pardieu, Olivrie, Pennequin & Pulford 2020), however older children take intention more into account than younger ones. Indeed, children aged 4 blame deliberate and accidental actions at the same level, while older children (5 and 6 y.o.) blame deliberate actions more than accidental actions. This effect increases with age. Contrary to the judgment of blame concerning intention, children aged 4 are able to differentiate between actions with and without consequences and blame actions with bad consequences more than actions without consequences.

Finally, the youngest children do not make any difference between the different levels of aggressiveness of the action and blame the "push" story as much as the "punch" story, while the older ones blame the "punch" story more than the "push" story. These results show that children before the age of 5 have difficulty in prioritizing the different levels of aggression.

**Moral Algebra According to Age**
The results of this experiment showed that moral algebra varied depending on the age of the children. While 4 y.o., place equal blame on intentional harm, attempted harm, and accidental harm, 5 and 6 y.o., place more blame on intentional harm. However, unlike 6 y.o. who blame both the "attempted harm" and "accidental harm" situation, 5 y.o. blame "accidental harm" more than "attempted harm". These results confirm previous results (Salvano-Pardieu et al. 2016; 2020) which show that the youngest children focus on the consequences more than on the actor's intention when it comes to judging social interactions. Finally, the analysis of the moral algebra of each age group for each level of aggression showed that the two youngest groups of children present, for each level of aggression, an additive algebra whereas the older children (6 y.o.), present multiplicative algebra. These results show that from the age of 6 y.o., children can modulate the weight of one factor in relation to another factor. In addition, our results show that the weight given to each "intention" and "consequence" factor increases with age. The older children

get, the more weight they assign to the "intention" and "consequence" factors. Finally, the two youngest groups of children assign a greater weight to the "consequences" factor than to the "intention" factor, while children aged 6 y.o., assign almost the same weight to each factor. These results show that advancing age allows children to take more and more account of intention in social interactions.

**The Theory of Mind According to Age** The results of this experiment showed an effect of age on performance in theory of mind and confirm previous results (Wellman and Liu, 2004). The oldest children (6 y.o.) are those who obtain the best theory of mind scores without however succeeding in obtaining all the points, especially for the question of hidden emotions (question of the Real-Apparent emotion) which is only succeeded by the half of them. This result shows that the affective component of ToM is mastered later than the cognitive component of ToM. This suggests that the two structures in the orbito-frontal lobe involved in the two components: the cognitive and the affective ToM are not developing in the same way: ventromedial prefrontal cortex involved in the affective component of ToM develops later than the dorso-lateral prefrontal cortex involved in the cognitive component of ToM.

This result also proves that children aged 6 have not yet mastered the first order theory of mind measured with the Wellman and Liu scale (2004). These results are in agreement with previous research which showed that the theory of mind, even if it appeared as early as 3 or 4 y.o. (Wimmer and Perner, 1983), developed gradually between 6 and 9 y.o. (Gweon et al., 2012) and continued to develop through social interactions during. Adolescence and early adulthood (Valle et al., 2015).

### 4.2   Moral Judgment and Theory of Mind

Comparing children with the highest theory of mind scores with those with the lowest scores, we found differences in judgment of blame and moral algebra. Children with the lowest theory of mind scores are the ones who, on average, blame actions the most severely in any situation. They perceive only weakly the difference between the two levels of aggressiveness of the action. Children with the lowest theory of mind scores are also those who least differentiate between deliberate and accidental actions when they have to blame these actions. Conversely, children with the highest theory of mind scores are the ones who differentiate between deliberate and accidental actions the most, blaming deliberate actions a lot more with a higher score for the intent value. These results, which confirm our hypothesis, show that the ability to take intention into account in the judgment of blame depends on the level of theory of mind. A link is observed between these two components: the higher the level of ToM, the more moral judgement is based on intention.

The more precise analysis of the moral algebra of the two groups of children confirms these first results. Indeed, children with the lowest theory of mind score are also those who blame accidental harm more severely than attempted harm, and this all the more so when the level of aggressiveness of the action is high. These results show that the children in this group rely mainly on the consequence of the action to make a judgment of blame and that they are sensitive to the seriousness of the consequence. When the consequence is as serious as a broken nose, children with the lowest level of theory of mind give an average of five times more weight to the consequence than to the intention. These results confirm those already carried out on the development of moral judgment (Piaget 1932; Monks, Smith, Swettenham., 2005; Salvano-Pardieu et al., 2016; Salvano-Pardieu, Olivrie, Pennequin, Pulford., 2020). Children with a low theory of mind score have more difficulty putting themselves in another person's shoes to judge the intention that leads to action. However, as we have observed, they are able to judge the action according to the presence or absence of negative consequences which are objective facts. As suggested by Fontaine et al., (2004); Salvano-Pardieu et al., (2016), to judge the consequences, children would use deontic reasoning based on knowledge of social rules, what is prohibited or authorized, (Manktelow and Over, 1991; Manktelow 1999; 2012) mastered around 2–3 y.o. with entry to the preschool and socialization. Smetana (1985) showed that children from the age of 3 knew how to differentiate between different types of social rules, for example: not storing one's belongings correctly and hitting a child, an action considered much more serious by children. These results were confirmed by other research (Smetana 1985; Tisak 1993) which showed that nursery and preschool children punished acts resulting in physical harm more severely than property damage.

The children in our study, all over the age of 3, therefore had a sufficient level of development to use deontic reasoning and judge an action according to the consequence it entails.

Contrary to the children in the first group, the children with a high score in the theory of mind, blame the situations of "accidental harm" as much as the situations of "attempted harm", which shows that they attribute less weight to the consequence and more weight to the intention than the children of the first group. Indeed, the weight they give to the two factors "Intent" and "Consequences" is equivalent. These results confirm that the most advanced children in theory of mind have a greater ability to "think the thoughts of others, develop a better understanding of social interactions (Deneault and Morin, 2007), and better social skills (Tourette, Recordon, Beard and Soares-Boucaud., 2000). Our results, in agreement with previous studies (Barisnikov, Van der Linder and Detraux., 2002; Killen et al., 2011; Fu et al., 2014) show that a better understanding of the mental states of others and therefore a higher level of theory of mind leads to better interpretation of actor's intent and social interactions.

# 5 Conclusion

This research shows that there is a link between children's level of theory of mind and their ability to judge social interactions taking into account the actor's intention. The higher the level of theory of mind, the more intention is taken into account in the judgment of blame. The distinction into two groups according to the level of theory of mind showed that children with a more developed and efficient theory of mind better understand the intentions of others and take them into account to blame an action.

The second aspect highlighted in this research is the level of theory of mind according to age. Our results in agreement with many authors confirmed that the theory of mind develops with age. The youngest children being the ones with less theory of mind skills. The theory of mind develops with the cognitive development of the child. In our study, although progress was visible with age between 4 and 7 years, on average, the oldest children had not reached the maximum of the theory of mind scale, and have more difficulties to understand emotional component of ToM than the cognitive component which confirms that the development of these two aspects of ToM are not simultaneous, and that the first-order theory of mind continues to develop beyond 7 years.

Finally, by comparing in this study the evolution of moral algebra as a function of age with the evolution of the theory of mind, we were able to confirm that the development of moral judgment depended on the development of the theory of mind.

The study of the development of these cognitive skills allows us to better understand, in young children, errors in the analysis of social interactions. New research could be interested in the educational tools that we could put in place to help them develop their theory of mind in order to regulate their behaviors.

## Annexes

### Scenarios Used in the Moral Judgement Task

**Low level of aggressiveness** "Push Story":
**Intentional harm.** "With Bad Intent, with bad Consequence".
Two girls, Marie and Fanny are playing together on the playground. Claire comes to play with them, but Fanny doesn't like Claire. She is angry and pushes Claire violently. Claire falls and hurts her knee.

How much blame do you assign to Fanny?

**Attempted harm.** "With bad Intent, without bad Consequence".
Two girls, Anna and Helen are playing together in the playground. Julia comes to join them, but Helen doesn't like Julia. She is angry and pushes Julia violently. Julia falls but she is not injured.

How much blame do you assign to Helen?

**Accidental harm.** "Without bad Intent, with bad Consequence".
Three girls who are friends: Alison, Suzanne and Dorothy are playing together on the playground. They run everywhere and suddenly Suzanne runs into Alison. Alison falls and hurts her knee.

How much blame do you assign to Suzanne?

**No Harm.** "Without bad Intent, without bad consequence".
Three girls who are friends: Marianne, Lily and Nathalie are playing together in the playground. They run everywhere and suddenly Lily runs into Nathalie. Nathalie falls to the floor, but she's not injured.

How much blame do you assign to Lily?

**High level of aggressiveness** "Punch Story":
**Intentional harm.** "With Bad Intent, with bad Consequence".

Two boys: Peter and Yan are quarrelling on the school playground. Very irritated by Peter, Yan gives a punch in Yan's face. Yan has his nose broken.

How much blame do you assign to Yan?

**Attempted harm**. "With bad Intent, without bad Consequence".
Two boys: George and Paul are quarrelling on the school playground. Very irritated by Paul, George tries to punch Paul in the face, but he misses him because Paul moves his head. Paul was not hurt.

How much blame do you assign to George?

**Accidental harm**. "Without bad Intent, with bad Consequence".
Two boys: Elliott and Tony are good friends. One day while they are playing one on either side of a punching ball, Elliott misses the punching ball and hits Tony in his face. Tony's nose is broken.

How much blame do you assign to Elliott?

**No Harm**. "Without bad Intent, without bad consequence".
Two boys: Tom and Steve are good friends. One day while they are playing one on either side of a punching ball, Tom misses the punching ball and nearly punches Steve in his face. Fortunately, Steve moves his head. Steve was not hurt.

How much blame do you assign to Tom?

# References

Anderson, N.H.: Foundations of Integration Theory. Academic, New York (1981)

Anderson, N.H.: A Functional Theory of Cognition. L. Erlbaum Associates, Mahwah, N.J (1996)

Anderson, N.H.: Contributions to Information Integration Theory, vol. I: Cognition. Psychology Press (2014)

Astington, J.W., Jenkins, J.M.: A longitudinal study of the relation between language and theory-of-mind development. Dev. Psychol. **35**(5), 1311–1320 (1999)

Astington, J.W.: Sometimes necessary, never sufficient: false belief understanding and social competence. In: Repacholi, B., Slaughter, V. (eds.), Individual differences in theory of mind: Implications for typical and atypical development, pp. 13–38. Psychology Press, New York (2003)

Barisnikov, K., Van der Linden, M., Detraux, J.J.: Cognition sociale, troubles du comportement social et émotionnel chez les personnes présentant une déficience mentale. Dans G. Petitpierre (dirs), *Enrichir les compétences*. Edition SPC (2002)

Baird, J.A., Astington, J.W.: The role of mental state understanding in the development of moral cognition and moral action. New Dir. Child Adolesc. Dev. **2004**(103), 37–49 (2004)

Buon, M., Seara-Cardoso, A., Viding, E.: Why (and how) should we study the interplay between emotional arousal, Theory of Mind, and inhibitory control to understand moral cognition?. Psychono. Bull. Rev. **23**, 1660–1680 (2016)

Coricelli, G.: Two-levels of mental states attribution: from automaticity to voluntariness. Neuropsychologia **43**(2), 294–300 (2005)

Cushman, F.: Action, outcome and value: a dual-system framework for morality. Pers. Soc. Psychol. Rev. **17**(3), 273–292 (2013)

Deneault, J., Morin, P.: La Théorie de l'esprit: ce que l'enfant comprend de l'univers psychologique. In: Larivée, S. (ed.) L'intelligence. Tome 1. Les approches biocognitives, développementales et contemporaines. ERPI, Montréal (2007)

Duval, C., Piolino, P., Bejanin, A., Laisney, M., Eustache, F., Desgranges, B.L.: Théorie de l'esprit: aspects conceptuels, évaluation et effets de l'âge. Rev. Neuropsychol. **1**(1), 41–51 (2011)

Flavell, J.H.: Cognitive development: children's knowledge about the mind. Annu. Rev. Psychol. **50**, 21–45 (1999). https://doi.org/10.1146/annurev.psych.50.1.21

Flavell, J.H., Flavell, E.R., Green, F.L., Moses, L.J.: Young children's understanding of fact beliefs versus value beliefs. Child Dev. **61**, 915–928 (1990)

Fontaine, R., Salvano-Pardieu, V., Renoux, P., Pulford, B.: Judgement of blame in Alzheimer's disease sufferers. Aging, Neuropsychol. Cogn. **11**(4), 379–394 (2004). https://doi.org/10.1080/13825580490521313

Fu, G., Xiao, W.S., Killen, M., Lee, K.: Moral judgment and its relation to second-order theory of mind. Dev. Psychol. **50**(8), 2085–2092 (2014)

Gweon, H., Dodell-Feder, D., Bedny, M., Saxe, R.: Theory of mind performance in children correlates with functional specialization of a brain region for thinking about thoughts. Child Dev. **83**(6), 1853–1868 (2012)

Hommers, W., Lee, W.Y.: Unifying Kohlberg with information integration: the moral algebra of recompense and of Kohlbergian moral informers. Psicológica **31**, 689–706 (2010)

Hommers, W., Lewand, M., Ehrmann, D.: Testing the moral algebra of two Kohlbergian informers. Psicológica **33**(3), 515–532 (2012)

Hughes, C., Leekam, S.: What are the links between theory of mind and social Relations? Review, reflections and new directions for studies of typical and a typical development. Soc. Dev. **13**(4), 590–619 (2004)

Hynes, C.A., Baird, A.A., Grafton, S.T.: Differential role of the orbital frontal lobe in emotional versus cognitive perspective-taking. Neuropsychologia **44**, 374–383 (2006)

Kahneman, D.: Thinking, Fast and Slow. macmillan (2011)

Kalbe, E., et al.: Dissociating cognitive from affective theory of mind: a TMS study. Cortex **46**(6), 769–780 (2010)

Killen, M., Mulvey, K.L., Richardson, C., Jampol, N., Woodward, A.: The accidental transgressor: morally-relevant theory of mind. Cognition **119**, 197–215 (2011)

Kohlberg, L.: Development of moral character and moral ideology. In: Review of Child Development Research, pp. 383–432. Russell Sage Foundation, New York, NY, US (1964)

Kohlberg, L., Kramer, R.: Continuities and discontinuities in childhood and adult moral development. Hum. Dev. **12**(2), 3–120 (1969)

Melot, A.M.: Les représentations du fonctionnement mental chez l'enfant d'âge préscolaire. In: Netchine-Grynberg, G. (ed.), Développement et fonctionnements cognitifs. Vers une intégration. Presses Universitaires de France (1999)

Manktelow, K.: Thinking and Reasoning: An Introduction to the Psychology of Reason, Judgment and Decision Making. Psychology Press (2012)

Manktelow, K.: Reasoning and Thinking. Psychology Press, Hove (1999)

Manktelow, K.I., Over, D.E.: Social roles and utilities in reasoning with deontic conditionals. Cognition **39**, 85–105 (1991)

Miller, S.A.: Theory of Mind: Beyond the Preschool Years. Psychology Press (2012)

Miller, S.E., Marcovitch, S.: How theory of mind and executive function co-develop. Rev. Philos. Psychol. **3**(4), 597–625 (2012)

Monks, C.P., Smith, P.K., Swettenham, J.: Psychological correlates of peer victimisation in preschool: social cognitive skills, executive function and attachment profiles. Aggressive Behav.: Official J. Int. Soc. Res. Aggression **31**(6), 571–588 (2005)

Perner, J., Wimmer, H.: "John thinks that Mary thinks that …" Attribution of second-order beliefs by 5- to 10-year-old children. J. Exp. Child Psychol. **39**, 437–471 (1985)

Piaget, J.L.: Jugement moral chez l'enfant. PUF, Paris (1932)

Przygotzki, N., Mullet, E.: Moral judgment and aging. Revue Européenne de Psychologie Appliquée **47**, 15–21 (1997)

Rogé, B., Mullet, E.: Blame and forgiveness judgements among children, adolescents and adults with autism. Autism **15**(6), 702–712 (2011)

Salvano-Pardieu, V., et al.: Judgement of blame in teenagers with Asperger's syndrome. Think. Reason. **22**, 251–273 (2016)

Salvano-Pardieu, V., Olivrie, M., Pennequin, V., Pulford, B.: Does Role Playing Improve Moral Reasoning's Structures in Young Children? In: Yama, H., Salvano-Pardieu, V. (eds.) Adapting Human Thinking and Moral Reasoning in Contemporary Society, pp. 199–222. IGI Global (2020). https://doi.org/10.4018/978-1-7998-1811-3.ch009

Smetana, J.G.: Preschool children's conceptions of transgressions: effects of varying moral and conventional domain-related attributes. Dev. Psychol. **21**(1), 18–29 (1985)

Tisak, M.S.: Preschool children's judgments of moral and personal events involving physical harm and property damage. Merrill-Palmer Q. **39**(3), 375–390 (1993)

Tourette, C., Recordon, S., Barbe, V., Soares-Boucaud, I.: Attention conjointe préverbale et Théorie de l'esprit à 5 ans: la relation supposée entre ces deux capacités peut-elle être démontrée ? Etude exploratoire chez des enfants non autistes. Dans V. Gerardin-Collet & C. Riboni. *Autisme: perspectives actuelles.* L'Harmattan (2000)

Valle, A., Massaro, D., Castelli, I., Marchetti, A.: Theory of mind development in adolescence and early adulthood: the growing complexity of recursive thinking ability. Europe's J. Psychol. **11**(1), 112–124 (2015)

Wellman, H.M.: Understanding the psychological world: Developing a theory of mind. In: Goswami, U. (ed.) Blackwell handbook of childhood cognitive development, pp. 167–187. Wiley (2002). https://doi.org/10.1002/9780470996652.ch8

Wellman, H.M., Liu, D.: Scaling of theory-of-mind tasks. Child Dev. **75**(2), 523–541 (2004)

Wimmer, H., Pemer, J.: Beliefs about beliefs: representation and constraining function of wrong beliefs in young children's under-standing of deception. Cognition **13**, 103–128 (1983)

# Trust in Algorithmic Advice Increases with Task Complexity

Mohammed Ali Tahtali[1,2]([⊠]) [iD], Chris Snijders[1] [iD], and Corné Dirne[2] [iD]

[1] Eindhoven University of Technology, 5600 MB Eindhoven, The Netherlands
m.a.tahtali@tue.nl
[2] Fontys University of Applied Sciences, 5600 AH Eindhoven, The Netherlands

**Abstract.** The use of algorithms in decision-making has increased in various fields, such as medicine, government, and business. Despite their proven accuracy, people often disregard algorithmic advice. When it comes to complex tasks, however, there is some evidence that people are more inclined to follow the advice of algorithms. This evidence is largely based on decision-making in rather artificial contexts, however, and studies in this field tend to rely on rather crude measures of complexity. We therefore investigate the effect of task complexity on trust in model-based advice in a realistic setting, measuring complexity in several standardized ways. We conducted an experiment with 151 participants, each assessing 20 real-life court cases of crimes (in the Dutch legal system). Participants were first asked to estimate the jailtime for each crime. They then received algorithmic advice, and were allowed to adjust their initial estimate. We measured task complexity in several ways. First, by simply counting the number of violations per case. Second, we focused on all the mitigating and aggravating circumstances associated with a specific case. Finally, we narrowed our focus to only the circumstances mentioned in the case text and concentrated on the violated sections of the law. We then used multi-level regression analysis (assessments within participants) on the target variable Weight on Advice (WOA) to assess the impact of complexity on trust in algorithmic advice. Our findings indicate that participants were more inclined to trust algorithmic advice as the complexity of tasks increased, for two of the three operationalizations of task complexity.

**Keywords:** algorithmic advice · trust · task complexity

## 1 Introduction

The use of algorithms in decision-making processes is on the rise in various domains [1–3]. While algorithms have demonstrated their potential to enhance outcomes and often outperform human experts, there are factors that lead people to, at times, disregard algorithmic advice. These factors include concerns about accuracy [4], transparency [5, 6], performance [7], and a preference for human expertise [8]. When facing complex tasks, human performance suffers [9, 10] and individuals may seek advice from others, including algorithms [11, 12]. However, individuals who are overconfident or underestimate the complexity of the task are unlikely to accept advice until they recognize their inability to independently perform the task [13].

J. Baratgin et al. (Eds.): HAR 2023, LNCS 14522, pp. 86–106, 2024.
https://doi.org/10.1007/978-3-031-55245-8_6

The impact of task complexity on trust in algorithmic advice remains understudied and shows mixed results. While some studies suggest that individuals rely more on algorithmic advice as tasks become more difficult [11, 14], others indicate that greater task difficulty can result in decreased trust in the automated system (cf. [15, 16]). Two issues hamper earlier studies on the effects of task complexity on the willingness to accept algorithmic advice. First, studies tend to use abstract tasks that do not align well with real-world tasks involving algorithmic advice. Second, earlier studies have varied substantially with respect to the operationalization of (increasing) complexity. Defining what it means for a decision to be complex is not straightforward and includes distinguishing between difficulty and complexity, as well as considering the subjective and objective aspects of complexity [17, 18]. It is unclear whether effects of complexity depend on the kind of complexity under consideration.

We seek to provide a comprehensive analysis of task complexity and its impact on decision making in a realistic task domain. Our main hypothesis is that individuals are more likely to trust algorithmic advice as task complexity increases. To test this hypothesis, we conducted an experiment in which 151 participants assess the imposed jailtime for 1,812 crime cases from the Dutch justice system, with and without the help of an algorithm. By examining the adoption of algorithmic advice in real-world cases, our study contributes to the ongoing discourse on the role of algorithmic advice in decision-making and highlights the need for careful consideration of the way in which the complexity of an assessment is evaluated.

## 2   Previous Literature and the Complicated Definition of Complexity

Research by Bogerts et al. [11] found that people rely more on algorithms than on advice "from the crowd" as a task becomes more complex. Participants were shown (real-world) photographs and asked to guess how many individuals were depicted. Each participant saw ten photographs, five with few people (about 15 individuals) and five with many (up to around 5,000 individuals). Subjects received advice from either an algorithm or the guesses of 5,000 people. Participants showed, in general, more trust in algorithmic advice, but they revised their answers more strongly when they received advice from an algorithm for the more complex photographs. A study by Wang and Du [13] examined the effect of task difficulty on participants' confidence in their initial estimation. They showed participants six photographs of a glass filled with coins. In the high complexity condition, coins were blurred, and for low-complexity tasks, pictures were clear. Participants (in their study two) were more likely to accept advice, deviating from their initial estimate, in the high complexity condition than in the low complexity condition, indicating lower confidence in their estimates for higher complexity tasks. However, the role of trust in mediating the effect of task difficulty on advice-taking was small and insignificant.

Schwark et al. [14] investigated how reliance and compliance rates were influenced by perceived task difficulty and the importance of the task. Participants were instructed to look for X's in a random range of 300 alphabetic characters. Before each trial, the level of difficulty was reported on a screen. The difficulty level was randomly assigned, and all trials were equally complex. During the experiment, the algorithm would at times show a message that it had detected an 'X' (or that it had detected that there was no 'X'). Subjects could agree or disagree with that assessment. The compliance rate was measured as the number of times the participant agreed with algorithmic advice for detecting a target, and the reliance rate was measured as the number of times the participant agreed with advice for not detecting a target. Reliance rates did not vary across conditions, but compliance rates increased as the trials varied from very easy to moderately easy tasks, and to very hard tasks. These results illustrate that just manipulating the perceptions of task difficulty (not the actual difficulty) was enough to affect trust in algorithmic advice. We found several other studies on the effect of task complexity on trust in algorithmic advice, the results of which we summarize in Table 1.

Previous studies (cf. Table 1) that have focused on the effects of task complexity on trust in algorithmic advice, have used the term 'complexity' in a rather diverse manner. It is useful to briefly reiterate parts of existing accounts of task complexity, as for instance proposed in Wood [19], Campbell [20], Bonner [21], Harvey and Koubek [22], Ham et al. [17], and Liu and Li [18]. A first characteristic of these accounts of complexity is that they distinguish between objective and subjective task complexity. Subjective complexity, sometimes also called 'perceived', 'experienced', or 'psychological' complexity [23, 24] refers to the extent to which the individual experiences the task as a complex one and is therefore related to both user characteristics and the relation between user characteristics and objective task characteristics. Objective complexity refers to characteristics of the task itself. This includes factors such as the number of elements involved in the task, their (number of) interactions, and the overall structure. A further issue is whether one should consider task 'complexity' as anything else than task 'difficulty'. Researchers disagree about this, ranging from those who argue that these concepts are identical, subsets of each other (in both ways), or completely dissimilar (cf. [18, p. 558]).

Properly defining complexity is a complex task, as others have argued before (cf. [25]. One way to grasp what task complexity is perceived to be, is by considering the conceptualizations that researchers have been using for the topic. Conceptualization involves determining indicators and dimensions of the concept [26] and changes vague and imprecise ideas to specific measurements and indicators, resulting in a clear and agreed-upon meaning for a concept (hopefully). Based on this approach, Liu and Li [18] accumulated 24 different operationalizations of the concept across different studies. In their paper, they ultimately reduce the complexity concept to 10 structural dimensions of complexity: Size, Variety, Ambiguity, Relationship, Variability, Unreliability, Novelty, Incongruity, Action complexity, and Temporal demand. This set of 10 allows us to categorize the literature on the effects of complexity on trust in algorithmic advice from Table 1 more clearly. In Table 2, we expand Table 1 to include the dimensions of complexity that we feel were manipulated in the different studies.

Upon examining Table 2, it becomes apparent that almost all studies manipulate size: the more there is to consider, the more complex the task is. Some studies combine multiple dimensions, such as size and unreliability. For instance, in the study by Bogerts et al. [11], task difficulty is manipulated by zooming out on a photo, which can make it impossible to count the individuals in the photo one-by-one, while size (the number of individuals on a photo varies from 15 to 5,000) is also manipulated. Another example with the same two dimensions is the study by Wang and Du [13]. In their study, participants were presented with the task of estimating the number of coins in a glass, where again both the number and picture blur were varied. It is not always completely obvious which other dimension besides size is being triggered (does adding blur to a picture change the 'uncertainty' or the 'unreliability'?), but what is clear is that size, the sheer number of objects or variants under consideration, is most often part of it, and that in some cases this is accompanied by something else that adds to the complexity. Previous studies, such as Wood [19], Bonner [21], Liu and Li [18], and Ham et al. [17], have in addition considered various related distinctions but orthogonal task components such as whether the complexity refers to Input, Process, or Output. Few or many coins in a glass is variation in the size dimension on the input, blurred pictures relate to process, and few or many options to choose from would be a variation in the size dimension of the output. All of this highlights the importance to at least be clear about the kind of complexity that is being considered, as the variations in complexity are substantial.

A second issue that is striking from Table 1 and 2 is that the tasks that are being used in the previous literature are quite abstract and somewhat artificial. Counting the number of coins in a glass, finding an X among 300 characters, or estimating the number of people on a picture are not tasks that people are likely confronted with in their daily lives. Taken together, this suggests two guidelines for a test of the effect of task complexity on trust in algorithmic advice: (1) consider carefully how to operationalize task complexity and evaluate complexity in terms of the dimensions suggested by Liu and Li [18] and (2) consider a task that is conceivable for lay-people in their everyday lives.

**Table 1.** Summaries of relevant research on task complexity and difficulty, and trust in algorithmic advice

| Conceptualization of complexity | Description of the task | Research | Variables | Conclusion of the paper |
|---|---|---|---|---|
| Number of acts | Four scenarios of robotic handling: fixture, handover, fetch and screwing. Task complexity increased with each scenario | [27] | Task complexity, usefulness and satisfaction | Scenarios with higher task complexity were perceived as more useful and satisfying compared to scenarios with lower complexity |
| Number of elements and similarity | Luggage scanning with varying complexity: low complexity (0–1 objects) and high complexity (2–3 objects). Coordinative complexity involved obscured, similar, both obscured and similar, or no such objects | [28] | Cognitive workload, task complexity and age | Task complexity interacts with automation in influencing object identification accuracy. Automation is more effective in high task complexity situations, but its impact varies depending on the specific complexity factors. Further research is needed to understand individuals' willingness to use automation in complex tasks |

*(continued)*

**Table 1.** (*continued*)

| Conceptualization of complexity | Description of the task | Research | Variables | Conclusion of the paper |
|---|---|---|---|---|
| Size of maze | Wayfinding task, larger mazes required more turns and were more difficult than smaller mazes | [15] | Task difficulty, accuracy and stress | Students' willingness to follow instructions from an algorithm changed over time. As the mazes became more difficult, participants showed a slight decrease in their likelihood to follow the system's instructions |
| Number of people in picture | Estimating the number of people in a picture, with easy images zooming in and containing 15 people, while difficult images zooming out and containing 5000 people | [11] | Task difficulty, low quality advice and social influence | Algorithmic advice had a positive impact on revising answers. Subjects were more likely to revise their responses when receiving advice from algorithms compared to advice from the crowd |
| Number of options | Participants were given a high complexity task involving developing an optimal investment strategy and a lower complexity task focused on finding the right savings account | [29] | Perceived complexity and adoption AI | Participants were more inclined to adopt algorithmic advice for complex tasks if they believed AI was more intelligent than humans |
| Identify the longest line among the options | Task was adapted from the Müller-Lyer illusion task. In simple tasks, the picture included eight vertical lines. In complex tasks, the picture included four vertical lines and four horizontal lines. The task is to select the longest line from the given options | [16] | Task complexity and risk | Participants showed greater trust in the robot during simple tasks than in complex ones |
| Depth of question | The task involves asking questions to an AI ChatBot. The easy question is "What should I do when my credit card expires?" and the complex question is "How do I get a credit card cash advance when I am abroad?" | [30] | Task complexity, problem solving ability and usage intention | Customers perceived AI as better for simple tasks but preferred human customer service for handling complex issues |

(*continued*)

**Table 1.**  (*continued*)

| Conceptualization of complexity | Description of the task | Research | Variables | Conclusion of the paper |
|---|---|---|---|---|
| Assessing with or without blur | Participants were shown six photographs of a glass filled with coins and were asked to estimate the number of coins in each photograph. In difficult tasks, the six photographs of the glass filled with coins were intentionally blurred | [13] | Confidence and trust | Task difficulty influenced participants' reliance on advice, with greater reliance observed in difficult tasks. Confidence mediated this effect |
| Announcement of task as simple or difficult | Looking for X's in a random array. Trials were all equally hard | [14] | Task difficulty and importance | Participants tended to rely on automation more when the task was perceived as difficult, regardless of feedback availability |
| Extent of consistency | Evaluate the internal control environment for a manufacturing firm. Task complexity was manipulated by varying the consistency of internal controls in cases | [31] | Task complexity, accuracy, feedback and task experience | Complex tasks led to increased reliance on decision aids by experienced participants. Providing feedback improved decision-making accuracy |

# 3  Method

## 3.1  Design

As the more real-life example that we were aiming for, participants in this study were presented with real-life court cases of crimes that have been committed in the Netherlands in the past decade. Participants were asked to estimate the jailtime sentence for the crime concerned. After giving their initial estimate, participants were provided with algorithmic advice and were then allowed to revise their initial judgment. The experiment was conducted in Dutch because the cases were based on the Dutch legal system. The online survey consisted of three parts. In the first part, we measured participants' levels of experience in the field of law (see Appendix 1) and introduced a court case example to help participants become familiar with the task.

In the second part, participants read and evaluated existing court cases that were divided over 20 cases in four blocks (4, 6, 6, and 4 cases). All participants completed all four blocks. Figure 1 shows an example case (the original example, including the Dutch text, can be found in Appendix 2). Using whole numbers between 0 and 30, participants estimated the number of years in the first box and the number of months in the second box. They then received algorithmic advice (for example: computer algorithm B has looked at this case and advices $x$ months jailtime. The algorithm took different factors into account, which included the use of violence, abuse of drugs and/or alcohol and possibly the criminal record of the offender) and were given the opportunity to revise their initial assessment.

**Table 2.** Summaries of relevant research including task complexity dimension on task complexity, difficulty, and trust in algorithmic advice

| Conceptualization of complexity | Research | Description of the task | Variables | Size | Variety | Ambiguity | Relationship | Variability | Unreliability | Novelty | Incongruity | Action complexity | Temporal demand |
|---|---|---|---|---|---|---|---|---|---|---|---|---|---|
| Number of acts | [27] | Four scenarios of robotic handling: fixture, handover, fetch and screwing. Task complexity increased with each scenario. | Task complexity, usefulness and satisfaction | ■ | | | | | | | | | |
| Number of elements and similarity | [28] | Luggage scanning with varying complexity: low complexity (0-1 objects) and high complexity (2-3 objects). Coordinative complexity involved obscured, similar, both obscured and similar, or no such objects. | Cognitive workload, task complexity and age | | | | | | | | | ■ | |
| Size of maze | [15] | Wayfinding task, larger mazes required more turns and were more difficult than smaller mazes. | Task difficulty, accuracy and stress | ■ | | | | | | | | | ■ |
| Number of people in picture | [11] | Estimating the number of people in a picture, with easy images zooming in and containing 15 people, while difficult images zooming out and containing 5000 people. | Task difficulty, low quality advice and social influence | | ■ | ■ | ■ | ■ | ■ | | | | |
| Number of options | [29] | Participants were given a high complexity task involving developing an optimal investment strategy and a lower complexity task focused on finding the right savings account. | Perceived complexity and adoption AI | | | | | | | | | ■ | |
| Identify the longest line among the options | [16] | Task was adapted from the Müller-Lyer illusion task. In simple tasks, the picture included eight vertical lines. In complex tasks, the picture included four vertical lines and four horizontal lines. The task is to select the longest line from the given options. | Task complexity and risk | | | | | | | | | | |
| Depth of question | [30] | The task involves asking questions to an AI ChatBot. The easy question is "What should I do when my credit card expires?" and the complex question is "How do I get a credit card cash advance when I am abroad?" | Task complexity, problem solving ability and usage intention | ■ | ■ | | | | | | | ■ | |
| Assessing with or without blur | [13] | Participants were shown six photographs of a glass filled with coins and were asked to estimate the number of coins in each photograph. In difficult tasks, the six photographs of the glass filled with coins were intentionally blurred. | Confidence and trust | | | | | | | | | | |
| Announcement of task as simple or difficult | [14] | Looking for X's in a random array. Trials were all equally hard. | Task difficulty and importance | | | | | | | | | | |
| Extent of consistency | [31] | Evaluate the internal control environment for a manufacturing firm. Task complexity was manipulated by varying the consistency of internal controls in cases. | Task complexity, accuracy, feedback and task experience | | | | | | | | ■ | | |

The third part consisted of a series of questions measuring, among others, the participant's expertise in the field of IT, experience with algorithms, and several demographic questions (see Appendix 1).

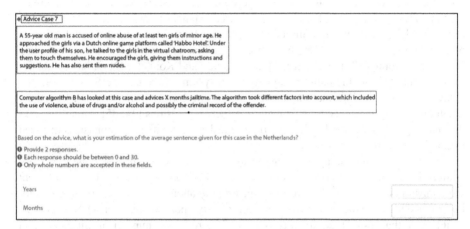

**Fig. 1.** Example of a case used in the experiment (translation from the original Dutch content).

## 3.2 Participants

In total, 151 participants completed the study, of which 126 (83%) completed the study at the psychology lab, and the other 25 (17%) completed the study online without the presence of experimenters. We initially invited the first group of participants to join our study through the JFS database of Eindhoven University of Technology. However, for the second group of participants, we contacted multiple universities in the Netherlands that offer law programs and asked for their cooperation in distributing our survey among their law students, in the hope of receiving input also from participants that are more knowledgeable of the law. Of all the participants, 59 (39%) were male, and 92 (61%) were female. The age of participants ranged from 18 to 55+. The participants were also divided into law students (4), former law students (19), and lay people who did not follow any law-related study (128). Most participants held a university degree as their highest education level, accounting for 47% (see Appendix 5 for the distribution details).

## 3.3 Target Variable

The dependent variable in our study is the Weight On Advice (WOA). This is measured by the extent to which participants modified their answers in the direction of the advice. The measurement is used more often in this line of research and based on the Judge Advisor System (JAS) model [32]. Individuals can choose to shift towards the given advice, to disregard it or, in theory though rarely in practice, to move further away from the advice. When the initial judgment and advice are equal, WOA is undefined.

$$Weight\ On\ Advice = \frac{revised\ judgment - initital\ judgment}{advice - initial\ judgement} \tag{1}$$

## 3.4  Task Complexity

In our experiment we focus on criminal law cases in The Netherlands. The final decision in the Dutch system is up to a single or multiple judges and no juries are used. Judges are provided with sentencing guidelines. In these guidelines, so-called 'orientation points' [33] include the sections of the law and the circumstances to be considered per section of the law. In our case it is not immediately obvious how to measure the size dimension of complexity. The Size dimension has been identified by several researchers, including Wood [19], Bonner [21], Simnett [34], and Harvey and Koubek [22]. A large quantity of input can overload the memory and attention system of the task performer [19, 21]. Conversely, too little information prevents the task performer from forming an accurate mental image for task execution or making effective decisions [34, 35]. This illustrates the challenge of assessing the alleged effect of task complexity and emphasises the need to carefully specify what Size entails.

When determining an appropriate jailtime, several additional aspects need consideration. First, one could argue that complexity is defined by the number of different violations: more violations lead to more uncertainty about how to sum up separate violations, and hence to higher complexity (Complexity operationalization 1). Second, one could argue that instead, complexity is defined more by the number of mitigating and aggravating circumstances that should be considered by the judge. For some violations there are few such circumstances, for some there are many (Complexity operationalization 2). The more circumstances need to be weighed against each other, the more complex a case could be. Finally, one could argue that, instead of focusing on the circumstances that are supposed to be considered, one should focus on how many of such circumstances are mentioned in the case text (assuming that assessors only consider the text they see; Complexity operationalization 3).

## 3.5  Negative Feedback

Between blocks 2 and 3, a message was displayed with feedback on the performance of the participant in the first two blocks. Participants received feedback (not based on their actual performance) suggesting that the algorithms had performed better than the participant. After this, participants were suggested to follow the advice of the algorithm more than they had done before. The feedback was provided at three different levels, and the level given to each participant was randomly varied. These three levels varied in the amount of negativity (see Appendix 4 for the literal text of the feedback).

## 3.6  Algorithm Transparency

Algorithmic advice was given in three different ways. In the experiment, the advice was explained as given by different algorithms: A, B, and C (see Appendix 3 for the literal text of the advice). Algorithm A gave advice without any further explanation, Algorithm B told the participant which factors were considered. Algorithm C would mention the relevant factors and present the formula used to compute the advice. These three gradations represent increasing levels of transparency. Human advice was manipulated by stating that the advice was given by a professional in criminal law. However, this study focuses on the cases where algorithmic advice was given.

## 3.7 Scales

We used five scales to assess participants' expertise in the field of law, their self-reported level of trust in human advice and in algorithms, their attitude towards feedback, and their expertise and experience in IT and algorithms. For each scale we employed existing (7-Point Likert) scales. Table 3 provides information on the construct of these scales, their sources, and the corresponding Cronbach's alpha scores.

**Table 3.** Scales used in the experiment

| Scale | Construct | Cronbach's alpha |
|---|---|---|
| 1 | Participants' expertise in the field of law [36, 37] | 0.88 |
| 2 | Participants self-reported trust in human advice [38] | 0.84 |
| 3 | Participants self-reported trust in algorithms [38] | 0.89 |
| 4 | Participants attitudes towards feedback | 0.78 |
| 5 | Participants IT expertise and experience, interest, and knowledge regarding algorithms | 0.83 |

# 4 Results

As mentioned above, we only consider the cases where participants received algorithmic advice (cases 5 through 16, see Appendix 6). We analyze the data in so-called "long format", where every row represents a jailtime assessment, leading to a multi-level data set of 12 rows of data for each of the 151 participants. From these 1,812 cases, we excluded 165 observations in which the advice was equal to the initial estimate as they provided no informative data, consistent with the study of Gino and Moore [12]. Additionally, we deleted 6 WOA values that were either smaller than -1 or larger than 3 to restrict the measure to a range consistent with its expected interpretation and practical application. The remaining 1,641 cases are used as a final data set for analysis.

First, we present descriptives of the data set. The sentence given by the participants' initial response averaged 49 (SD = 40.4) months. After receiving advice, which averaged 62 months, the response increased to an average 58 (SD = 67.7) months. Figure 2 gives an overview of the different jailtime guesses throughout the study for those cases where algorithmic advice was given.

After receiving algorithmic advice, participants changed their initial answers 1,399 times (82%) and 302 times (18%) participants adhered to their initial answers. The mean WOA was 0.57 (SD = 0.41) and Median 0.66.

## 4.1 Main analysis

To test our hypothesis that people trust algorithmic advice more when tasks become more complex, we use multi-level regression analysis (assessments within participants) on the

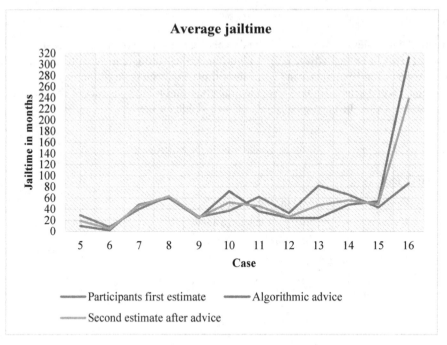

**Fig. 2.** Average jailtime estimations for cases 5 to 16.

target variable WOA. We conducted our data analysis using Stata 17 software, exclusively, without the use of any external source packages. Our data exhibits a hierarchical structure with two levels: 20 cases are nested within participants. Conventional regression techniques cannot account for this dependency. Therefore, we adopted multi-level regression analysis to model the data.

As mentioned in the Method section, we defined three different complexity variables. Per complexity variable, we show three models: a model with just the complexity variable as a predictor, a model with transparency and feedback included (the two factors in the experiment), and a model that also includes other participant characteristics (Table 4). The purpose of the models is, per used complexity measurement, to assess whether the size of effects (if existing at all) is similar across different implementations of the model (Models 1–3 for measure 1, Models 4–6 for measure 2, and Models 7–9 for measure 3).Let us first consider the effect of the complexity variables as the sole predictor in models 1, 4, and 7. For two of the complexity operationalizations, complexity focuses on the number of mitigating and aggravating circumstances (0.004, $p < .001$) and the number of law sections violated in the case (0.069, $p < .001$) we find that with increasing complexity, the weight on advice increases, which we interpret as increased trust in the algorithm. We find no significant effect for complexity that focuses on the number of mitigating and aggravating circumstances based on the case text provided (0.003 $p = .41$). Models 2, 3, 5, 6, 8, and 9 show that this occurs irrespective of the inclusion of the additional covariates. Hence, the results in Table 4 show that whether we find an effect of complexity on the extent to which algorithmic advice is incorporated does not depend (much) on whether different additional covariates are incorporated, but do depend on the kind of operationalization of complexity.

Differences in transparency refer to the different ways in which the algorithms that participants faced were introduced (no transparency, medium, high). Compared to no transparency, we find a difference with medium ($-0.120\,p < .001$) and high transparency ($-0.059\,p < .001$) only for the complexity measurement that focuses on the number of mitigating and aggravating circumstances. More transparency leads to lower WOA and hence less trust in the algorithm in this case. This implies that the notion that trust in advice increases with increased transparency about the algorithm is refuted. For negative feedback we do find results that are in line with the general intuition: medium ($0.118, p < .001$) and high negative ($0.071\,p < .010$) feedback lead to more trust in the algorithmic advice, presumably because they emphasize that the participant is less competent in these assessments than initially considered.

The participant characteristics that were added are the "trust in algorithms" scale, age, male versus female, and education (as an interval score), mainly as control conditions and guided by the data. The "trust in algorithms" scale was the only scale that led to some significant results ($0.037, p < .001$). Those who score higher on the scale trust the algorithm more. We examined the effects of age more closely and found that when we distinguished between those above and below 35, we saw evidence that older participants trusted the algorithm less ($-0.070, p < .10$). We also found some evidence that males are less trusting than females ($-0.059, < .10$), but find no effect of education ($0.001$, p $= .90$). We further considered work status and law expertise, but neither of them reached significance or influenced the other results in a substantial way.

**Table 4.** Multi-level logistic regression models for trust in algorithmic advice and task complexity Cell entries represent the coefficients (and standard errors).

| | Complexity measure: Number of mitigating and aggravating circumstances | | | Complexity measure: Number of mitigating and aggravating circumstances based on case text | | | Complexity measure: Number of law sections violated | | |
|---|---|---|---|---|---|---|---|---|---|
| | (1) | (2) | (3) | (4) | (5) | (6) | (7) | (8) | (9) |
| Target variable | WOA | WOA | WOA | WOA | WOA | WOA | WOA | WOA | WOA |
| Complexity | 0.00*** | 0.01*** | 0.01*** | 0.00 | 0.01 | 0.01 | 0.07*** | 0.07*** | 0.07*** |
| | (0.00) | (0.00) | (0.00) | (0.00) | (0.01) | (0.01) | (0.01) | (0.01) | (0.01) |
| Algorithm A | | | | | | | | | |
| No transparency | | 0.00 | 0.00 | | 0.00 | 0.00 | | 0.00 | 0.00 |
| | | (.) | (.) | | (.) | (.) | | (.) | (.) |
| Algorithm B Medium transparency | | −0.12*** | −0.12*** | | −0.02 | −0.02 | | 0.02 | 0.02 |

(*continued*)

**Table 4.** (*continued*)

| | Complexity measure: Number of mitigating and aggravating circumstances | | | Complexity measure: Number of mitigating and aggravating circumstances based on case text | | | Complexity measure: Number of law sections violated | | |
|---|---|---|---|---|---|---|---|---|---|
| | (1) | (2) | (3) | (4) | (5) | (6) | (7) | (8) | (9) |
| | | (0.03) | (0.03) | | (0.03) | (0.03) | | (0.02) | (0.02) |
| Algorithm C High transparency | | $-0.06^{**}$ | $-0.06^{**}$ | | $-0.00$ | $-0.00$ | | 0.01 | 0.01 |
| | | (0.03) | (0.03) | | (0.03) | (0.03) | | (0.02) | (0.02) |
| Low negative feedback | | 0.00 | 0.00 | | 0.00 | 0.00 | | 0.00 | 0.00 |
| | | (.) | (.) | | (.) | (.) | | (.) | (.) |
| Medium negative feedback | | $0.12^{***}$ | $0.09^{**}$ | | $0.12^{***}$ | $0.09^{**}$ | | $0.12^{***}$ | $0.09^{**}$ |
| | | (0.04) | (0.04) | | (0.04) | (0.04) | | (0.04) | (0.04) |
| High negative feedback | | $0.07^{*}$ | $0.07^{*}$ | | $0.07^{*}$ | $0.07^{*}$ | | $0.07^{*}$ | $0.07^{*}$ |
| | | (0.04) | (0.04) | | (0.04) | (0.04) | | (0.04) | (0.04) |
| Scale trust in algorithms | | | $0.04^{***}$ | | | $0.04^{***}$ | | | $0.04^{***}$ |
| | | | (0.01) | | | (0.01) | | | (0.01) |
| Age 35+ | | | $-0.07^{*}$ | | | $-0.07^{*}$ | | | $-0.07^{*}$ |
| | | | (0.04) | | | (0.04) | | | (0.04) |
| Male | | | $-0.06^{*}$ | | | $-0.06^{*}$ | | | $-0.06^{*}$ |
| | | | (0.03) | | | (0.03) | | | (0.03) |
| Education level | | | 0.00 | | | 0.00 | | | 0.00 |
| | | | (0.02) | | | (0.02) | | | (0.02) |
| Constant | $0.53^{***}$ | $0.47^{***}$ | $0.39^{***}$ | $0.56^{***}$ | $0.50^{***}$ | $0.41^{***}$ | $0.46^{***}$ | $0.39^{***}$ | $0.30^{***}$ |
| | (0.02) | (0.03) | (0.10) | (0.02) | (0.03) | (0.10) | (0.02) | (0.04) | (0.11) |
| N | 1641.00 | 1641.00 | 1641.00 | 1641.00 | 1641.00 | 1641.00 | 1641.00 | 1641.00 | 1641.00 |

*Standard errors in parentheses $*p < 0.1$, $**p < 0.05$, $***p < 0.01$.*

## 5 Conclusion and Discussion

This research considered the effect of task complexity on the willingness to accept algorithmic advice in a real-world task domain. Our hypothesis was that individuals are more likely to trust algorithmic advice as the task complexity increases. Defining task complexity proved challenging, in part because earlier studies on the topic have used (reasonable but) ad-hoc operationalizations rather than definitive guidelines for what complexity is or could be. Based on the 10 dimensions of task complexity of Liu and Li [18], we positioned earlier studies with respect to the kinds of complexity they considered. We observe that most studies primarily concentrate on the dimension of Size, while the remaining dimensions receive limited attention. We also discovered that several studies, such as Frazier et al. [28], and Bogert et al. [11], concentrated on multiple dimensions.

Sticking to the size dimension of complexity, our findings show that participants demonstrated a greater inclination to trust algorithmic advice as the complexity of tasks increased for two out of the three ways in which complexity was operationalized. The operationalizations that focused on the number of sections of the law and the number of mitigating and aggravating circumstances demonstrated a positive and significant effect of task complexity, which aligns with some findings from previous studies in the field [11, 13, 14, 27, 29]. However, the operationalization focusing on only those circumstances that were visible in the case text did not exhibit such a significant effect. These results in and of itself show that relatively subtle differences in the operationalization of complexity, even within a given dimension (size) of complexity, can lead to different results. Future research could investigate task complexity's impact on trust in algorithmic advice in other real-world task domains, to provide insights into whether the relationship is similar across different contexts. By examining the role of complexity in diverse domains, researchers can uncover commonalities or variations in how individuals trust algorithmic advice based on task complexity. Additionally, further exploration of additional dimensions of task complexity (such as Variety or Uncertainty) is warranted. However, our findings also highlight the importance of following up this research with attempts to try and investigate the underlying mechanism that is involved in more detail. Apparently, effects are subtly dependent on the kind of task complexity and the specific context in which assessments take place, and a better understanding of these effects can only occur by figuring out what happens underneath the surface.

Our study also investigated the impact of negative feedback and algorithm transparency. We found that for the complexity variable focusing on mitigating and aggravating circumstances, medium and high levels of transparency negatively affected trust, compared to no transparency. One possible explanation for this could be that in medium transparency, we have not always shown the appropriate mitigating and aggravating circumstances that match with the case text. Even attempts to further enhance transparency using a formula did not yield a positive trust outcome. Höök [39] emphasised that this approach may further alienate inexperienced users, thus hampering the intended goal of fostering trust. Instead, an alternative approach to explain the algorithm's decision to laypeople (or even experts) could be to come up with less precise but more straightforward explanations of algorithmic suggestions, for instance through tree-based models or localized regression methods. Furthermore, exposing inexperienced individuals with negative feedback about their own decisions increased the likelihood of accepting algorithmic advice, a finding consistent with De Vries et al. [40]. Similarly, our study found that medium and high levels of negative feedback resulted in greater trust in algorithmic advice.

It is crucial to interpret the results of our study with caution due to several limitations we encountered. For instance, participants in our study were not provided with an explanation of the complexity level of the cases they were evaluating. Previous research has indicated that people approach advice differently when tackling complex tasks, so what we consider complex might be different from how participants have perceived this. Participants, most of whom did not have a lot of knowledge about the law, were not informed about the framework or 'orientation points' that the law prescribes to consider.

This lack of awareness about the specific criteria used to determine complexity could have affected their perceptions and decision-making during the study.

**Acknowledgments.** This publication is part of the project Vertrouwen in algoritmes with project number 023.017.109 of the research programme Promotiebeurs voor Leraren which is (partly) financed by the Dutch Research Council (NWO).

# Appendix 1

1.Statements item scale human advice

1. When an important issue or problem arises, I would feel comfortable depending on the advice of other people.
2. I can always rely on the advice of other people in a tough situation.
3. I feel that I could count on the advice of other people to help me with a crucial problem.
4. Faced with a difficult situation, I would feel comfortable using the advice of other people.

2. Statements item scale algorithmic advice

1. When an important issue or problem arises, I would feel comfortable depending on the advice of a computer.
2. I can always rely on the advice of a computer in a tough situation.
3. I feel that I could count on the advice of a computer to help me with a crucial problem.
4. Faced with a difficult situation, I would feel comfortable using the advice of a computer.

3. Statements item scale IT experience

1. I often watch series and/or movies concerning artificial intelligence.
2. During my study I have gained knowledge about computer algorithms.
3. I have a lot of knowledge about IT.
4. I have experience with computer algorithms.
5. People in my social circle work in the IT sector.
6. I often use a Google Home, Alexa, or similar technology.
7. I encounter computer algorithms/artificial intelligence on a daily basis.
8. I follow the news with regard to the most recent developments of new technologies.

4. Statements item scale law expertise.

1. I have experience with the Dutch law system, for example through my work environment, study, internship or other courses.
2. I come into contact with criminal law on a daily basis.
3. I often watch series or movies concerning crime.
4. People in my social circle work in the field of law.
5. During my study I have gained knowledge about the Dutch law system.

8. I often read, watch, or listen news items with regard to crime.
7. I have a lot of knowledge about criminal law.
8. I am interested in anything related to the Dutch law system.

5. Statements item scale feedback.

1. When I make a mistake, I do not mind getting negative feedback on my performance.
2. I can handle negative feedback well.
3. When I get feedback, I use this to my advantage.
4. Feedback helps me to perform better.

## Appendix 2

Fig. 3.

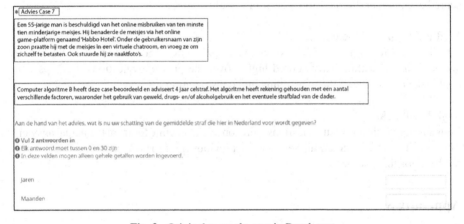

**Fig. 3.** Original example case in Dutch text.

## Appendix 3

Textual algorithm descriptions.

**Algorithm A**

Computer algorithm A has looked at this case and advices $x$ months jailtime.

**Algorithm B**

Computer algorithm B has looked at this case and advices $x$ months jailtime. The algorithm took different factors into account, which include the use of violence, abuse of drugs and/or alcohol and possibly the criminal record of the offender.

**Algorithm C**

Computer algorithm C has looked at this case and advices $x$ months jailtime. The algorithm calculated this advice based on the following formula: $T = x1 * w1 + x2 * w2 + \ldots + xn * wn$ where $T$ indicates a range of jailtimes, $xi$ indicates a relevant factor, and $wi$ the adequate weight. The values of the factors can vary from 0 to 9, where 0 means that the factor does not apply here and 9 means that the factor is really relevant. The weights vary between 0 and 1, and will sum to a total of 1. The value of $T$ will give a value that is linked to a specific range of jail times. The algorithm will take the median value and gives this as output.

## Appendix 4

Textual feedback received.

### Low Negative Feedback
Your answers to the previous questions were, on average, close to the actual answer. This means that, until now, you have performed reasonably well. However, the algorithms have scored even higher. Perhaps the algorithms could help you.

### Medium Negative Feedback.
Your answers to the previous questions were, on average, not that close to the actual answer. The algorithms have scored higher over the previous questions. Perhaps you could take the advice of the algorithms into consideration.

### High Feedback.
Your answers to the previous questions were, on average, quite far from the actual answer. The algorithms have scored higher over the previous questions. You should focus more on the algorithms advice.

## Appendix 5

Representation of demographical variables (Tables 5 and 6)

**Table 5.** Distribution of participants by age group

| Age group | Number of participants |
|-----------|------------------------|
| 18–24 | 88 |
| 25–34 | 13 |
| 34–44 | 8 |
| 45–54 | 9 |
| > 55 | 33 |

**Table 6.** Distribution of participants by highest education level

| Education level | Number of participants |
| --- | --- |
| Elementary Education | 1 |
| High School | 7 |
| Secondary vocational education | 9 |
| Bachelor's Degree (Applied sciences) | 17 |
| Bachelor's Degree | 71 |
| Master's Degree | 46 |

# Appendix 6

Case 5

On an outside parking lot in the Netherlands, a victim has been beat on his head by a man using a crowbar. Victim 1 lost consciousness immediately and fell to the ground. When a bystander saw this happening, he tried to stop the attacker, resulting in a slap on his cheek. Victim 2 fell on the ground, breaking his cheek bone on multiple places. Both victims must live with sustaining injuries. The attacker carried the crowbar illegally.

Case 6

A Chinese woman is charged with the possession of a false Chinese national passport. Judges are convinced that she was aware of the falsification, or that she at least could have had presumed that it was falsified. She claims to be illiterate, because of which she allegedly was not aware of the falsification.

Case 7.

A 55-year old man is accused of online abuse of at least ten girls of minor age. He approached the girls via a Dutch online game-platform called 'Habbo Hotel'. Under the user profile of his son, he talked to the girls in the virtual chatroom, asking them to touch themselves. He encouraged the girls, giving them instructions and suggestions. He has also sent them nudes.

Case 8

A 50-year old man is accused of sexual abusing four women between 82 and 94 years old. Over a timespan of three years, he approached the elderly women at their apartments, acting as their physiotherapist, physician or caregiver. Once inside their apartments, he asked the women to undress and to let him massage them. Subsequently, he touched them. He even raped one of the women.

Case 9

A 23-year-old man has used a firearm within a nightlife area. The man was under influence of a significant amount of alcohol. He shot multiple times in the air and towards the ground. Luckily, no one was hit by the man. On top of this, he was keeping another weapon at home, accompanied by enough ammunition for both guns.

Case 10

Two brothers have committed an armed robbery on a value transport. They have stolen expensive jewelry and highly exclusive watches that were carried in the van. The driver (one of the brothers) is suspected to be part of the robbery. After a phone call, the driver differed from his usual route. Minutes later, a fire weapon was aimed at him, after which the other brother entered the van. When they reached a remote zone, the van stopped and was completely emptied by a group of accomplices. After this, the van was set on fire and the police was called.

Case 11

The defendant was driving at a speed of 79 km/h where the speed limit was 50 km/h while under the influence of alcohol and drugs. As a result, he crashed his car against a bus. The two other passengers in the car were killed in the accident. The defendant was known by the police since he had been caught speeding under the influence of alcohol before.

Case 12

Three drunk men decided spontaneously that they would break into a house, assuming nobody was home. When they got into the house, they encountered the resident which resulted in a struggle. The resident was beaten by the defendants. The main defendant had a leading role and was the one using the most violence.

Case 13

Five men were involved in the preparation of a terrorist attack. They provided a large batch of ammunition for two terrorism suspects. The police found 45 kg of ammunition as well as 3700 patterns for Kalashnikov-rifles.

Case 14

A 76-year old men abused 2 girls on a camping. He was known as the friendly granddad of the camping and was therefore trusted by the girls. The abuse of one of the girls, who was mentally deficient, was going on for about 10 years. The other girl was abused during the summer at the age of 9. He had received a sentence of 1 year of jailtime before, for abuse of his 9-year old granddaughter.

Case 15

The father of a 14-year old girl beat the 46-year old stalker of his daughter with a snow shovel. The stalker had been trying to meet up with the girl and left flowers and chocolate in the garden of the girl. The father believed that the stalker had been accused of sexual abuse before, but this was not the case. The actions of the father were seen as an attempt to murder, since he could have reasonably foreseen that the stalker would possibly die as a result of getting beaten with the snow shovel.

Case 16

Two men shot an innocent guy in a so-called 'mistaken murder'. They shot the guy in the porch of his own flat. The men believed the innocent guy to be another man who lived in the same flat and was their actual target.

# References

1. Wang, S., et al.: A deep learning algorithm using CT images to screen for Corona virus disease (COVID-19). Eur. Radiol. **31**(8), 6096–6104 (2021). https://doi.org/10.1007/s00330-021-07715-1
2. Karadimas, N.V., Papatzelou, K., Loumos, V.G.: Optimal solid waste collection routes identified by the ant colony system algorithm. Waste Manag. Res. **25**(2), 139–147 (2007). https://doi.org/10.1177/0734242X07071312

3. Engin, Z., Treleaven, P.: Algorithmic government: automating public services and supporting civil servants in using data science technologies. Comput. J. **62**(3), 448–460 (2019). https://doi.org/10.1093/comjnl/bxy082

4. Yin, M., Vaughan, J.W., Wallach, H.: Understanding the effect of accuracy on trust in machine learning models. In: Conference Human Factors Computer System – Proceedings, pp. 1–12 (2019). https://doi.org/10.1145/3290605.3300509

5. Kizilcec, R.F.: How much information? Effects of transparency on trust in an algorithmic interface. In: Conference Human Factors Computing System – Proceedings, pp. 2390–2395 (2016). https://doi.org/10.1145/2858036.2858402

6. Zhang, Y., Vera Liao, Q., Bellamy, R.K.E.: Effect of confidence and explanation on accuracy and trust calibration in AI-assisted decision making. In: FAT* 2020 – Proceedings of the 2020 Conference Fairness, Accountability, Transparency, pp. 295–305 (2020). https://doi.org/10.1145/3351095.3372852

7. Dietvorst, B.J., Simmons, J.P., Massey, C.: Overcoming algorithm aversion: People will use imperfect algorithms if they can (even slightly) modify them. Manage. Sci. **64**(3), 1155–1170 (2018). https://doi.org/10.1287/mnsc.2016.2643

8. Önkal, D., Goodwin, P., Thomson, M., Gönül, S., Pollock, A.: The relative influence of advice from human experts and statistical methods on forecast adjustments. J. Behav. Decis. Mak. **22**(4), 390–409 (2009). https://doi.org/10.1002/bdm.637

9. Jacko, J.A., Ward, K.G.: Toward establishing a link between psychomotor task complexity and human information processing. Comput. Ind. Eng. **31**(1–2), 533–536 (1996)

10. Zhao, B.: A structured Analysis and Quantitative Measurement of Task Complexity in Human-Computer Interaction. Purdue University (1992)

11. Bogert, E., Schecter, A., Watson, R.T.: Humans rely more on algorithms than social influence as a task becomes more difficult. Sci. Rep. **11**, 8028 (2021). https://doi.org/10.1038/s41598-021-87480-9

12. Gino, F., Moore, D.A.: Effects of task difficulty on use of advice. J. Behav. Decis. Mak. **20**(1), 21–35 (2007). https://doi.org/10.1002/bdm.539

13. Wang, X., Du, X.: Why does advice discounting occur? The combined roles of confidence and trust. Front. Psychol. **9**, 2381 (2018). https://doi.org/10.3389/fpsyg.2018.02381

14. Schwark, J., Dolgov, I., Graves, W., Hor, D.: The influence of perceived task difficulty and importance on automation use. Proc. Hum. Factors Ergon. Soc. **2**, 1503–1507 (2010). https://doi.org/10.1518/107118110X12829370088561

15. Monroe, S., Vangsness, L.: The effects of task difficulty and stress on trust in an automated navigation aid. Proc. Hum. Factors Ergon. Soc. Annu. Meet. **66**(1), 1080–1084 (2022). https://doi.org/10.1177/1071181322661406

16. Huang, H., Rau, P.L.P., Ma, L.: Will you listen to a robot? Effects of robot ability, task complexity, and risk on human decision-making. Adv. Robot. **35**(19), 1156–1166 (2021). https://doi.org/10.1080/01691864.2021.1974940

17. Ham, D.H., Park, J., Jung, W.: Model-based identification and use of task complexity factors of human integrated systems. Reliab. Eng. Syst. Saf. **100**, 33–47 (2012). https://doi.org/10.1016/j.ress.2011.12.019

18. Liu, P., Li, Z.: Task complexity: a review and conceptualization framework. Int. J. Ind. Ergon. **42**(6), 553–568 (2012). https://doi.org/10.1016/j.ergon.2012.09.001

19. Wood, R.E.: Task complexity: definition of the construct. Organ. Behav. Hum. Decis. Processes **37**, 60–82 (1986). https://doi.org/10.1016/0749-5978(86)90044-0

20. Campbell, D.J.: Task complexity: a review and analysis. Acad. Manag. Rev. **13**(1), 40–52 (1988). https://doi.org/10.5465/amr.1988.4306775

21. Bonner, S.E.: A model of the effects of audit task complexity. OAccount. Organ. Soc. **19**(3), 213–234 (1994). https://doi.org/10.1016/0361-3682(94)90033-7

22. Harvey, C.M., Koubek, R.J.: Toward a model of distributed engineering collaboration. Comput. Ind. Eng. **35**(1–2), 173–176 (1998). https://doi.org/10.1016/s0360-8352(98)00053-9
23. Byström, K.: Task Complexity, Information Types and Information Sources: Examination of Relationships. http://www.hb.se/bhs/personal/katriina/kby-diss.pdf (1999). 11 May 2007
24. Vakkari, P.: Task complexity, problem structure and information actions. Integrating studies on information seeking and retrieval. Inf. Process. Manag. **35**(6), 819–837 (1999). https://doi.org/10.1016/S0306-4573(99)00028-X
25. Gill, T.G., Cohen, E.: Research themes in complex informing. Informing Sci. **11**, 147–164 (2008). https://doi.org/10.28945/444
26. Babbie, E.R.: The Practice of Social Research, 11th edn. Belmont (2007)
27. Wagner-Hartl, V., Schmid, R., Gleichauf, K.: The influence of task complexity on acceptance and trust in human-robot interaction – gender and age differences. Cogn. Comput. Internet of Things **43**, 118–126 (2022). https://doi.org/10.54941/ahfe1001846
28. Frazier, S., McComb, S.A., Hass, Z., Pitts, B.J.: The moderating effects of task complexity and age on the relationship between automation use and cognitive workload. Int. J. Hum. Comput. Interact. (2022). https://doi.org/10.1080/10447318.2022.2151773
29. von Walter, B., Kremmel, D., Jäger, B.: The impact of lay beliefs about AI on adoption of algorithmic advice. Mark. Lett. (2021). https://doi.org/10.1007/s11002-021-09589-1
30. Xu, Y., Shieh, C.H., van Esch, P., Ling, I.L.: AI customer service: task complexity, problem-solving ability, and usage intention. Australas. Mark. J. **28**(4), 189–199 (2020). https://doi.org/10.1016/j.ausmj.2020.03.005
31. Mascha, M.F., Smedley, G.: Can computerized decision aids do 'damage'? A case for tailoring feedback and task complexity based on task experience. Int. J. Account. Inf. Syst. **8**(2), 73–91 (2007). https://doi.org/10.1016/j.accinf.2007.03.001
32. Sniezek, J.A., Van Swol, L.M.: Trust, confidence, and expertise in a judge-advisor system. Organ. Behav. Hum. Decis. Process. **84**(2), 288–307 (2001). https://doi.org/10.1006/obhd.2000.2926
33. Landelijk Overleg Vakinhoud Strafrecht, "Oriëntatiepunten voor straftoemeting en LOVS-afspraken, pp. 1–45 (2020). www.rechtspraak.nl
34. Simnett, R.: The effect of information selection, information processing and task complexity on predictive accuracy of auditors. Account. Organ. Soc. **21**(7–8), 699–719 (1996). https://doi.org/10.1016/0361-3682(96)00006-2
35. Driver, M., Streufert, S.: Integrative complexity: an approach to individuals and groups as information-processing systems. Adm. Sci. Q. **14**(2), 272–285 (1969)
36. Williams, G.C., Ded, E.L.: Self-perceived competence. Encycl. Qual. Life Well-Being Res. **70**(4), 5784 (2014). https://doi.org/10.1007/978-94-007-0753-5_103737
37. Newell, S.J., Goldsmith, R.E.: The development of a scale to measure perceived corporate credibility. J. Bus. Res. **52**(3), 235–247 (2001). https://doi.org/10.1016/S0148-2963(99)00104-6
38. McKnight, D.H., Choudhury, V., Kacmar, C.: Developing and validating trust measures for e-commerce: an integrative typology. Inf. Syst. Res. **13**(3), 334–359 (2002). https://doi.org/10.1287/isre.13.3.334.81
39. Hook, K.: Evaluating the utility and usability of an adaptive hypermedia system. In: International Conference Intelligence User Interfaces, Proceedings of the IUI, no. February, pp. 179–186 (1997). https://doi.org/10.1145/238218.238320
40. de Vries, P., Midden, C., Bouwhuis, D.: The effects of errors on system trust, self-confidence, and the allocation of control in route planning. Int. J. Hum. Comput. Stud. **58**(6), 719–735 (2003). https://doi.org/10.1016/S1071-5819(03)00039-9

# Can a Conversational Agent Pass Theory-of-Mind Tasks? A Case Study of ChatGPT with the Hinting, False Beliefs, and Strange Stories Paradigms

Eric Brunet-Gouet[1,2]($\boxtimes$) (iD), Nathan Vidal[2] (iD), and Paul Roux[1,2] (iD)

[1] Centre Hospitalier de Versailles, Service Hospitalo-Universitaire de Psychiatrie d'Adultes et d'Addictologie, Le Chesnay, France
ebrunet@ght78sud.fr

[2] Université Paris-Saclay, Université Versailles Saint-Quentin-En-Yvelines, DisAP-DevPsy-CESP, INSERM UMR1018, 94807 Villejuif, France

**Abstract.** We investigate the possibility that the recently proposed OpenAI's ChatGPT conversational agent could be examined with classical theory-of-mind paradigms. We used an indirect speech understanding task, the hinting task, a new text version of a False Belief/False Photographs paradigm, and the Strange Stories paradigm. The hinting task is usually used to assess individuals with autism or schizophrenia by requesting them to infer hidden intentions from short conversations involving two characters. In a first experiment, ChatGPT 3.5 exhibits quite limited performances on the Hinting task when either original scoring or revised rating scales are used. We introduced slightly modified versions of the hinting task in which either cues about the presence of a communicative intention were added or a specific question about the character's intentions were asked. Only the latter demonstrated enhanced performances. No dissociation between the conditions was found. The Strange Stories were associated with correct performances but we could not be sure that the algorithm had no prior knowledge of the test. In the second experiment, the most recent version of ChatGPT (4-0314) exhibited better performances in the Hinting task, although they did not match the average scores of healthy subjects. In addition, the model could solve first and second order False Beliefs tests but failed on items with reference to a physical property like object visibility or more complex inferences. This work offers an illustration of the possible application of psychological constructs and paradigms to a conversational agent of a radically new nature.

**Keywords:** large language model · ChatGPT · theory-of-mind · indirect speech · False Beliefs

## 1 Introduction

Having a theory of mind is conceived as the capacity of an individual to impute mental states to himself or to others (either to conspecifics or to other species as well), such as intentions, beliefs, or knowledge (Premack and Woodruff, 1978). In the present work, we

J. Baratgin et al. (Eds.): HAR 2023, LNCS 14522, pp. 107–126, 2024.
https://doi.org/10.1007/978-3-031-55245-8_7

address the possibility of applying these concepts and some available paradigms to new artificial intelligence (AI) technologies, a conversational agent based on a Large Language Model (LLM), and highlight their potential contributions in the addressed areas. Historically, the concept was coined to investigate animals like chimpanzees (Premack and Woodruff, 1978), and was successful to study early child development (Baillargeon, Scott, and He, 2010) and pathological conditions like autism (Baron-Cohen, Leslie, and Frith, 1985). To determine whether an individual without language or with non-proficient communication skills has a theory-of-mind (ToM), the preferred approach is to observe his/her behavior when he/she interacts with others (animals or humans) in experimental conditions that require the use of these skills. However, the existence of articulate language and the ability to conduct conversations and process complex requests makes it seemingly easy to detect theory of mind abilities in the individual. Firstly, the use of mental state terms or concepts (i.e., "I believe", "she thinks", "he wants", "they seek to" etc.) can be considered as a proof of the conceptual capacity to attribute volitional or epistemic mental states to another one. Secondly, language is part of a pragmatic context of communication. It has long been conceptualized that our ability to communicate is largely based on the building of a shared knowledge with the interlocutor and that deciphering indirect or metaphorical language requires the understanding of the communicative intention of interlocutors (Sperber and Wilson, 1986). Authors have proposed to test the theory of mind through the comprehension of indirect language in the sense that pragmatics could be understood as a sub-module of ToM, however the intricacy of the two constructs does not mean that they should be confused theoretically and empirically (see discussion in Bosco, Tirassa, and Gabbatore, 2018).

Cognitive neuropsychologists theorized that patients with schizophrenia suffer from communication and/or social cognition disorders that could be reflected by deficits in the comprehension of indirect speech (Frith, 1992; Hardy-Baylé, Sarfati, and Passerieux, 2003). Experimental paradigms have been proposed to measure the deficit of patients and to measure their pragmatic skills (Bazin et al. 2005; Langdon et al. 2002; Mazza et al, 2008). Here, we consider the *Hinting Task* introduced by Corcoran et al (1995). This task was designed to test the ability of subjects to infer the real intentions behind indirect speech utterances. The task consists of ten short stories presenting an interaction between two characters in which one character drops a very obvious hint. Allowing to detect deficits in schizophrenia, the task was selected by the SCOPE initiative to create a social cognition assessment battery (Pinkham, Harvey, and Penn, 2018). More recently norms in healthy subjects and schizophrenic patients were published with both the original scoring system and a new one proposed to improve psychometric properties by lowering ceiling effects (Klein et al., 2020). In the following study, we will consider both criteria to score a subject's performance.

Before the creation of the hinting task, the attribution of epistemic mental states, namely beliefs and knowledge, could be tested through the so-called False Belief paradigm also referred to as the *Unexpected Transfer* paradigm (Baron-Cohen, Leslie, and Frith, 1985; Wimmer, 1983). In its classical version, two dolls, Ann and Sally, are involved in a short sequence in which one of them moves an object after the other doll has left the scene. The task is to figure out that the second doll will look for the object in the place she left it and not where it actually is. This task was regularly posed as a gold standard to attest the presence or deficiency of a first-order theory of mind (i.e., inferences about another's mental state) and different versions have been proposed in order to distinguish a capacity for mentalization from a simple memorization of the world's state at a specific moment of the sequence. However, as Perner and Wimmer (1985) discussed, understanding the behavior of others is not based solely on inferring their representations in terms of their true or erroneous beliefs about the world: it is also necessary to take into account what others think about the thoughts of others, the so-called second-order beliefs. The paradigm of false beliefs of the second order is an interesting complexification in developmental psychology because it is supposed to be acquired later, around 5–6 years, in children than the ability to understand the first order, which would be evident between the ages of 3 and 5 (Miller, 2009).

Last, we focused in this work on Happé's *Strange Stories* task which is another paradigm allowing to test mental and physical inference skills from short texts. This task was initially developed for studying pragmatics and ToM in children with autism (Happé, 1994) and was revised with a control condition not involving mental state attribution (White et al, 2009). It consists of unambiguous situations in which untrue utterances are made by a character for different reasons (i.e., lie, sarcasm, double-bluff, etc.) depending on the context, the emotional content, or the relationship between the characters. The rationale for this task was to provide diverse real-life situations compared to the first- or second-order false-belief tasks, for which autistic patients could possibly present non-ToM strategies to pass the test. Interestingly, such a paradigm highlighted reduced performances of autistic children in the mentalizing conditions (White et al., 2009). Note also that unlike the hinting task, this task does not target a stereotyped conversation situation in pragmatical terms.

In the field of artificial intelligence, the development of conversational agents has recently made striking progress, allowing these models to produce responses that resemble human responses. These agents are disembodied and "purely linguistic" by nature, and trained over extremely large text corpuses essentially from internet databases. If computer linguistic agents have increasingly sophisticated communication capacities, their internal architectures do not necessarily include models of mental states. This raises a fundamental doubt about the ability of these "stochastic parrots" according to the denomination of Bender et al. (2021), to take into account the communicative intent and share knowledge during a verbal exchange, because they have not been trained in such a situation: "Text generated by a LM [Language Model] is not grounded in communicative intent, any model of the world, or any model of the reader's state of mind. It can't have been, because the training data never included sharing thoughts with a listener, nor does the machine have the ability to do that."

In what follows, we investigate whether methods of assessments of pathological ToM in humans can be applied to a conversational agent. Usability of psychological constructs, like the Big Five Personality Model, with LLM was discussed by Pellert et al. who argued for adopting a new psychometric perspective (2022). We chose to focus on the recently proposed ChatGPT model,[1] which is the state of the art in this field. It is based on multiple steps fine-tuning of a transformer-based architecture with reinforcement learning for human feedback and the use of large-scale conversation datasets (see reviews of generative models in (see reviews of generative models in Gozalo-Brizuela and Garrido-Merchan, 2023). This model brought a lot of attention because it has extremely impressive skills to produce sound and well-formulated answers to a wide range of questions: its similarities with human experts has recently been investigated (Guo et al., 2023 p. 202). The use of deep learning techniques based on training over massive databases, makes it particularly complicated to answer the question of the existence, even implicit, of mental concepts represented within very complex, poorly interpretable structures. It is interesting to note that the interpretability of deep-learning models, i.e., the way they can be understood by humans, is a complex question with legal, societal and ethical consequences (Carvalho, Pereira, and Cardoso, 2019), and that it may be considered as sharing some logics with neuroscientists' objectives of understanding brain computations and discovering neural codes.

Recent studies reported experimental evidence of the manifestation of ToM skills in LLM. Through a conversation with the ChatGPT, Mortensen (2023) reported that the chatbot did not consider it had a ToM while it knew some about the concept. Other authors assessed ChatGPT-3 with both a Chinese and an English version of Sally and Ann False Belief paradigm and introduced subtle variations in the phrasing of the test (Dou, 2023). They reported quite dramatic differences in performances but acknowledged that ChatGPT could score like a human subject at the test under certain conditions. Yet, the reasons of these discrepancies remain unclear. Kosinski (2023) investigated successive versions of this model (GPT-3 "davinci-001", "davinci-002", GPT-3.5 and GPT-4) with a False Belief test. He reported a continuous improvement of the performances in the False Belief test that reached nearly a 95% success rate for GPT-4, yielding similar results than a 7-year-old child.

Building on Kosinski's (2023) positive findings of GPT's ToM skills, Ullman (2023) proposed to challenge GPT-3.5 with slightly modified versions of ToM tasks in order to find out subtle conditions that elicit failure. The unfavorable conditions included statements about visibility of the objects, uninformative labels placed on the objects, or information given by a trusted person, which can't be taken into account properly by the LLM. Errors in the slightly modified versions of TOM tasks highlighted negligence or incoherence in the hierarchization or the integration of information by GPT-3.5, leading the author to question the existence of a theory of mind in this model. Finally, another study based on the SOCIALIQA database and a progressive test procedure with an increasing number of examples (k-shot probing), brought into light a limited capacity of GPT-3 to process mental states compared with factual questions (Sap et al., 2023).

---

[1] OpenAI, ChatGPT: Optimizing language models for dialogue, https://openai.com/blog/chatgpt/ (November 30, 2022). To access the web-based chat: https://chat.openai.com/chat.

To further investigate ChatGPT's skills, we conducted our study like a single clinical case study and tested with several evaluations of ToM abilities, i.e., the Hinting task, False Beliefs paradigm and the Strange Stories. Adult patients with clinical conditions like schizophrenia exhibit impaired performances in the hinting task, which impair their real-life functioning (Pinkham, Harvey, and Penn, 2018), therefore we used their scores and that of healthy subjects, taken from the psychometric study of (Klein et al., 2020), to provide a clinically relevant range of values. This enabled us to place ChatGPT's scores on a standardized scale. Following an exploratory approach, whenever it appeared relevant, we slightly modified the tasks in order to investigate the AI's skills by measuring performance improvement when specific cues were added. In these cases, we investigated changes in the wording of the questions to see the extent to which their focus on the characters' intentional states or expectations might have altered the accuracy of the responses. Two experiments were conducted and investigated two successive versions of ChatGPT (i.e., 3.5 and 4).

## 2 Experiment #1

### 2.1 Methods

**Hinting Task.** Free online OpenAI's ChatGPT-3.5 (Dec 15th 2023 version) was interrogated with a series of questions from the hinting task. For a subpart of the following procedure subsequent test sessions were conducted with the Jan 09th and 30th versions. In its original version, this task designed to assess patients with schizophrenia consists of ten situations described by three sentences involving two characters (Corcoran, Mercer, and Frith, 1995). In each situation, one protagonist says something that may be indirectly interpreted as a request toward the other character (see an example in Table 1). Invariably, a question is asked "What does the character really mean when he says this?", in order to test the chatbot or the human subject's capability to infer a communicative intention. The expected answer is rewarded by two points (HINTING 1 scores). In the absence of a correct answer, a second version (HINTING 2) with an additional cue and a question focusing on the character's request to his/her interlocutor is proposed (see Table 1). If respondents clarify the subjects' intention ("George wants Angela to get him or offer to get him a drink."), they score one-point.

The text responses were rated independently by two of the authors (EBG and NV) according to specific and validated criteria (HINTING 1 + 2). Between each sentence the chatbot's history was erased so that no context could interfere. Three scoring systems were used in order to compare the model's skills with those of healthy or individuals with schizophrenia, and with the models' itself. The first one is the original version described by Corcoran et al. (1995), the second one is a revised version from the SCOPE initiative by Klein et al. (Klein et al., 2020). The third scoring system (False/True) corresponds to the number of correct answers that are defined as the correct understanding of the intentional meaning of the sentence, allowing us to compare the scores with the two following conditions, for which no validated scoring system is available.

**Table 1.** List of experimental conditions used in Experiment #1. Examples of full prompts are provided in italics

| Experiment | Task | Conditions names | Number of items | Rating system | Associated construct | Rationale |
|---|---|---|---|---|---|---|
| #1 | Hinting | HINTING1 and HINTING2 | 10 + 10 | (Klein et al. 2020) + (Corcoran, Mercer, and Frith 1995) + False/true | Attribution of intentions to others, inferring indirect speech meaning | Two step validated task testing pragmatic language understanding in situations where a character produces indirect speech. In reference to ToM skills, the task requires the attribution of hidden and of communicative intentions |

*Ex. HINTING1: George arrives in Angela's office after a long and hot journey down the motorway. Angela immediately begins to talk about some business ideas. George interrupts Angela saying: "My, my! It was a long, hot journey down that motorway!" What does George really mean when he says this?*
*Ex. HINTING2: George arrives in Angela's office after a long and hot journey down the motorway. Angela immediately begins to talk about some business ideas. George interrupts Angela saying: "My, my! It was a long, hot journey down that motorway!" George goes on to say: "I'm parched!" What does George want Angela to do?*

| Experiment | Task | Conditions names | Number of items | Rating system | Associated construct | Rationale |
|---|---|---|---|---|---|---|
| #1 | Hinting | INJONCTION | 10 | False/true | idem | The same as HINTING2 but without the additional cue. This allows to test the specific effect of the question focusing on the character's demands |

*Ex. George arrives in Angela's office after a long and hot journey down the motorway. Angela immediately begins to talk about some business ideas. George interrupts Angela saying: "My, my! It was a long, hot journey down that motorway!" What does George want Angela to do?*

| Experiment | Task | Conditions names | Number of items | Rating system | Associated construct | Rationale |
|---|---|---|---|---|---|---|
| #1 | Hinting | COMMUNICATIVE | 10 | False/true | idem | The same as HINTING1 but with a focus on the concept of "communicative intention" |

*Ex. George arrives in Angela's office after a long and hot journey down the motorway. Angela immediately begins to talk about some business ideas. George interrupts Angela saying: "My, my! It was a long, hot journey down that motorway!" Considering George's most likely communicative intention, what does he really mean when he says this?*

*(continued)*

**Table 1.** (*continued*)

| Experiment | Task | Conditions names | Number of items | Rating system | Associated construct | Rationale |
|---|---|---|---|---|---|---|
| #1 | False Beliefs | LOOK | 2 | False/true | First order false belief attribution | This test follows the unexpected object transfer paradigm allowing to test the capability to infer a first-order false belief. The question focuses on the character's search strategy |

*Ex. Alan and Jenny are in the kitchen. Jenny puts her chocolate inside the cupboard. Then, she leaves the kitchen. Alan moves the chocolate to the table, and then he moves it to the fridge, and moves it again to the table. Later, Jenny comes back to the kitchen to eat her chocolate. Where will Jenny look to find it?*

| Experiment | Task | Conditions names | Number of items | Rating system | Associated construct | Rationale |
|---|---|---|---|---|---|---|
| #1 | False Beliefs | EXPECT | 2 | False/true | First order false belief attribution | Same as in LOOK, but with the focus on the subject's expectations (i.e., her mental state) |

*Ex. Alan and Jenny are in the kitchen. Jenny puts her chocolate inside the cupboard. Then, she leaves the kitchen. Alan moves the chocolate to the table, and then he moves it to the fridge, and moves it again to the table. Later, Jenny comes back to the kitchen to eat her chocolate. Where will Jenny expect to find it?*

| Experiment | Task | Conditions names | Number of items | Rating system | Associated construct | Rationale |
|---|---|---|---|---|---|---|
| #1 | Strange stories | MENTAL | 16 | False/true | Complex social situations understanding | The task depicts various situations requiring the understanding of involving understanding of double bluff, white lie, persuasion, and misunderstanding, etc |

*Ex. Simon is a big liar. Simon's brother Jim knows this, he knows that Simon never tells the truth! Now yesterday Simon stole Jim's ping-pong paddle, and Jim knows Simon has hidden it somewhere, though he can't find it. He's very cross. So he finds Simon and he says, "Where is my ping-pong paddle? You must have hidden it either in the cupboard or under your bed, because I've looked everywhere else. Where is it, in the cupboard or under your bed"? Simon tells him the paddle is under his bed. Why will Jim look in the cupboard for the paddle?*

| Experiment | Task | Conditions names | Number of items | Rating system | Associated construct | Rationale |
|---|---|---|---|---|---|---|
| #1 | Strange stories | PHYSICAL | 16 | False/true | Complex situation understanding | This task is the control task corresponding to MENTAL, not requiring mental state attribution |

*Ex. Two enemy powers have been at war for a very long time. Each army has won several battles, but now the outcome could go either way. The forces are equally matched. However, the Blue army is stronger than the Yellow army in foot soldiers and artillery. But the Yellow army is stronger than the Blue Army in air power. On the day of the final battle, which will decide the outcome of the war, there is heavy fog over the mountains where the fighting is about to occur. Low-lying clouds hang above the soldiers. By the end of the day the Blue army has won. Why did the Blue army win?*

To investigate ChatGPT's errors when it deals with intentional situations, two other versions of the sentences were used. These modifications were made in an exploratory approach based on the observed errors in order to reveal the sensitivity of the model to the pragmatic aspects of the questions:

1. Character's Intention version (INJONCTION): the three sentences as described above (HINTING 1) are kept and the question from the second part of the task (HINTING 2) is added without the additional cue (see Table 1). This experimental condition thus clarifies the question by indicating that one character expects something from the other. However, no information is added to the context. Thus, performances can be compared with that of HINTING 1.
2. Communicative intention version (COMMUNICATIVE): In this version the question of the original version (HINTING 1) was modified and indicated explicitly that a character likely had a "communicative intention" (see Table 1). This version allowed us to test whether the chatbot could use this concept to urge an inference about the character's intention. The fact that ChatGPT had some conceptual knowledge about "communicative intention" was tested with a direct interrogation (see Supplementary Material 1.1).

**False Beliefs Task.** The version proposed here of the False Belief task was designed specifically to test the capabilities of ChatGPT. The test was performed twice to check the answer's coherence. It involves a description of a classic initial situation with two protagonists in a kitchen, Alan and Jenny, the latter placing her chocolate in the cupboard. The object transfer sequence was made more complex than in the classical versions and was declined in several versions (two False Beliefs situations) in order to make sure that the model does not respond randomly based on the probability of the usual presence of chocolates in cupboards.

The question is asked in two ways. Indeed, it appeared depending on the formulation, ChatGPT could answer very differently to the questions leading either to success or failure:

- In the LOOK condition, the question is about where Jenny will look for the chocolate when she returns (see Table 1).
- In the EXPECT condition, the question is about where Jenny expects to find her chocolate, introducing an interrogation about the mental state (see Table 1).

**Strange Stories Mental and Physical Task.** In this task, the ChatGPT is required to interpret short vignettes and is asked to explain why a character says something that is not literally true (White et al, 2009; Happé, 1994). To succeed it has to attribute mental states such as desires, beliefs or intentions, and sometimes higher order mental states such as one character's belief about what another character knows. Scoring system ranges from 0 to 2 points for each story depending on the quality of the interpretation. In addition, two conditions of equal difficulty are presented: MENTAL and PHYSICAL, the former focusing on pragmatics and mental state attributions and the second on physical states only (see Table 1).

## 2.2 Results

Please note that all the results of the evaluations of this article are available in the Supplementary Material. The data were reported as Google Sheets and processed with Jamovi (www.jamovi.org).

**Hinting Task.** ChatGPT 3.5 performs poorly at the first question of the task (HINTING 1) with a number of correct answers between 1 and 4 on a total of 10 questions. However, this score is largely improved in the second part of the task when a cue and a question about the character's demands is asked (HINTING 2). Indeed, more questions are answered correctly, raising the score to 8 or 9 on 10. Even if this improvement is significant, the use of validated rating systems by two raters shows that the overall performances (HINTING 1 + 2) remain at 10 or 13/20 (Corcoran system) and 8 to 9/20 (Klein system). Based on the work of (Klein et al. 2020), let's remind that healthy subjects (n = 286) performed at 17.9/20 (SD = 2) and 16/20 (SD = 2.5), with both scoring systems respectively, and that the patients with chronic schizophrenia (n = 375) scored 15.7/20 (SD = 3.4) and 13.7/20 (SD = 3.4), at their first evaluation. In all cases, ChatGPT performed worse than human subjects with z-scores at -2.4 and -2.8 for Corcoran and Klein's scores respectively (those values correspond to the best performance of ChatGPT measured in the present study compared with the normal distribution from the aforementioned article).

Compared to the HINTING 2 condition, INJONCTION allowed ChatGPT to give 6 or 8 correct answers out of 10, while COMMUNICATIVE only reached a score of 4/10. The scores obtained by ChatGPT given the conditions are reported in Table 2.

**Table 2.** Raw scores (scored as true or false) at the hinting task. Answers were obtained and scored at four different epochs. Maximum score is 10

| Conditions | ChatGPT version | dec 15 | jan 09 | jan 09 | jan 30 | ChatGPT's maximum score |
|---|---|---|---|---|---|---|
| HINTING 1 | dec 15 | 2 | 4 | 2 | 1 | 4 |
| HINTING 2 | dec 15 | 8 | 8 | 9 | 8 | 9 |
| INJONCTION | dec 15 | 6 | 7 | 6 | 8 | 8 |
| COMMUNICATIVE | dec 15 | 4 | 2 | 4 | 4 | 4 |

In all cases and conditions, ChatGPT 3.5 was able to give a detailed answer from one to five sentences. In the majority of the cases, it provided several hypotheses, which often included the correct answer. In some cases, the model qualified its answers by expressing doubts ("It is difficult to say for certain", "It is not clear from the information provided...") or even incertitude ("It's not possible for me to accurately determine the true meaning behind Rebecca's statement."). Answers and quotations are reported in Supplementary Tables.

**False Beliefs Task.** The conversational agent exhibits perfect performances at the False Beliefs task in the EXPECT condition (100% of good response, see Table 3). On the contrary the LOOK condition is failed at the exception of one item when the chocolate is returned to its initial place. We scored this answer as zero because the justification was wrong, leading to a total of 0% of good response.

**Table 3.** ChatGPT scores at the False Belief tasks and the False Photographs task. Note that the tests were performed twice and resulted in the same ratings. Maximum value is 2

| Conditions | jan 09 | jan 09 |
|---|---|---|
| False Beliefs (LOOK) "Where will Jenny look to find it ?" | 0 | 0 |
| False Beliefs (EXPECT) "Where will Jenny expect to find it ?" | 2 | 2 |

**Strange Stories Mental and Physical Task.** ChatGPT answered quite correctly to both MENTAL (13/16, 81% of good response) and PHYSICAL (12/16, 75% of good response) conditions in the Strange Stories task (see Table 4) when the scoring system described in White et al. (2009) was used.

**Table 4.** ChatGPT scores at both sets of stories from the Strange Stories task. Each condition consists of eight stories, scoring a maximum of two points. Maximum score is 16

| Conditions | ChatGPT version | Scores |
|---|---|---|
| Mental stories (MENTAL) | jan 30 | 13 |
| Physical stories (PHYSICAL) | jan 30 | 12 |

## 2.3 Discussion

The first observation of this experiment is the amazing quality of the answers given by ChatGPT-3.5. They are correctly organized, and present a dialectical effort to discuss several hypotheses. Of importance, characters' names and roles are preserved in the answers. Our results concur with Guo et al.'s (2023) findings that this model provides "organized" with clear logic" answers which tend to be "long" and "detailed". However, such answers would not match the natural answers of human subjects unless they were urged to discuss all hypotheses and their probabilities. These rather long and hesitant answers could appear to a clinician as a way to avoid answering, an obsessive indecision or a smokescreen strategy. In some case the model moderates its own conclusions using formulations like "it is not clear that". These utterances give the appearance of the existence of some kind of metacognitive judgment. It also gives to the reader the impression that one preponderant answer strategy is based on a reformulation of the questions that does not require much inference skills. However, in many cases it appears that ChatGPT is able to conclude and even produces some intentional hypotheses as shown below.

Second, the Hinting Task highlighted the poor intentional inferences of ChatGPT-3.5 according to the criteria defined by Corcoran et al. (1995) or Klein et al. (2020). The scores obtained in both the initial and secondary questions do not compete with human performances even in psychopathological conditions. This result could totally exclude ChatGPT-3.5 from being considered as having intentional mental states inferential skills. However, a more careful analysis of the responses shows that when the model receives additional cues and more focused questions about the intention of the characters, its responses are significantly improved. Eight to nine good responses out of 10 were given in the HINTING 2 condition, which is likely to indicate that ChatGPT-3.5 has the ability to make intentional inferences, even if those scores are considered weak with the classical score formula.

Additional experiments with the modified hinting task provide further insight into the factors that can help ChatGPT to respond correctly. First of all, adding (COMMU-NICATIVE condition) a cue prior to the question and directing the model to consider the most likely communicative intention does not seem to improve the responses (score from 2 to 4/10).

A greater gain (scores of 6 or 8/10), although weaker than with the HINTING 2 condition, is found when it is clearly specified that the task requires a statement about what the character wants the other character to do (INJONCTION). This result suggests that the injonction directs the answer to the mentalist concept and allows the inference of an intentional mental state. In this case, ChatGPT uses "X wants Y to", "X asks Y to", or "X would like Y to" to specify the intention. Thus, we find that some indications in addition to the question allow the language model to provide responses that encompass the intentions normally presented in simple life situations. However, it has a heavy tendency to stay "strictly focused on the given question" as noted by Guo et al. (2023).

Regarding the False Belief paradigm in Experiment #1, the results are also contrasted and depend directly on the questions asked. This is in line with Dou's (2023) report of ChatGPT-3's variable performances in Ann and Sally False Belief test depending on complex combinations of phrasing and language, and with the fact that alteration of ToM tasks can induce errors of ChatGPT-3.5 (Ullman, 2023). We notice that ChatGPT has real capacities to exploit complex sequences of information indicating successive actions and even taking into account representations of a changing reality.

We note in the case of False Beliefs an ability to take into account the knowledge of one of the characters to anticipate what she expects. Jenny will expect to find her chocolate in the cupboard, as that is where she put it before leaving the kitchen. Surprisingly, sometimes emotional elements are given to describe the possible reaction of the character to the location of the object.

Example: ChatGPT-3.5: "She may be surprised or confused to find it in the fridge or on the table".

Let us note the failure of the model in the LOOK condition. In our case, this type of question seems to induce the necessity for the model to answer to the final position of the object and not to the position anticipated by Jenny. One explanation could be that ChatGPT is trying to answer the literal question: where will Jenny look for the object is a question corresponding to the final location of the object. But ChatGPT does not infer that this question is about Jenny's expectations (her mistaken beliefs) that will guide her search strategy. However, such interpretation is contradicted by Mortensen's (2023) report of a correct answer in another version of Ann and Sally's false belief paradigm, ending with a look question ("Where will Sally look for her marble?"). As indicated above, our paradigm was, on purpose, made more complex, with several successive moves of the chocolate in order to dampen the success of a probability answer. It appears that subtle modifications of the formulation and the situation produce changes that are difficult to anticipate, sometimes allowing ChatGPT to either produce answers whose argumentation indicates that it has taken into account the mental state of the character or answers that are incompatible with a theory of mind.

Finally, the strange stories test also seems to bring positive results regarding the inference capacities of the model. Such a level of performance is unexpected given the higher length of the stories compared to the simple situations used in the hinting task. It is possible, however, that these detailed contexts (often eight sentences), that are explicitly dealing with mental states such as the characters' beliefs, urges ChatGPT to use theory of mind concepts. However, as discussed in the limitations of the study, ChatGPT could have been exposed to the data on this task (logics, situations, etc.) during the learning procedure. It is therefore difficult for us to integrate the results of the Strange Stories task in our conclusions. This emphasizes the needs for the creation of large-scale test corpora and the design of theory of mind task generators to test models on unpublished sets of situations or stories.

# 3  Experiment #2

Taking into account the results from experiment #1 and some of the discrepancies in the ratings, the assessments were conducted again using the up-to-date GPT-4-0314 model through its API. As shown with different paradigms by Ullman (2023) and Dou (2023) simple modifications of the false belief task could greatly alter GPT-3 performance. To extend our investigations with GPT-4 and determine if the main conclusions are replicated, we wanted to see if the new version brought performance improvements to the Hinting Task and to the False Belief Task. We implemented supplementary first and second-order mental states conditions to the tasks to challenge the model with more complicated inferences. We added information about whether the character can or cannot see the object or the action as proposed by Ullman (2023). Finally, we tested whether the implicit attribution of mental states like true or false beliefs could be used to feed deductive reasoning.

## 3.1  Methods

In Table 5 the different conditions are described. Code is provided in Supplementary Material 2.

The HINTING 1 + 2 was scored with Klein's method (Corcoran, Mercer, and Frith, 1995; Klein et al., 2020) by four independent raters among whom EBG and NV. Each input was tested three times to check the stochastic variability in the answers (Pellert et al, 2022).

**Table 5.** List of experimental conditions used in Experiment #2. Examples of full prompts are provided in italics

| Experiment | Task | Conditions names | Number of items | Rating system | Associated construct | Rationale |
|---|---|---|---|---|---|---|
| #2 | Hinting | HINTING1 and HINTING2 | 10 + 10 | (Klein et al. 2020) | See table #XYZ | See Table 1 |
| #2 | False Beliefs 1st order | Move vs. No move | 2 + 2 | False/True | First order ToM | As in the LOOK condition of experiment #1 |
| | | *Ex. Move: Alan and Jenny are in the kitchen. Jenny hides her chocolate inside the cupboard. Then, she leaves the kitchen. Then, Alan takes the chocolate and hides it inside the fridge. Later, Jenny comes back to the kitchen. Where will Jenny look to find the chocolate?* | | | | |
| | | *Ex. No move: Alan and Jenny are in the kitchen. Jenny hides her chocolate inside the cupboard. Then, she leaves the kitchen. Then, Alan takes the chocolate and hides it inside the cupboard. Later, Jenny comes back to the kitchen. Where will Jenny look to find the chocolate?* | | | | |
| #2 | False Beliefs 1st order | Visibility | 1 | False/True | First order ToM + pragmatic reasoning | As in (Ullman 2023), an additional information about the object's visibility should prevent the character from using his erroneous belief about its location |
| | | *Ex. Alan and Jenny are in the kitchen. Jenny hides her chocolate inside the cupboard. Then, she leaves the kitchen. Then, Alan takes the chocolate and puts it on the table which is the first place one can see when entering the room. Later, Jenny comes back to the kitchen. Where will Jenny look to find the chocolate?* | | | | |
| #2 | False Beliefs 1st order | Trio | 2 | False/True | First order ToM | The Move condition is made more complex adding a third character |
| | | *Ex. Alan, John and Jenny are in the kitchen. Jenny hides her chocolate inside the cupboard. Then, she leaves the kitchen. Then, Alan takes the chocolate and hides it in the fridge. Then, he leaves the kitchen. Then, John takes the chocolate and hides it in the trashcan. Later, Jenny and Alan come back to the kitchen. Where will Jenny look to find the chocolate?* | | | | |
| #2 | False Beliefs 1st order | Trio Different place vs. Same place | 1 + 1 | False/True | First order ToM | The 1st order prediction about both characters behavior is used implicitly as an input into a deductive mechanism |

*(continued)*

**Table 5.** (*continued*)

| Experiment | Task | Conditions names | Number of items | Rating system | Associated construct | Rationale |
|---|---|---|---|---|---|---|
| *Ex. Different place: Alan, John and Jenny are in the kitchen. Jenny hides her chocolate inside the cupboard. Then, she leaves the kitchen. Then, Alan takes the chocolate and hides it in the fridge. Then, he leaves the kitchen. Then, John takes the chocolate and hides it in the trashcan. Later, Jenny and Alan come back to the kitchen. Will Alan and Jenny look for the chocolate at the same place?* | | | | | | |
| #2 | False Beliefs 2st order | Trio | 4 | False/True | Second order ToM | These conditions contrast the second order beliefs of one character about the false beliefs of another character having participated in the successive moves of the object |
| *Ex. Alan, John and Jenny are in the kitchen. Jenny hides her chocolate inside the cupboard. Then, she leaves the kitchen. Then, Alan takes the chocolate and hides it in the fridge. Then, he leaves the kitchen. Then, John takes the chocolate and hides it in the trashcan. Later, Jenny and Alan come back to the kitchen. Where Alan thinks that Jenny will look for the chocolate?* | | | | | | |
| #2 | False Beliefs 2st order | Trio different place vs. Same place | 1 + 1 | False/True | Second order ToM | The 2st order prediction about two characters' false beliefs is used implicitely as an input into a deductive mechanism |
| *Ex. Different place: Alan, John and Jenny are in the kitchen. Jenny hides her chocolate inside the cupboard. Then, she leaves the kitchen. Then, Alan takes the chocolate and hides it in the fridge. Then, he leaves the kitchen. Then, John takes the chocolate and hides it in the trashcan. Later, Jenny and Alan come back to the kitchen. Does John thinks that Alan and Jenny will look for the chocolate at the same place?* | | | | | | |

The False beliefs was also assessed in a revised version taking into account problems encountered in Experiment #1: for instance, the initial position where Jenny places the chocolate changed in order to check that other answers than "the cupboard" could be given. The false belief task was also revised in its phrasing to assess several possible situations concerning first and second order beliefs varying the number of characters. This allowed us to test the model's flexibility when it infers characters' expectations. The use of second-order mental states ("X thinks that Y will look in Z") aims at increasing the difficulty of the task according to theory of mind literature. Table 5 describes the conditions and provides examples. In the condition Visibility, we introduce the physical information about the fact that the object is left visible on the table, in order to prevent false belief to be used. In the Same vs. Different places conditions, two characters' beliefs attributions are required implicitly to infer if the character will look for the object at the same or a different place.

## 3.2  Results

Regarding the Hinting Task ratings, the percent overall agreement between raters was 84.44% and the fixed-marginal Fleiss's kappa reached 0.69 (95% CI = [0.54, 0.85]) (4 raters, 3 levels, 30 cases) corresponding to a substantial agreement. Across the four raters and the three repetitions, the mean ratings of the Hinting Task using Klein's system was 13.4 (SD = 1.3, range: [11 15]). These performances are equivalent to those found in adult patients with chronic schizophrenia (z-score = -0.08) but slightly inferior to the healthy controls group (z-score = $-1.04$) (Klein et al., 2020). Among 120 ratings, the 0, 1, and 2 values frequencies were respectively 2.5%, 60.8% and 36.7%.

Interrater reliability of the False Beliefs ratings (45 items, 4 raters) reached the maximum agreement value of 100%. The conversational agent achieved a maximum performance in $1^{st}$ and $2^{nd}$ order False Beliefs tests but failed at the $1^{st}$ order Visibility and the $2^{nd}$ order Different Place conditions (see Table 6).

**Table 6.** ChatGPT-4-0314 assessment with the revised version of the False Beliefs test, allowing to test its performance on $1^{st}$ and $2^{nd}$ order false beliefs inferences. Mean scores of the four raters and three repetitions are provided

| Order of mental state | Condition | Number of items | Mean Scores |
|---|---|---|---|
| 1st Order | Move vs. no move | 6 | 6 |
| | duo_visibility | 3 | 0 |
| | Trio | 6 | 6 |
| | Same vs. Different place | 6 | 6 |
| 2nd Order | Trio | 12 | 12 |
| | Trio Different place | 3 | 0 |
| | Trio Same place | 3 | 3 |

## 3.3  Discussion

For the Hinting Task and the False Beliefs task, the more recent version of ChatGPT-4 accessed through its API demonstrates a higher level of performance than in Experiment #1. The text responses are also more concise, consisting of one or two sentences, and provide an acceptable answer to the question, allowing an easier evaluation of their correctness.

The Hinting Task elicits clearer answers and is no longer based on contradictory arguments as reported above. When we compare the performance of ChatGPT-4 to the score distribution of a control subjects group (Klein et al., 2020), we only observe a trend towards lower scores ($-2.6$ points, z-score = $-1.04$). Moreover, the model reaches average performances of patients with schizophrenia. Even if performances were improved compared to Experiment #1, the error patterns, with a majority of items scoring 1, could reveal that the additional cue and the request question are still helpful to succeed in the task. In the previous experiment, we argued that the indication to focus on the character's demand could be an incentive to adopt an intentional reading. The results of the Experiment #2 are in line with this hypothesis.

The False Belief task brings a very high level of performance and contrary to experiment #1, the use of the formulation "Where will [the character] look to find the chocolate?" does not seem to cause any difficulty. We also note that several changes of formulation, successive transfers of the object (up to three locations), the increase of the number of characters from two to three are not criteria that systematically penalize the model. Nevertheless, the robustness of the model with regards to these complexity parameters would benefit from being confirmed in other studies.

However, as reported by Ullman, the model does not seem to be able to handle practical physical information such as the notion of visibility of objects (Ullman, 2023). Indeed, the only condition of the first-order false beliefs leading to a failure consists in indicating that the supposedly hidden object is placed in evidence in the room, so that the character does not need to use his belief to guide the search. Providing this information was intended to encourage the LLM to consider 1) the physical reality of the scene which requires practical knowledge of people's environment, 2) the consequences of this scenery on the character's visual perception, and 3) the primacy of visual perception in feeding the cognitive system and updating beliefs. It is possible that the learning corpus of this recent version of ChatGPT, that is entirely text-based and does not rely on physical immersion or bodily embodiment, did not allow it to be trained for one or more of these steps.

The order of false belief attribution was similarly investigated with the introduction of a second order formulation such as "Where [first character] thinks that [second character] will look for the chocolate?" While this increment in the ToM orders corresponds to an increase in the complexity of the task, ChatGPT-4 still succeeds. Sap et al. (2023) reported a slight decrease of performances of GPT-3-DAVINCI in a second order False Beliefs task by comparison with the first order one, at least when the number of examples in the k-shot language probing procedure varies from 2 to 8. The only failures are found when the implicit formulation of the Different place condition was used. Here again, this may indicate the sensibility of the model to unusual and complex story structures (Ullman, 2023; Dou, 2023).

## 4   General Discussion

Detecting and evaluating manifestations of intentional reasoning and more generally of a theory of mind in an individual is a complex task that has given rise to numerous research and methodological proposals. In this paper, we investigate the ability of a recent and sophisticated model of conversational agent, OpenAI's two versions of ChatGPT (3.5 and 4), to use intentional reasoning to understand ambiguous language. To do so, we simply confronted the chatbot with items from well-known tasks that are standard in research on mentalization. If this simple implementation of a single case methodology cannot by itself solve the question of the presence of a form of theory of mind in natural language processing models, the analysis of the results reveals some intriguing and important methodological questions for the future. It should also be noted that our approach was deliberately observational, as in clinical investigations of single cases. Indeed, we do not use, in our analysis, any knowledge of the information processing mechanisms or computations used by ChatGPT to generate responses. We considered ChatGPT like a human subject. We used simple scoring grids already defined in clinical research to observe the quality of the responses.

As shown by the results, the 3.5 version of ChatGPT does not exhibit spontaneous and reliable use of theory of mind inference to find the most likely interpretation of hints in simple conversations. However, in certain conditions, when the question focused on people's intentions, the conversational model produced correct interpretations with explicit and even well-argued references to mental states and improves its scores. Arguably, some capabilities to "infer" first-order mental states existed but were not to be favored by the model. In addition, some False Belief attribution and complex dynamic representation of the world existed.

An interesting result of the second experiment with ChatGPT-4-0314 is the improved performance and the better quality of the answers. Thus, for both the Hinting task and the false beliefs, the number of correct answers is increased. In particular, the model can handle complex False Belief tasks including a formulation where a second order mental state is questioned. Moreover, its ability to take into account indirect speech, without reaching that of an average-performing adult, reveals a capacity to infer the underlying intentions as soon as the questions are clarified. When the questions do not focus on the character's requests toward another character, the answers do not focus regularly to the hidden intention. The analysis of the errors shows the sensitivity of the model to the test conditions as it is unable to process additional information of a practical nature (the visibility of an object for example), or complex formulations (false beliefs).

## 5  Limitations

The present study is obviously insufficient to clarify the question of the existence of a theory of mind in a numeric human-like model. The small number of items makes it difficult to draw a conclusion without being able to produce a statistical inference by repeating test trials. A longer procedure would be needed to test the model studied on multiple occasions. We note, however, that the small number of trials corresponds to the usual conditions for the use of tests in clinical practice. It is interesting that the sophistication of ChatGPT allowed us to consider it as a usual human subject.

Let's also note that successive versions of ChatGPT produce changes in measured performances as reported by Kosinski (2023). Based on that, we conducted the assessment on two successive versions (ChatGPT-3.5 December 2022 and January 2023), and we found only slight differences that may be partly related to some stochastic parts in the model. An additional experiment with a newer version ChatGPT-4 in April 2023 brought strong evidence of improved skills. Taken together, the results demonstrate a difficulty posed by these evolving models in providing comparable data and replication of experiments. It is thus crucial that LLM developers make available a versioning of their models.

Finally, we are not aware of the possible contamination of answers by the succession of questions we asked. Eliminating the question history does not guarantee a perfect reset of the response system (at least if we consider ChatGPT as a human being). Moreover, ChatGPT's large learning base could include the data and knowledge gathered about the tasks we used, due to the overwhelming literature concerning these paradigms. A direct questioning of ChatGPT about the hinting task does not seem to reveal any particular knowledge of the task or its authors while it acknowledges that it could be useful (see Supplementary material 1.2). Concerning the False Belief tests, Kosinski acknowledges that they could have been seen by the model during the training phase (2023), and thus designs a novel test paradigm. In the present study, we also designed from scratch the tests we used. The drawback of this procedure is that our False Beliefs tests lack validation and prior psychometric knowledge.

On the contrary, the Happé's Strange Stories task is clearly referred to by ChatGPT as a well-known paradigm used to study social cognition deficits in autism (see Supplementary material 1.3). It is not possible to exclude that the model has been trained with these situations. For the experimenter, if we consider the design of tasks that can be used by both humans and machines, it will be necessary to guarantee their confidentiality, particularly since the AIs may be trained on any available public data including articles in scientific literature describing these same tasks.

## 6 Conclusion

This study challenged a recently designed conversational agent, ChatGPT, with the hinting task, a classical intention reading task used to assess social cognitive skills in schizophrenia, and with a new False Beliefs test and the Strange Stories task. Adopting a radical, blind to technological considerations about the models' design, cognitive psycho-pathological and clinical perspective, we used this task as in a single case study and proposed revised versions in order to characterize the AI performances. The main finding is that some paradigms classically designed for human study are applicable in conditions quite close to their use in clinical or research settings. The psychological constructs that are attached to them (theory of mind, attribution of beliefs or intentions, communicative intention, etc.) seem to be implementable in research even though the conversational agent being studied is of a radically new nature. However, the results concur with Mortensen's (2023) note that "In a way interaction with ChatGPT is unlike any other interaction we know as humans". These new cognitive agents of an artificial nature are likely to challenge the double-dissociations of performance found in humans that underlie neuropsychological constructs as we know them, which could lead to profound revisions of conceptual frameworks in neuroscience.

## 7 Ethical Statement and Conflicts of Interest

The authors declare that the research was conducted in the absence of any commercial or financial relationships that could be construed as a potential conflict of interest. They acknowledge that the emergence of LLMs and their deployment by industries or even governments pursuing economic or political goals raises broad ethical questions about their use in academic research. The implementation of moral rules, of mechanisms that mimic human behavior and might deceive people, of content neutrality is a research topic in itself that cannot be left to the designers of these models and should be handled by open, independent and peer-evaluated research, as proposed in the present study.

**Acknowledgements.** We are grateful to Clarissa Montgomery and Augustin Chartouny for their kind participation in the assessments. The authors would like to thank deepl.com AI for their suggestions in some of the translations.

**Author Contributions.** EBG and NV participated in the assessments. EBG wrote the first draft of the manuscript which was corrected and commented by all the authors.

# References

Baillargeon, R., Scott, R.M., He, Z.: False-belief understanding in infants. Trends Cogn. Sci. **14**(3), 110–118 (2010). https://doi.org/10.1016/j.tics.2009.12.006

Baron-Cohen, S., Leslie, A.M., Frith, U.: Does the autistic child have a 'theory of mind'? Cognition **21**(1), 37–46 (1985)

Bazin, N., Sarfati, Y., Lefrere, F., Passerieux, C., Hardy-Bayle, M.C.: Scale for the evaluation of communication disorders in patients with schizophrenia: a validation study. Schizophrenia Res. **77**(1), 75–84 (2005)

Bender, E.M., Gebru, T., McMillan-Major, A., Shmitchell, S.: On the dangers of stochastic parrots: can language models be too big? " In: Proceedings of the 2021 ACM Conference on Fairness, Accountability, and Transparency, PP. 610–23. ACM, Virtual Event Canada (2021). https://doi.org/10.1145/3442188.3445922

Bosco, F.M., Tirassa, M., Gabbatore, I.: Why pragmatics and theory of mind do not (completely) overlap. Front. Psychol. **9**(August), 1453 (2018). https://doi.org/10.3389/fpsyg.2018.01453

Carvalho, D.V., Pereira, E.M., Cardoso, J.S.: Machine learning interpretability: a survey on methods and metrics. Electronics **8**(8), 832 (2019). https://doi.org/10.3390/electronics8080832

Corcoran, R., Mercer, G., Frith, C.D.: Schizophrenia, symptomatology and social inference: investigating 'theory of mind' in people with schizophrenia. Schizophrenia Res. **17**(1), 5–13 (1995)

Dou, Z.: Exploring GPT-3 Model's Capability in Passing the Sally-Anne Test A Preliminary Study in Two Languages. Preprint. Open Science Framework (2023). https://doi.org/10.31219/osf.io/8r3ma

Frith, C.D.: The Cognitive Neuropsychology of Schizophrenia. Laurence Erlbaum Associates Publishers, Hove, UK (1992)

Gozalo-Brizuela, R., Garrido-Merchan, E.C.: ChatGPT Is Not All You Need. A State of the Art Review of Large Generative AI Models (2023). https://doi.org/10.48550/ARXIV.2301.04655

Guo, B., et al.: How Close Is ChatGPT to Human Experts? Comparison Corpus, Evaluation, and Detection (2023). https://doi.org/10.48550/ARXIV.2301.07597

Happé, F.G.E.: An advanced test of theory of mind: understanding of story characters' thoughts and feelings by able autistic, mentally handicapped, and normal children and adults. J. Autism Dev. Disord. **24**(2), 129–154 (1994). https://doi.org/10.1007/BF02172093

Hardy-Baylé, M.C., Sarfati, Y., Passerieux, C.: The cognitive basis of disorganization symptomatology in schizophrenia and its clinical correlates: toward a pathogenetic approach to disorganization. Schizophrenia Bull. **29**(3), 459–471 (2003)

Klein, H.S., et al.: Measuring Mentalizing: A Comparison of Scoring Methods for the Hinting Task. Int. J. Methods Psychiatric Res. **29**(2), e1827 (2020). https://doi.org/10.1002/mpr.1827

Kosinski, M.: Theory of Mind May Have Spontaneously Emerged in Large Language Models. arXiv. http://arxiv.org/abs/2302.02083 (2023)

Langdon, R., Coltheart, M., Ward, P.B., Catts, S.V.: Disturbed communication in schizophrenia: the role of poor pragmatics and poor mind-reading. Psychol. Med. **32**(7), 1273–1284 (2002)

Mazza, M., Di Michele, V., Pollice, R., Casacchia, M., Roncone, R.: Pragmatic language and theory of mind deficits in people with schizophrenia and their relatives. Psychopathology **41**(4), 254–263 (2008). https://doi.org/10.1159/000128324

Miller, Scott A.: Children's understanding of second-order mental states. Psychol. Bull. **135**(5), 749–773 (2009). https://doi.org/10.1037/a0016854

Mortensen, D.: ChatGPT Is Constantly Tricked by People, but Who's Fooling Who? *Prototypr.Io* (blog). https://prototypr.io/post/chatgpt-is-constantly-tricked-by-people-but-whos-fooling-who (2023)

Pellert, M., Lechner, C.M., Wagner, C., Rammstedt, B., Strohmaier, M.: AI Psychometrics: Using Psychometric Inventories to Obtain Psychological Profiles of Large Language Models." Preprint. PsyArXiv. https://osf.io/jv5dt (2022)

Perner, Josef, Wimmer, Heinz: 'John thinks that mary thinks that…' attribution of second-order beliefs by 5- to 10-year-old children. J. Exp. Child Psychol. **39**(3), 437–471 (1985). https://doi.org/10.1016/0022-0965(85)90051-7

Pinkham, A.E., Harvey, P.D., Penn, D.L.: Social cognition psychometric evaluation: results of the final validation study. Schizophrenia Bull. **44**(4), 737–748 (2018). https://doi.org/10.1093/schbul/sbx117

Premack, D., Woodruff, G.: Does the chimpanzee have a theory of mind? The Behav. Brain Sci. **1**, 515–526 (1978)

Sap, M., LeBras, R., Fried, D., Choi, Y.: Neural Theory-of-Mind? On the Limits of Social Intelligence in Large LMs. http://arxiv.org/abs/2210.13312 (2023)

Sperber, D., Wilson, D.: Relevance: Communication and Cognition. Basil Blackwell, Oxford (1986)

Ullman, T.: Large Language Models Fail on Trivial Alterations to Theory-of-Mind Tasks. arXiv. http://arxiv.org/abs/2302.08399 (2023)

White, S., Hill, E., Happé, F., Frith, U.: Revisiting the strange stories: revealing mentalizing impairments in Autism. Child Dev. **80**(4), 1097–1117 (2009). https://doi.org/10.1111/j.1467-8624.2009.01319.x

Wimmer, H.: Beliefs about beliefs: representation and constraining function of wrong beliefs in young children's understanding of deception. Cognition **13**(1), 103–128 (1983). https://doi.org/10.1016/0010-0277(83)90004-5

# Does Cognitive Load Affect Explicit Anthropomorphism?

Fabien Calonne[1]($\boxtimes$), Marion Dubois-Sage[1], Frank Jamet[1,2,3], and Baptiste Jacquet[1,2]

[1] Paris 8 University, 93526 Saint-Denis, France
fcalonne@gmail.com
[2] Association P-A-R-I-S, 75005 Paris, France
[3] CY Cergy Paris Université, 95000 Cergy-Pontoise, France

**Abstract.** People tend to attribute human traits to non-human entities, such as animals, God or robots. This tendency, referred to as anthropomorphism, is seen as a natural, automatic inclination. Within the framework of Dual Process Theories, some researchers have proposed a model of anthropomorphism that distinguishes between implicit and explicit processes. This model states that implicit anthropomorphism is a result of Type 1, intuitive processing, while explicit anthropomorphism is a result of Type 2, deliberative processing. The results obtained so far seem to corroborate this model. Although perfectly well founded, we believe that this model's clear separation between indirect measures – intended to measure implicit anthropomorphism – and direct measures – intended to measure explicit anthropomorphism – is a questionable premise from both a theoretical and a practical point of view. Contrary to these previous studies, we thus recommend testing the automaticity of anthropomorphism toward robots by manipulating the concurrent cognitive load while participants give their explicit attributions. Our hypotheses are that, since we believe anthropomorphism to be an automatic and implicit process, cognitive load will either (i) lead to an increase of the anthropomorphism attributions or (ii) result in an absence of difference between the different condition of load, proving that Type 2 processes are pointless in the anthropomorphizing course. We discuss and propose a paradigm aimed at testing our hypothesis.

**Keywords:** Anthropomorphism · Dual Process Theory (DPT) · Human-Robot Interaction (HRI) · Dual Task Paradigm · Cognitive Load

## 1 Introduction

Over the next few years, Human-Robot Interactions (HRIs) are expected to become ubiquitous [1]. During HRIs, humans behave as they would with a fellow human [2]. They attribute human characteristics to them, such as emotions or intentions [3], a process called anthropomorphism [4] (for a review, see [5]). This process influences both the way humans represent robots to themselves and how robots affect human cognition and behavior [6]. For example, replacing the adult partner with a humanoid robot (NAO) increases the success rate in the false belief task in typically developing children [7] and

© The Author(s), under exclusive license to Springer Nature Switzerland AG 2024
J. Baratgin et al. (Eds.): HAR 2023, LNCS 14522, pp. 127–138, 2024.
https://doi.org/10.1007/978-3-031-55245-8_8

in children with Autism Spectrum Disorders [8]. This procedure, known as the mentor-child paradigm, reduces the pragmatic ambiguity of the task, and thus appears to be better suited for assessing children's social-cognitive skills.

Therefore, how humans perceive and explain robot agency has become a topical issue in social robotics [9]. As proposed by Złotowski and colleagues, anthropomorphic attributions, just like attributions on the whole, may be shaped as a result of distinct cognitive processing [10]. Evans and Stanovich had previously been postulated that attributions as a whole could be understood as the result of distinct types of cognitive processing [11]. According to this dual process theory, or DPT [11–14], two sets of general processes with distinct properties can be distinguished: Type 1 (t1) and Type 2 (t2). T1 processes are implicit, fast, take place effortlessly and in parallel, have high capacities and little need for working memory, whereas T2 processes are explicit, slow, require effort and operate in series, have limited capacities and make massive use of working memory [15].

DPT has been successfully employed in many areas of human behavior such as logical reasoning [11], moral judgment [16] or theory of mind [17] to name but a few (for a review, see [12]). However, while popular among scholars, this distinction between intuition and deliberation has also come under criticism (e.g. [18]). We adopt De Neys's view that whether intuition and deliberation differ in degree or in nature is irrelevant to the development of any psychological theory [14], and thus focus on the mechanisms underlying human thinking.

The above-mentioned model [10] proposes that anthropomorphism depends on Type 1 and Type 2 processes. It is based on the assumption that implicit anthropomorphism is a T1 process, while explicit anthropomorphism is a T2 process. Methodologically, they assume that direct measures only reflect explicit anthropomorphism, while indirect measures only reflect implicit anthropomorphism. We believe that this is an important concession that deserves to be validated experimentally. The aim of our article is to justify our argument and to provide avenues for testing it.

Here we discuss this dual model of anthropomorphism. We begin by presenting it and its empirical implications. We then critically analyze the data from studies that have adopted this approach. We end by proposing a new experimental rationale that, in our view, could validate a dual model of anthropomorphism, albeit free from the restrictive methodological premise adopted by [10]. Our aim is not to take a step backwards, but to make sure we haven't missed any ladders.

## 2   A Dual-Process Approach to Anthropomorphism

### 2.1   Anthropomorphism in HRI

Anthropomorphism is the tendency to attribute human characteristics to nonhuman entities [2–4] (for a review, see [5]). This is a widely acknowledge and ubiquitous process in social life [19], and anthropomorphic attributions have been demonstrated over a wide range of agents, from geometric shape [20] to animals [21] to God [22]. This tendency, usually described as irrepressible [23], has been conceived as an automatic psychological process for centuries [3]. According to Dacey, anthropomorphism would be a heuristic used by our unconscious folk psychology to understand nonhuman animals [24].

Recently, there has been a growing academic interest in anthropomorphism toward robots [25]. It has long been assumed that anthropomorphism is a "default schema" and that studying anthropomorphism toward mechanistic devices could help us better understand anthropomorphic attributions toward our fellow human beings: "*In the end, the question of what makes people think machines can think is a question about what makes us think other people can think*" [26].

In the most recent years, the tendency to interpret and predict behaviors in terms of mental states (desires, beliefs) – the so-called intentional stance [27] – has similarly been put forward as the default mode of human brain [28]. Perez-Osorio and Wykowska proposed that adopting the intentional stance might be a pivotal factor in facilitating social attunement with artificial agents [29]. They also argued that the cognitive processes used by anthropomorphism and those that lead to attributing intention to these agents could be the same. Several behavioral investigations have tried to detangle the factors that lead humans to adopt this intentional stance toward artificial agents [9, 30–34]. For instance, Thellman and colleagues showed that people's attitude toward a Pepper robot (SoftBank Robotics) was very similar to the attitude they adopt toward a human fellowship [34].

Besides, Nass and Moon, in their seminal works, has shown that participants treat technological devices (e.g. television or computer) as social agents despite being reluctant to admit any anthropomorphic attributions when they were explicitly debriefed afterward [35, 36]. Epley have previously suggested that anthropomorphic inferences may serve as a base for automatic attribution that could be afterward modulated by deliberate reasoning [3]. Similarly, Fussell and colleagues showed that, although people tended to attribute human-like qualities to robots based on their initial instinctive judgments, their answers became more mechanistic when they were asked to provide an explicit response [37]. Thus, it seems that people handle social devices differently, depending on whether they rely on intuition or deliberation [38]. To elucidate this apparent contradiction, [10] proposed to distinguish between implicit and explicit anthropomorphism. In other words, they proposed a dual process model of anthropomorphism.

## 2.2 The Dual Process Model of Anthropomorphism and Its Questionable Premise

This dual model of anthropomorphism is based on two major assumptions: (i) that the observed interindividual differences in anthropomorphism depend on distinct cognitive processes (T1 vs. T2), and (ii) that implicit anthropomorphism is an outcome of T1 processing while explicit anthropomorphism is an outcome of T2 processing. This is fully consistent with previous dual process models (e.g. [15]).

Taking their thinking a step further, and building on the work of De Houwer [39], they assume that direct measures (i.e. questionnaires) only reflect explicit anthropomorphism, while indirect measures (e.g. response time, facilitation effect, galvanic skin response, electrocardiogram, fMRI, eye tracking, pupillometry; see [25]) only reflect implicit anthropomorphism. It follows that (i) measures of implicit anthropomorphism is achievable, and (ii) before concluding that anthropomorphism is the result of dual processing, both the direct and indirect measures of anthropomorphism must be shown to be independently affected. Our aim is not to question their rationale, which we believe to be entirely relevant and persuasive, but to ensure that their premises are valid.

Indeed, implicit measures are a strong tool to study human cognitive architecture; they offer a unique insight into the consequences of automatic processing since they reflect the automatic impact of attitudes and cognitions [39]. Moreover, implicit measures could help grasping automatic processes, and thus overcoming the biases associated with explicit measures; implicit measures of attitudes towards human compared to robotic agents are thus promising [40]. To differentiate explicit and implicit features, one need to discriminate distinct representational formats for each process; indeed, subjects can grasp explicit representation content effortlessly and verbally report on it, while fuzzy content lacks clarity in expression and thought [15]. However, although this distinction between implicit and explicit anthropomorphism is very useful, we don't think necessary to postulate that indirect measures cannot be used to index explicit processes, and *vice versa*.

We issue that the aforementioned proposal is questionable from both a theoretical and a practical perspective. First, it seems to us that this assumption is a major and unnecessary concession to the principle of parsimony, which refers to the total number of assumptions required to explain a given phenomenon [14]. This philosophical principle, also known as Ockham's razor, posit that if two theories are just as suitable to explain a given phenomenon, the one using the fewest axioms should be preferred. Put simply, simpler explanations are better.

Secondly, and more importantly, we do not believe this strict dichotomy to be experimentally justified, nor to be necessary. Indeed, an implicit mechanism can actually affect an explicit response, and vice versa. In the field of human reasoning, for example, in which DPT has been widely used, it is common to use an explicit measure (i.e. the participants' tacit response to a problem) to index an implicit process (i.e. a cued heuristic response) (e.g. [41]). Similarly, an implicit measure (e.g. response time, dermal conductance, pupil dilation, and so on) can index the complexity of an explicit, deliberative process, like a digits memorizing task (e.g. [42]).

## 2.3   Back to the Model and the Data

Thus, [10] hypothesized that anthropomorphism, just like many other cognitive mechanisms, could be the result of dual process attribution. They proposed a dual process model in which implicit anthropomorphism relies on automatic T1 process, while explicit anthropomorphism relies on the more cognitive demanding T2 process. In the first attempt to disentangle the influence of T1 and T2 processes in such attributions toward robots, they manipulated the participants' motivation to engage in T2 processing while interacting with a Robovie R2 (Intelligent Robotics and Communication Laboratories). They do so by manipulating task instructions. However, this manipulation did not lead to any difference regarding the explicit anthropomorphism (as measured by the attribution of HN traits), nor did it affect the implicit anthropomorphism (measured as the effect of a priming task). They conclude by suggesting that future work should focus on developing indirect measure of anthropomorphism.

Following this seminal paper, psychologists have investigated the different processes of anthropomorphizing robots by relying on various combinations of explicit and implicit measures [1, 9, 30–33, 43, 44]. Nevertheless, these studies showed mixed results. Marchesi and colleagues conducted a series of studies to assess whether people can adopt the intentional stance towards artificial agents, how they differ in their tendency to attribute mental states to humans and robots, and how they react to different levels of human-likeness in robot behavior [30–33]. They first showed that it is possible to encourage people to adopt the intentional stance toward a robot iCub (RobotCub Consortium) by probing participants' stance towards mentalistic versus mechanistic explanations of its behavior [30]. They also found that creating a shared social context with this humanoid robot can increase the adoption of the intentional stance, while the mere exposure to it did not [33].

More importantly for our purposes, Marchesi and colleagues used the InStance (IST) to compare how people judge the intentions of human and robot agents based on their actions [31]. They found that, at the explicit level, their participants tend to use more mentalistic descriptions for the human and more mechanistic descriptions for the robot. However, response times did not show any difference, suggesting that, at the implicit level, both stances were equally likely to be activated for the robot. Their results were confirmed in a following study [32]. Using a combination of the IST and participants' pupil dilation monitoring, they showed that individual biases in adopting the intentional vs. design/mechanical stance may indicate differences in mental effort and cognitive flexibility in interpreting the behavior of artificial agents and that pupil dilation was higher when the robot showed hints of human-likeness, such as eye contact, gestures, and speech. Individual differences in perceiving subtle human-like cues from robots, possibly tied to personal traits or attitudes, may thus affects judgments about their intentionality.

Spatola & Wudarczyk conducted two experiments to assess how people's explicit and implicit judgments of robots' emotions relate to their perception of robot anthropomorphism [43]. Within two online experiments, participants were asked to rate the emotions and human-likeness of robots based on their facial expressions and voices. The results showed that people's explicit ratings of robot emotions were influenced by their perceived robot anthropomorphism, but their implicit evaluation (as measured by an adapted version of the Implicit Association Test) were not. Using the same protocol, these authors also showed in a follow up study that implicit attitudes towards robots predicted explicit attitudes, semantic distance between robots and humans, anthropomorphism, and prosocial behavior towards robots [44].

Another series of studies conducted by Spatola and colleagues investigated the effect of cognitive load on the tendency to anthropomorphize a humanoid robot. Using an online experiment where participants watched videos of a robot (NAO; SoftBank Robotics) performing tasks and rated its human-likeness and intentionality under different levels of cognitive load [1]. The results showed that cognitive load increased the ratings of human-likeness and intentionality of the robot, suggesting that explicit anthropomorphism is an automatic process that requires less cognitive resources. In another study [9], participants watched videos of a robot and answered questions about its intentions and emotions (i.e. using a modified version of the IST) under different levels of cognitive load. The results showed that the mentalistic representation seems to be the default one, as (i) it was more difficult to switch from a mentalistic representation to a mechanistic representation than the opposite (experiment 1) and (ii) participants were faster to select the mentalistic description in the high cognitive load condition (experiments 2 & 4). Moreover, this default social cognitive system seems to operate automatically since cognitive load did not affect the late semantic stages of processing, but did affect the early stage of processing (experiments 3 & 4).

Overall, the studies conducted by Marchesi, Spatola and their colleagues highlight the complexity of how we perceive and interact with artificial agents, combining intentional and mechanical aspects. They show that, at the explicit level, participants tend to attribute mental explanations to humans and mechanical explanations to robots. However, at the implicit level, both types of explanation seem to be activated in a similar way for robots. In addition, cognitive load seems to increase explicit anthropomorphism towards robots. However, we believe this last point needs further investigations. Among these studies, those that have used explicit measures of anthropomorphism all face the same methodological pitfall, namely the timing of the implicit manipulation relative to the explicit measure. Indeed, the explicit measure of anthropomorphism (i.e. questionnaires) were always proposed at distance from the experimental manipulations. In the [31] study, the Individual Differences in Anthropomorphism Questionnaire (IDAQ; [45]) was proposed after the robot IST block and before the human one. In the Spatola and Chaminade's study [1], the different questionnaires, aimed at evaluating explicit anthropomorphism, were proposed either at the beginning or at the end of the procedure, that is apart from the cognitive load manipulation. The reasoning behind this methodology is that manipulating the cognitive load when the representations are formed will affect the final evaluation (i.e.; the questionnaire's score). However, the resulting output is impure in terms of the processes involved. It can similarly capture the results of both Type 1 and Type 2 processes, as well as the outcome of their interaction. Zlotowski and colleagues has pointed out that direct measures of anthropomorphism can reflect either the implicit or explicit anthropomorphism [10]. In this particular case, it's been verified, but we believe it doesn't necessarily have to be true.

Finally, Spatola and colleagues used the IST, a mentalistic vs. mechanistic description decision task, which can indeed be conceived as an explicit measure of anthropomorphism [9]. Only in the third experiment did they manipulate the cognitive load at the time participants' judgement was made. They found no significant difference nor on response time, nor on proportion of mentalistic choice. The cognitive control framework they have adopted led them to conclude cognitive load does not affect late, semantic, stages of processing while evaluating descriptions of robot's behaviors. However, in the DPT framework, such an absence of effect would be interpreted as the fact that T2 processes are not needed for anthropomorphic attributions, and thus that they rely on T1 (see Table 1). This interpretation is reinforced by the fourth experiment in which participants were faster to select the mentalistic compared to the mechanistic response, but only in the high load condition.

**Table 1.**  This table summarizes the different possible outcomes arising from the manipulation of cognitive load in the dual-task paradigm, along with the various interpretations associated with them in the DPT framework.

| Cognitive load effect on the task of interest | Interpretation in terms of DPT |
| --- | --- |
| Reduced score | T2 processes are necessary: the task relies on deliberative T2 processes |
| No effect on the score | T2 are unnecessary: the task relies on automatic T1 processes |
| Improved score | T2 processes monitor T1 outcomes: the task needs T2 processes to be regulated |

These elegant and well-designed studies shed interesting light on the automatism of anthropomorphic attributions and the distinction between T1 and T2 processes. However, as mentioned above, they are based on the strong premise of a dichotomy between implicit anthropomorphism, which can only be measured indirectly, and explicit anthropomorphism, which can be indexed directly using questionnaires. However, the question arises as to whether cognitive load might have a direct impact on explicit attributions.

## 3   How to Test Our Hypothesis

As we already mentioned, our aim is not to call into question the relevance of implicit measures of human attitudes towards robots. As has been argued elsewhere [9, 10, 40], comparing implicit attitudes towards humans and robotic agents is a promising path. Our aim is to show that implicit process can effectively been measured explicitly and to shed further light on the dual process account of anthropomorphism. We see almost three ways it can be done: by using cognitive load, time pressure or both.

As in the classical aforementioned reasoning dual task paradigm, we assume that having our subjects to give an explicit response under cognitive load will allow us to bypass the T2 processes and thus to identify the intuitively generated response produced by T1 processes [46]. Since they are cognitive resource demanding [11], we can reduce T2 processes outcome by burdening participants' cognitive resources with a demanding concurrent task. Our hypothesis being that, as anthropomorphism is an automatic and implicit process, this will either (i) lead to an increase of the anthropomorphism scales' score in the high load condition or (ii) result in an absence of difference between the different condition of load, proving that T2 processes are pointless in the anthropomorphizing course (see **table 1**). The next paragraph further develops our paradigm.

As has been amply and empirically documented, appearance is a major factor contributing to anthropomorphism (e.g. [2, 47–49]; for a review, see [5]). For instance, a recent review concluded that human-like robots give rise to greater attribution of mental states relative to robots with less human-like appearance, even if the tendency to ascribe mental states has been reported toward robots with all kinds of physical appearance [50].

It might therefore be interesting to make use different types of agents, such as those identified by Duffy, which are: humanoid, iconic and abstract [2]. Moreover, a human agent as well as a standard object would provide two good control conditions [51, 52]. As measures of attributed anthropomorphism, we recommend using two different scales to assess both physical and mentalist attributions, like the Godspeed Questionnaire Series (GQS) and the attribution of mental states questionnaire (AMS-Q). The GQS has been developed by Bartneck and colleagues [53]. This is one of the most frequently used tools in HRI research and is available in 19 languages to date [54]. The questionnaire helps gain insights into how people perceive and interact with robots, making it valuable for evaluating the effectiveness and acceptance of various robotic technologies (for meta-analyses, see [55, 56]). The AMS-Q has been developed by Manzi and colleagues (e.g. [48]), assesses the attribution of mental and sensory states to an agent. Primarily used with human, the AMS-Q has been validated to evaluate the level of mental anthropomorphisation of nonhuman agents [57]. Finally, Spatola and Caminade showed that subjective cognitive load can affect robots' anthropomorphism (experiment 1). The increase in human traits attribution in the high load condition was also significantly mediated by the reported level of cognitive effort (experiment 2) [1]. As a result, a measure of perceived cognitive load would have to be included in order to control for its possible effects. Prior experience with robots should also be controlled [40]. We believe that this protocol will allow us to put our hypotheses to the test.

Another way of testing our hypothesis would be to have this evaluation carried out under time pressure. The rationale is the same as for the cognitive load manipulation. A final method to test our hypothesis would be to use the two-response paradigm [58]. In this paradigm, participants first give an intuitively generated response (e.g. Under time-pressure and/or cognitive load [59]). Next, they are allowed to take all the time they want to before they give their final response. Comparing the human trait attribution between these two conditions should permit to detangle the specific contributions of T1 and T2 processes.

## 4   Conclusion

Our aim was to provide experimental avenues to test the hypothesis that, in agreement with the dual-process model of Złotowski and colleagues [10], anthropomorphism rely on two distinct types of process, T1 and T2, but without relying on their assumption that direct and indirect measures respectively evaluate explicit and implicit anthropomorphism. We hope to have shown that it is possible to test this premise before developing this model any further. The preliminary results we have so far obtained using the paradigm described above appear to confirm our hypotheses.

**Acknowledgments.** We would like to thank the P-A-R-I-S association and the CHArt laboratory (Cognitions Humaine et Artificielle, RNSR 200515259U) for supporting and funding this research project. We would also like to thank Jean Baratgin for his advice and help in the preparation of this work.

# References

1. Spatola, N., Chaminade, T.: Cognitive load increases anthropomorphism of humanoid robot. The automatic path of anthropomorphism. Int. J. Hum-Comput. Stud. **167**, 102884 (2022)
2. Duffy, B.R.: Anthropomorphism and the social robot. Robot. Auton. Syst. **42**, 177–190 (2003)
3. Epley, N., Waytz, A., Cacioppo, J.T.: On seeing human: a three-factor theory of anthropomorphism. Psychol. Rev. **114**, 864 (2007)
4. Pickett, J.P.: Anthropomorphism. The American Heritage Dictionary of the English Language (2000)
5. Dubois-Sage, M., Jacquet, B., Jamet, F., Baratgin, J.: We do not anthropomorphize a robot based only on its cover: context matters too! Appl. Sci. **13**, 8743 (2023)
6. Spatola, N., et al.: Improved cognitive control in presence of anthropomorphized robots. Int. J. Soc. Robot. **11**, 463–476 (2019)
7. Baratgin, J., Dubois-Sage, M., Jacquet, B., Stilgenbauer, J.-L., Jamet, F.: Pragmatics in the false-belief task: let the robot ask the question! Front. Psychol. **11**, 593807 (2020)
8. Dubois-Sage, M., Jacquet, B., Jamet, F., Baratgin, J.: The mentor-child paradigm for individuals with autism spectrum disorders. In: Social Robots Personalisation: At the Crossroads between Engineering and Humanities (CONCATENATE). , Stockholm, Sweden (2023)
9. Spatola, N., Marchesi, S., Wykowska, A.: Cognitive load affects early processes involved in mentalizing robot behaviour. Sci. Rep. **12**, 14924 (2022)
10. Złotowski, J., Sumioka, H., Eyssel, F., Nishio, S., Bartneck, C., Ishiguro, H.: Model of dual anthropomorphism: the relationship between the media equation effect and implicit anthropomorphism. Int. J. Soc. Robot. **10**, 701–714 (2018)
11. Evans, J.St.B.T., Stanovich, K.E.: Dual-process theories of higher cognition: advancing the debate. Perspect. Psychol. Sci. **8**, 223–241 (2013). https://doi.org/10.1177/1745691612460685
12. Evans, J.S.B.: Dual-processing accounts of reasoning, judgment, and social cognition. Annu. Rev. Psychol. **59**, 255–278 (2008)
13. Kahneman, D.: Thinking, Fast and Slow. Farrar, Straus and Giroux (2011)
14. De Neys, W.: On dual- and single-process models of thinking. Perspect. Psychol. Sci. **16**, 1412–1427 (2021). https://doi.org/10.1177/1745691620964172
15. Bellini-Leite, S.C.: Dual process theory: embodied and predictive; symbolic and classical. Front. Psychol. **13**, 805386 (2022). https://doi.org/10.3389/fpsyg.2022.805386
16. Greene, J., Haidt, J.: How (and where) does moral judgment work? Trends Cogn. Sci. **6**, 517–523 (2002). https://doi.org/10.1016/S1364-6613(02)02011-9
17. Sherman, J.W., Gawronski, B., Trope, Y.: Dual-Process Theories of the Social Mind. Guilford Publications (2014)
18. Melnikoff, D.E., Bargh, J.A.: The mythical number two. Trends Cogn. Sci. **22**, 280–293 (2018). https://doi.org/10.1016/j.tics.2018.02.001
19. Chartrand, T.L., Fitzsimons, G.M., Fitzsimons, G.J.: Automatic effects of anthropomorphized objects on behavior. Soc. Cogn. **26**, 198–209 (2008). https://doi.org/10.1521/soco.2008.26.2.198
20. Heider, F., Simmel, M.: An experimental study of apparent behavior. Am. J. Psychol. **57**, 243–259 (1944). https://doi.org/10.2307/1416950
21. Eddy, T.J., Gallup, G.G., Povinelli, D.J.: Attribution of cognitive states to animals: anthropomorphism in comparative perspective. J. Soc. Issues **49**, 87–101 (1993). https://doi.org/10.1111/j.1540-4560.1993.tb00910.x
22. Barrett, J.L., Keil, F.C.: Conceptualizing a nonnatural entity: anthropomorphism in god concepts. Cogn. Psychol. **31**, 219–247 (1996). https://doi.org/10.1006/cogp.1996.0017

23. Epley, N.: A mind like mine: the exceptionally ordinary underpinnings of anthropomorphism. J. Assoc. Consumer Res. **3**, 591–598 (2018)

24. Dacey, M.: Anthropomorphism as cognitive bias. Philos. of Sci. **84**, 1152–1164 (2017). https://doi.org/10.1086/694039

25. Dacey, M., Coane, J.H.: Implicit measures of anthropomorphism: affective priming and recognition of apparent animal emotions. Front. Psychol. **14**, 1149444 (2023). https://doi.org/10.3389/fpsyg.2023.1149444

26. Caporael, L.R.: Anthropomorphism and mechanomorphism: two faces of the human machine. Comput. Hum. Behav. **2**, 215–234 (1986). https://doi.org/10.1016/0747-5632(86)90004-X

27. Dennett, D.C.: Intentional systems. J. Philos. **68**, 87–106 (1971). https://doi.org/10.2307/2025382

28. Spunt, R.P., Meyer, M.L., Lieberman, M.D.: The default mode of human brain function primes the intentional stance. J. Cogn. Neurosci. **27**, 1116–1124 (2015). https://doi.org/10.1162/jocn_a_00785

29. Perez-Osorio, J., Wykowska, A.: Adopting the intentional stance toward natural and artificial agents. Philos. Psychol. **33**, 369–395 (2020). https://doi.org/10.1080/09515089.2019.1688778

30. Marchesi, S., Ghiglino, D., Ciardo, F., Perez-Osorio, J., Baykara, E., Wykowska, A.: Do we adopt the intentional stance toward humanoid robots? Front. Psychol. **10**, 450 (2019)

31. Marchesi, S., Spatola, N., Perez-Osorio, J., Wykowska, A.: Human vs humanoid. A behavioral investigation of the individual tendency to adopt the intentional stance. In: Proceedings of the 2021 ACM/IEEE International Conference on Human-Robot Interaction, pp. 332–340 (2021)

32. Marchesi, S., Bossi, F., Ghiglino, D., De Tommaso, D., Wykowska, A.: I am looking for your mind: pupil dilation predicts individual differences in sensitivity to hints of human-likeness in robot behavior. Front. Robot. AI. **8**, 653537 (2021). https://doi.org/10.3389/frobt.2021.653537

33. Marchesi, S., De Tommaso, D., Perez-Osorio, J., Wykowska, A.: Belief in sharing the same phenomenological experience increases the likelihood of adopting the intentional stance toward a humanoid robot. Technol. Mind Behav. **3**, 11 (2022). https://doi.org/10.1037/tmb0000072

34. Thellman, S., Silvervarg, A., Ziemke, T.: Folk-psychological interpretation of human vs. humanoid robot behavior: exploring the intentional stance toward robots. Front. Psychol. **8**, 1962 (2017). https://doi.org/10.3389/fpsyg.2017.01962

35. Nass, C., Steuer, J., Tauber, E., Reeder, H.: Anthropomorphism, agency, and ethopoeia: computers as social actors. In: INTERACT '93 and CHI '93 conference companion on Human factors in computing systems – CHI '93, pp. 111–112. ACM Press, Amsterdam, The Netherlands (1993). https://doi.org/10.1145/259964.260137

36. Nass, C., Moon, Y.: Machines and mindlessness: social responses to computers. J. Soc. Issues **56**, 81–103 (2000). https://doi.org/10.1111/0022-4537.00153

37. Fussell, S.R., Kiesler, S., Setlock, L.D., Yew, V.: How people anthropomorphize robots. In: Proceedings of the 3rd ACM/IEEE international conference on Human robot interaction, pp. 145–152. ACM, Amsterdam The Netherlands (2008). https://doi.org/10.1145/1349822.1349842

38. Guzman, A.L., McEwen, R., Jones, S.: The SAGE Handbook of Human–Machine Communication. SAGE Publications (2023)

39. De Houwer, J.: What are implicit measures and why are we using them? In: Handbook of implicit cognition and addiction, pp. 11–28. Sage Publications, Inc, Thousand Oaks, CA, US (2006). https://doi.org/10.4135/9781412976237.n2

40. Li, Z., Terfurth, L., Woller, J.P., Wiese, E.: Mind the machines: applying implicit measures of mind perception to social robotics. In: 2022 17th ACM/IEEE International Conference on Human-Robot Interaction (HRI), pp. 236–245 (2022). https://doi.org/10.1109/HRI53351.2022.9889356

41. Tversky, A., Kahneman, D.: Extensional versus intuitive reasoning: the conjunction fallacy in probability judgment. Psychol. Rev. **90**, 293–315 (1983). https://doi.org/10.1037/0033-295X.90.4.293

42. Beatty, J., Kahneman, D.: Pupillary changes in two memory tasks. Psychonomic Sci. **5**, 371–372 (1966). https://doi.org/10.3758/BF03328444

43. Spatola, N., Wudarczyk, O.A.: Ascribing emotions to robots: explicit and implicit attribution of emotions and perceived robot anthropomorphism. Comput. Hum. Behav. **124**, 106934 (2021)

44. Spatola, N., Wudarczyk, O.A.: Implicit attitudes towards robots predict explicit attitudes, semantic distance between robots and humans, anthropomorphism, and prosocial behavior: From attitudes to human–robot interaction. Int. J. Soc. Robot. **13**, 1149–1159 (2021)

45. Waytz, A., Epley, N., Cacioppo, J.T.: Social cognition unbound: insights into anthropomorphism and dehumanization. Curr. Dir. Psychol. Sci. **19**, 58–62 (2010)

46. Bago, B., De Neys, W.: Advancing the specification of dual process models of higher cognition: a critical test of the hybrid model view. Think. Reason. **26**, 1–30 (2020). https://doi.org/10.1080/13546783.2018.1552194

47. Broadbent, E., et al.: Robots with display screens: a robot with a more humanlike face display is perceived to have more mind and a better personality. PLoS ONE **8**, e72589 (2013). https://doi.org/10.1371/journal.pone.0072589

48. Manzi, F., et al.: A robot is not worth another: exploring children's mental state attribution to different humanoid robots. Front. Psychol. **11**, 2011 (2020). https://doi.org/10.3389/fpsyg.2020.02011

49. Zhao, X., Malle, B.F.: Spontaneous perspective taking toward robots: the unique impact of humanlike appearance. Cognition **224**, 105076 (2022). https://doi.org/10.1016/j.cognition.2022.105076

50. Thellman, S., de Graaf, M., Ziemke, T.: Mental state attribution to robots: a systematic review of conceptions, methods, and findings. ACM Trans. Human-Robot Interact. **11** (2022). https://doi.org/10.1145/3526112

51. Sacino, A., et al.: Human- or object-like? Cognitive anthropomorphism of humanoid robots. PLoS ONE **17**, e0270787 (2022). https://doi.org/10.1371/journal.pone.0270787

52. Złotowski, J., Bartneck, C.: The inversion effect in HRI: are robots perceived more like humans or objects? In: 2013 8th ACM/IEEE International Conference on Human-Robot Interaction (HRI), pp. 365–372. IEEE (2013)

53. Bartneck, C., Kulić, D., Croft, E., Zoghbi, S.: Measurement instruments for the anthropomorphism, animacy, likeability, perceived intelligence, and perceived safety of robots. Int. J. Soc. Robot. **1**, 71–81 (2009). https://doi.org/10.1007/s12369-008-0001-3

54. Bartneck, C.: Godspeed questionnaire series: translations and usage. In: Krägeloh, C.U., Alyami, M., Medvedev, O.N. (eds.) International Handbook of Behavioral Health Assessment, pp. 1–35. Springer International Publishing, Cham (2023). https://doi.org/10.1007/978-3-030-89738-3_24-1

55. Mara, M., Appel, M., Gnambs, T.: Human-Like robots and the uncanny valley. Zeitschrift für Psychologie. **230**, 33–46 (2022). https://doi.org/10.1027/2151-2604/a000486

56. Weiss, A., Bartneck, C.: Meta analysis of the usage of the Godspeed Questionnaire Series. In: 2015 24th IEEE International Symposium on Robot and Human Interactive Communication (RO-MAN), pp. 381–388 (2015). https://doi.org/10.1109/ROMAN.2015.7333568

57. Miraglia, L., Peretti, G., Manzi, F., Di Dio, C., Massaro, D., Marchetti, A.: Development and validation of the Attribution of Mental States Questionnaire (AMS-Q): a reference tool for assessing anthropomorphism. Front. Psychol. **14** (2023)
58. Thompson, V.A., Prowse Turner, J.A., Pennycook, G.: Intuition, reason, and metacognition. Cogn. Psychol. **63**, 107–140 (2011). https://doi.org/10.1016/j.cogpsych.2011.06.001
59. Bago, B., De Neys, W.: Fast logic?: examining the time course assumption of dual process theory. Cognition **158**, 90–109 (2017). https://doi.org/10.1016/j.cognition.2016.10.014

# Human Thinking and Reasoning

# Narrative Empowerment for Intelligence: Structural Study Applied to Strategic Notes

François Marès(✉) ⓘD, Bruno Bachimont, and Olivier Gapenne

Université de Technologie de Compiègne, Costech (EA2223),
60203 Compiègne, France
francois.mares@alten.com

**Abstract.** This article presents the initial results of the research project titled "Narrative Empowerment for Analysis: Epistemic Challenges in the Technical Construction of Strategic Narratives." To instrument the construction of strategic narratives, it is necessary to characterize their technical features. Gerard Genette's modal narratology, as developed in "Figures" III, allows us to describe the relationship between the pseudo-time of the narrative – the succession of sentences and pages – and the diegetic time – the time of the story being told – as well as phenomena of mode and voice. We conducted a micronarrative analysis of DGSE notes related to the Rwandan crisis in the 1990s and are now conducting a macronarrative analysis of the entire set of documents, considering their chronological succession as a single narrative for the recipients. We present the challenges encountered during the decomposition process and provide a description of the analysis output, which will serve as material for reflecting on an appropriate digital medium for the construction of strategic narratives.

**Keywords:** Empowerment · Storytelling · Narrative · Pseudo-time · Intelligence

## 1 Empowerment for Strategic Analysis

Strategic analysis, as an information exploitation process for action, is, in practice, the site of numerous interactions between human and machine. The innovations that enable new relationships, sometimes involving assistance, support or substitution, are most often the corollaries of technological progress. Our work, on the other hand, consists of defining a new relationship of empowerment[1] (or *supplementation*) on the basis of the study of analysts' capacity to manipulate

---

[1] The concept of empowerment used in this research stems directly from work devoted to perception and its instrumentation. Substitution means a relationship between an agent and an instrument and indicates that this relationship enables and constrains a new power, for example perceptive power. Typically, it is possible for blind people to experience objects at a distance without contact via the active use of a dedicated

textual material aimed at producing narratives and auguring a new cognitive power of synthesis. To achieve this objective, we first need to find a technical description of the narrative produced by strategic analysis, and in particular of the temporal reconfiguration it brings about. The structural approach thus enables us to identify the constraints on its design. This article presents the theoretical contribution of Gérard Genette's narratology to the description of the finished product of military intelligence exploitation. If this structural description proves to be relevant and effective, it will be possible to apply the categories of this structure to the informational material used in strategic analysis, in order to facilitate the production of notes for decision-makers.

## 1.1   Storytelling at the Heart of Military Intelligence Exploitation

Among the places where strategic analyses are carried out, the military intelligence process is a prime case study.

Its function is to manipulate and produce intelligence, which is described as a state of coherence of data (the *knowledge*), for decision-making (*strategic*). This *strategic knowledge* can be raw or elaborated, and is derived from data which can be factual or event-based, primary or secondary and more or less truthful [5]. Analysts, on the one hand, receive and direct the production of raw intelligence and, on the other hand, are the interlocutors of decision-makers, receiving their abundant questions and responding as quickly as possible with elaborated intelligence.

The rational study of the constraints of this activity and its organisation is not new[2]. However, in the current context of massification of the heterogeneous information that intelligence analysts process in a limited timeframe and of strong innovation in knowledge technologies, multi-source exploitation is seeking to integrate scientific paradigms [11]. This dynamic, natural at the level of data production, encounters more difficulties at the level of the construction of elaborated intelligence.

The textbook by Olivier Chopin and Benjamin Oudet [6], which quotes a definition of the analyst's work by the American researcher Gregory F. Treverton, sums up the fundamental place that narrative occupies in the exploitation process:

Intelligence ultimately is storytelling. It is helping policy makers build or adjust stories in light of new or additional information or arguments. [17]

---

instrument [12]. By extension, we propose to mobilise this concept to evoke the conquest of a new capacity for strategic analysis linked to the synthetic power of the narrative resulting itself from the analyst's relationship with a textual instrument possessing a manipulable structure.

[2] Particularly with reference to the visionary and very comprehensive study entitled *Tactique des renseignements* (Intelligence tactics), published in two volumes (1881 [13] and 1883 [14]) by Major General Jules-Louis Lewal, to whom Franck Bulinge pays tribute in his textbook which we quote in this article.

This statement is the fundamental hypothesis of our research project: the development of intelligence is fundamentally narrative. We therefore seek to understand how this narrative work plays its part in the exploitation, and how it makes it possible to combine technological and human faculties for analysis.

In general, a work of writing can be the site of semiotic freedom: Barthes [4] speaks of the "forces of freedom" in literature against the power inscribed in language[3]. However, the writing work we are studying is not arbitrary; it composes a narrative made up of a series of adjustments to answers to strategic questions. We can thus study the narrative structure of the notes issued by the analysts, which responds on the one hand to the combination of constraints of the operating process, and on the other hand to the needs of the decision-makers.

We thus present our attempt to use Genette's theorisation to describe the modal specificities of the finished product of exploitation. In particular, the constraints of the temporal reconfiguration carried out. The Sect. (4) then succinctly presents the contribution of Barthes' definition of codes to our question.

These constraints will later have to be linked to the exploitation methods, in particular structured [16].

### 1.2   Corpus of Notes Issued by the DGSE Relating to Rwanda (1993-1994)

An event will ensure our access to a set of documents representative of the product of the exploitation of intelligence of military interest. The order of 6 April 2021 provides for the opening of archives relating to Rwanda between 1990 and 1994, including copies of the documents cited in the report of the Commission de recherche sur les archives françaises relatives au Rwanda et au génocide des Tutsi (1990–1994) (art.3). In this last batch, a file combines documents issued by the DGSE [1]. It consists of around 300 sheets (268 for 1993 and 1994 combined), divided between situation reports and thematic analyses, plus a few maps and mission reports.

The product of the analysis reflects the issues of those commissioning it. Thus, its content is totally dependent on the area and period concerned, the issues at stake in the crisis, etc. But its narrative construction constraints are inherent in its strategic dimension – with a view to action, whether political, humanitarian or military. But the constraints of its narrative construction are inherent in its strategic dimension – with a view to action, whether political, humanitarian or military. The relative proximity in time of the Rwandan crisis of the 1990s allows us to be closer, not in terms of data collection techniques, which have largely changed, but in terms of the use of narrative in strategic dialogue with decision-makers. The distance and singularity of the *story* being

---

[3] In his inaugural lecture at the Collège de France, Barthes distinguished three forces of freedom in literature: *mathesis*, its capacity to engage in a dramatic (as opposed to epistemological) discourse on life; *mimésis*, its capacity to grasp reality through stubbornness and displacement; and *sémiosis*, freedom in the heteronymy of the things of language.

told is thus not an obstacle to the generalisation of the study of the *narrative* – according to the distinction made by narratology.

## 2    Genette's Structural Analysis

If we distinguish, as Genette does [8, p. 27], *story* (the signified), *narrative* (the signifier) and *narrating* (the producing narrative action), we can identify two branches in the literary structuralist work of the $XX^{th}$ century: those who tend to study the story – first Propp, then Bremond, then Greimas – and those who tend to study narrative (or "narrative mode"), of which Genette is one. Barthes, like Todorov, is seen by Genette as being between the two approaches [9, p. 17]. His *Introduction à l'analyse structurale des récits* first places him in a functional approach [2], then his critique of Balzac [3], as we shall see later, can be seen as complementary to Genette's – while initiating Barthes's "conversion to post-structuralism"[15, p. 6]. We will not elaborate here a structural analysis of the *stories* narrated by the exploiters of intelligence, but will limit ourselves to the modal approach of the *narratives* produced by the exploitation of intelligence.

The work of theorising, based on Genette's critique of Proust's *Recherche*, published in 1972 under the title *Figures* III [7][4], will thus be our reference point. Although this narratology is restricted to literary texts, it is nevertheless intended to pave the way for a study of factual narrative, of the "eternal report" (Genette uses Mallarmé's expression)[10, p. 9]: the daily gossip, the press account, the police report and, as far as we are concerned, the intelligence note.

In *Figures* III, Genette successively explores the order, duration, frequency, mode and voice of narrative discourse. Following the numerous discussions that these chapters gave rise to, Genette published *Nouveau discours du récit* [9] in 1983, which clarified or developed certain points.

### 2.1    Presentation of the Main Concepts for Our Study

The first three chapters, on order, duration and frequency, describe the relationships between the *pseudo-time of the narrative*[5] and the *diegetic time* – assimilated here to real time, as in historical narrative. They are briefly presented below, and their application to the corpus of notes is described in the Sect. (3). Genette's considerations on mode and voice are also succinctly presented and directly commented on.

**Order.** Anachronies are passages which take place in a diegetic time earlier – *analepses* –, or later – *prolepses* – than that of the *first* narrative – or *primary* narrative[6]. The latter is defined arbitrarily: it is the narrative in relation to which

---

[4] The English translation of *Figures* III which will serve as a reference for this article is that of Lewin published in 1980 [8].

[5] The narrative *pseudo-time* corresponds to the sequence of words and sentences, whereas the *diegetic time* refers to the time in the world of the story being told.

[6] This second name is a proposal from *Nouveau discours du récit* [9], p. 29 and 88.

we are situating ourselves, and its definition can vary according to the aim of the analysis or whether we are carrying out a micronarrative or macronarrative analysis[7].

Anachronies are defined by their *reach*, their distance in diegetic time with the interruption of the primary narrative; and their *extent*. The difficulty, or even impossibility, of defining the reach or extent of an anachrony makes it difficult to understand and even more difficult to represent.

Depending on their temporal position relative to the primary narrative, anachronies are said to be *internal*, *mixed* or *external*, and analepses which join the moment of interruption of the primary narrative are said to be *complete* – they *overlap* the primary narrative if they go beyond the interruption.

Anachronies can be *heterodiegetic*, i.e. concern a line of action distinct from that of the primary narrative, or *homodiegetic*. Among the latter, a distinction should be made between completive anachronies, which fill in an ellipsis or paralipse[8], i.e. which communicates information hitherto omitted; and repetitive anachronies relating to questions of frequency.

**Duration.** Following Müller (1948) and Barthes (1967), Genette defines the speed of a narrative as "the relationship between a duration (that of the story [...]), and a length (that of the text, measured in lines and pages)." [8, p. 88]. For the macroscopic study of rhythmic effects, Genette defines four *narrative movements*: the *ellipse*, the *summary narrative*, the *scene* and the *pause*[9].

The duration of the ellipsis is sometimes difficult to define. It can be *explicit*, expressed by a short proposition or by simple elision; or *implicit*, i.e. the solution to a reasoning on chronology; or *hypothetical*, when it cannot be located until, perhaps, it is revealed by a complétive analepsis.

**Frequency.** The repetition of an event is always *abstract*, in the sense that an event is localised in time and two similar events are distinct. What can be repeated in history is a virtual event, a class of events. However, the narrative of an event can be repeated by the author. We thus obtain four cases by combination: the *singulative* narrative or narratives (only one story per event), the *repetitive* narrative (multiple accounts of a single event) and the *iterative* narrative (a single narrative for several events). Genette adds a final case, typical of *la Recherche*, which he calls *pseudo-iterative* and which consists of initiating

---

[7] To break down the temporal structure of a narrative, we can choose to situate ourselves on an arbitrary scale. Genette calls *micronarrative* the level of the shortest anachronies, and *macronarrative* the overall level, which neglects them. A primary narrative at the micro level can thus belong to an anachrony at the macro level.

[8] A paralipse is a "lateral omission" [7, p. 290], which consists in not giving information that one possessed (in the sense of focus) and which is therefore missing from the understanding.

[9] In *Nouveau discours du récit*, Genette specifies that the pause is not necessarily descriptive – it can be a digression –, and that a description is not necessarily a pause (p. 35).

an iteration from a singular narrative. It differs from the iterative narrative in that the inaugural narrative gives a description of the event in its singularity, rather than limiting itself to its abstraction, which is the only one that is really repeated.

Iterations can be *internal* or *external* to the segment under consideration (they are called *synthesizing* or *generalizing* iterations respectively).

**Mood.** Genette proposes to distinguish three diegetic levels, discourse *narratized, transposed* or *reported*[10], and then establishes the notion of *focalization*. The dominant focus of a narrative can be *external* – the narrator says less than the characters know –, *internal* – fixed, variable or polymodal –, or *nonfocalized* – the narrator is omniscient. Focalisation is potentially accompanied by *alterations, paralipses* and *paralepses*, which correspond respectively to cases where the narrator gives less and more information than his position enables him to give.

In the case of intelligence, the focus most often oscillates between *external* and *nonfocalized*. The actors in the conflict sometimes know more than the Service, or vice versa.

Alterations are difficult to identify, since the sources are hidden. Only an audit with access to all the DGSE's sources would be able to classify certain passages as paralepses, in other words, for intelligence purposes, as assertions that were disseminated but could not be justified. The equivalent of a paralipse for the sequence of files produced by the operation corresponds to information that is possessed but not disseminated, whether voluntarily or not. It can be guessed at or raised as a question by the reader. An ideal note answers all the reader's questions to which the analysts know the answer.

**Voice.** In his chapter on voice, Genette distinguishes several diegetic levels: the *intradiegetic;*, or *diegetic*, narrative is the referential; the *extradiegetic* level is that of the recitation of the diegetic narrative; the *metadiegetic* level is that of a narrative within a considered diegetic narrative. He then draws up a typology of level changes which he extends in *Nouveau discours du récit* (p. 90).

*Figures* III also distinguishes the levels of presence of the author according to whether he is absent from the universe of the story (*extradiegetic*), present to varying degrees (*homodiegetic*) or the hero of the story (*autodiegetic*)[11]. The intelligence services can be considered to be more or less involved in the world of the crisis that is the subject of the analysis, so we can speak of a homodiegetic narrative. But in our case, the DGSE almost never refers to itself in the description of the conflict, except when it formulates hypotheses and thus provides an account of the process it is implementing. This is the case with the files

---

[10] Genette, in *Nouveau discours du récit*, completes his rejection of the opposition *mimesis/ diégésis* by replacing *mimesis* by *rhésis*, transcription.

[11] In *Nouveau discours du récit*, Genette insists on the continuum between homodiegetic and heterodiegetic levels (p. 102).

that present the service's hypotheses concerning the attack of 6 April 1994 ( [1] 18615/N - "Responsibilities for the attack"; 19404/N - "Service hypothesis on responsibility for the attack on President Habyarimana's plane").

Finally, although Genette's work is analytical and not functionalist, he distinguishes five functions of the narrator: one narrative and four extra-narrative (governing, communicative, testimonial and ideological). In *Nouveau discours du récit*, he specifies that "the extra-narrative functions are more active in the 'narratorial' type, that is, in our terms, unfocused" [9, p. 130]. We could list the occurrences of these functions, but our work here is limited to an analytical approach. We will only note that most of the notes have a final 'commentary' section which allows the narrator to exercise his ideological and sometimes testimonial function (although this latter function is exercised throughout the body of the note).

# 3 Structural Analysis of Notes

We carried out a micronarrative and macronarrative analyses of part of the corpus presented above (1.2), in order to test Genette's decomposition methodology and identify the structural properties of the notes. The first observation is that these decompositions work. Their implementation can incorporate specific features linked to the strategic nature of the narratives.

We naturally identify the micro- and macro-narrative levels at the note and corpus levels respectively.

## 3.1 Micronarrative Analysis of Notes

The notes in the file are almost all structured as follows: 1) heading 2) main points 3) facts 4) commentary 5) appendices.

**Choice of First Narrative.** The first step, before determining the anachronies, is to choose a primary narrative. This can be chosen according to various criteria: the longest story, the slowest story, the first story to be read, the story closest to the date on which the note was issued, the story closest to the time of reading, or the story that appears to be the most important in relation to a strategic issue. Often, the title or the box (known as "for the reader in a hurry") indicates which story is the most important in the memo.

In the case of a systematic micronarrative analysis that would prepare a macronarrative analysis, we can detail the consequences of the last two criteria:

- By selecting the most recent story, there can no longer be any external prolepses that predate the moment when the note was issued. Thus prolepses, even at the micronarrative level, will necessarily be either internal (ignored by the macro analysis) or predictive (probably to be retained for the macro analysis).

- By selecting the most important story in relation to a strategic question, in the context of an analysis which answers a question, we facilitate the use of decomposition.

Let us take the example of note 18974/N, "Involvement of Uganda and Libya", dated 8 November 1990 [1]:

> The policy of the government in Kigali was to accuse President Museveni from the outset of a deliberate attack in order to attract international support, achieve national unity and provide an explanation for his initial military setbacks. [...]
> Uganda's involvement
> The Ugandan president was aware of the Tutsis' preparations. However, he has drawn the attention of his Rwandan counterpart to the problem three times this year, but the latter has never paid any attention and has always opposed the opening of negotiations concerning the return of external refugees. It is unlikely that the Ugandan Head of State was aware of the date of the launch of the operation, otherwise he would have had every reason to oppose it, particularly at a time when he held the presidency of the OAU.

We could choose as the first narrative the description of Rwanda's current politics (at the date of the note). Then the second paragraph on knowledge is an analepsis (with an iterative internal analepsis on the refugee negotiations). This choice applies the criterion of the most recent narrative. Or the second paragraph is the first narrative and the first a prolepse. The choice of either option may slightly modify the transition to macronarrative decomposition in the case of an algorithmic process, if the analepses and prolepses are treated differently.

**Difficulty in Determining Reaches and Extents.** As we said earlier, the temporal location of identified ellipses is sometimes difficult. This is also the case for many analepses and prolepses, or even primary narratives. If we take the example of note 19205/N, "Collaboration between former FAR officers and the RPF", of 5 August 1994 [1]:

> The Rwandan Patriotic Front (RPF) is said to have commissioned several officers from the former Rwandan Armed Forces (FAR) to visit refugees in Goma (Zaire) in order to encourage them to return to Rwanda.
> With the approval of the Prefect of Gisenyi, these officers – who were not involved in the atrocities perpetrated against the Tutsis – have been authorised by the RPF to return to their country since the end of July.

The RPF order for the officers of the former FAR cannot be precisely located. The uncertainty obviously extends to the explicit ellipsis of the execution of their mission.

## 3.2   Macronarrative Analysis of the Corpus

Macro analysis first raises the question of scale: which anachronisms should be preserved? Two types of criteria can be used.

- a criterion on meaning; in the context of a strategic issue, importance can still be the criterion.
- a temporal criterion; for example on the scope, integrating the closest analepses and prolepses. A particular regime may concern prolepses whose predictions have not yet been verified (as well as uncertain portions with no order considerations). When using scope as a criterion, the heterodiegetic character of micronarrative anachronies may be more important than their anachronistic character.

For macronarrative analysis, the thematic succession of notes is almost anarchic, but their temporal succession is usually linear: each note concerns the latest facts for the *situation points*, but also for the thematic sheets, although their narratives are often more extensive and contain more anachronisms.

**Table 1.** Sequence of notes from 29 and 30 June 1994 in the corpus [1].

| Date | N° AN | N° DGSE | Title |
|------|-------|---------|-------|
| 1994/06/29 | 389 | 18908/N | Tensions in northern Burundi |
| 1994/06/29 | 391 | 18915/N | Information on the Zero network |
| 1994/06/29 | 94 | 18918/N | Refugee movements towards Gisenyi |
| 1994/06/30 | 388 | 18919/N | RPF state of mind |
| 1994/06/30 | 384 | 18920/N | Situation in the Butare region |
| 1994/06/30 | 387 | 18921/N | About Radio des mille collines |
| 1994/06/30 | 383 | 18922/N | Use of NGOs by the RPF |
| 1994/06/30 | 386 | 18926/N | Update at 1:00 pm |
| 1994/06/30 | 385 | 18927/N | Opinion of President Museveni |
| 1994/06/30 | 384 | 18930/N | Tanzanian position |

The Table 1 shows a set of ten notes issued over two days. Notes 18918 and 18920 relate certain facts in common with the "Situation update at 13:00", but the essential content does not overlap, while the diegetic time of the ten notes extends in each case up to the issue of the note concerned.

A few notes that shed light on the interpretation of events of great strategic importance, such as the presidential bombing of 6 April 1994 or the signing of treaties, relate entirely to events that took place some time after their date of issue. But the other notes mainly concern recent events, with some anachronisms.

Ellipses, within a complete corpus - in the sense that it represents the entirety of narratives available to a decision-maker or analyst - are significant sources of temporal distortion. They are logically quite numerous in our corpus, as the analysis work primarily involves selecting important information, contrary to the search for narrative continuity in classical literature. Implicit and hypothetical ellipses also pose problems in reconstructing the link between the pseudo-time of the narrative and the diegetic time, as their location can be uncertain or even undefined.

In our context, repetitive narratives clearly indicate that the concerned story holds high strategic value, as long as the repetition continues. Iterative and pseudo-iterative narratives with a *generalizing* nature must therefore receive special attention. In the case of a series of intelligence notes, an external iteration or pseudo-iteration means that a note initiates a series of events that will persist long after the initial note has been disseminated. This often applies to population movements: a note reports an initial displacement, and the next note that discusses it will summarize subsequent movements.

## 4    Barthes' Proairetic and Hermeneutic Codes

Roland Barthes, whose contribution pertains to both the structural approach to history and narrative, provides us with a method of sense decomposition in his work *S/Z* [3], which aligns well with Genette's structural approach.

In this work, Barthes dissects Balzac's *Sarrasine* into *lexias*. He states that this division "will be as arbitrary as possible; it will not imply any methodological responsibility since it concerns the signifier, while the analysis proposed here focuses solely on the signified" [3, p. 20]. This arbitrariness can also serve as the basis for methodological choices specific to strategic analysis.

Barthes identifies five codes: the *hermeneutic, semantic, symbolic, proairetic,* and *cultural* codes. However, the hermeneutic code - consisting of morphemes around which an enigma is structured - and the proairetic code - comprising actions organized into sequences - limit the multivalence of the narrative, as they "constrain by their connection to truth and empiricism" [3, p. 37]. Therefore, Barthes himself points out that proairetic sequences are "subjected to a logical-temporal order, [...] both syntagmatic and orderly, and they can form the privileged material for a certain structural analysis of the narrative" [3, p. 209]. This latter analysis will be provided by Genette in *Figures* III.

In strategic analysis, events (proairetic code) and the treatment of uncertainty (hermeneutic code) are the focal points of the narrative. This justifies, on the one hand, our choice of Genette's analysis, and on the other hand, it prompts us to distinguish between these two codes.

## 5    Conclusion and Future Work

The analytical elements of *Figures* III allow us to effectively describe the microstructure and macro-temporal structure of the DGSE intelligence notes corpus.

The need for clarity and concision in the notes reduces the diversity of codes used, but implies a high speed of narrative and a large number of ellipses, which are sources of difficulties in representation. The numerous anachronisms, whatever the scale of observation, attest to the strong temporal distortions operated by the strategic narrative, rarely reversible for the recipient.

We have identified some methodological variations stemming from the arbitrary nature, particularly in the choice of primary narratives. The strategic stakes, underlying the purpose for which we intend to use the decomposition, become evident in these choices. A comprehensive analysis of the entire corpus, using an appropriate tool, will enable us to more accurately compare the implications of these methodological choices. We also need to experiment with integrating a distinction between Barthes' hermeneutic and proairetic codes.

**Acknowledgements.** This research was supported by Alten SA (CIFRE). We would also like to thank the students of the *Recitech* group at the Université de Technologie de Compiègne for their assistance in applying Genette's method to the corpus.

# References

1. Archives Nationales: Direction générale de la sécurité extérieure (dgse). No. 20210031/10 in Copies des documents cités dans le rapport "La France, le Rwanda et le génocide des Tutsi (1990–1994)" remis au Président de la République le 26 mars 2021, dites Cartons sources

2. Barthes, R.: Introduction à l'analyse structurale des récits. Communications 8(1), 1–27 (1966). https://doi.org/10.3406/comm.1966.1113

3. Barthes, R.: S/Z : essai. Collection Tel Quel, Éd. du Seuil, Paris (1970)

4. Barthes, R.: Leçon: leçon inaugurale de la chaire de sémiologie littéraire du Collège de France, prononcée le 7 janvier 1977. No. 205 in Points, Éditions Points (2015)

5. Bulinge, F.: Maîtriser l'information stratégique: méthodes et techniques d'analyse. Information & stratégie, De Boeck supérieur, Bruxelles [Paris], 2nd edn. (2022)

6. Chopin, O., Oudet, B.: Renseignement et sécurité, 2nd edn. Science politique, Armand Colin (2019)

7. Genette, G.: Figures III. Éditions Points, Paris (1972)

8. Genette, G.: Narrative Discourse: An Essay in Method. Cornell University Press, Ithaca, N.Y (1980)

9. Genette, G.: Nouveau discours du récit. Editions du Seuil, Paris (1983)

10. Genette, G.: Récit fictionnel, récit factuel. Protée 19(1) (1991). 1re pub. 1979

11. Laurent, S.: Promouvoir une authentique communauté épistémique d'analystes du renseignement (cedar): étude comparée (États-unis, france, grande-bretagne). Consultance au profit de la das, Délégation aux affaires stratégiques (January 2014)

12. Lenay, C., et al. (eds.): Toucher pour connaître. Psychologie cognitive de la perception tactile manuelle, chap. La substitution sensorielle: Limites et perspectives, pp. 287–306. PUF (2000)

13. Lewal, J.L.: Etudes de guerre: tactique des renseignements, vol. 1. Baudoin & Ce (1881)

14. Lewal, J.L.: Études de guerre. Tactique des renseignements, vol. 2. Baudoin & Ce (1883)

15. Messager, M.: Barthes, de A à (S/)Z. Acta Fabula 13(5) (2012). https://doi.org/10.58282/acta.7024

16. Pherson, R.H., Heuer, R.J. (eds.): Structured Analytic Techniques for Intelligence Analysis. SAGE, CQ Press, third edition edn. (2021)

17. Treverton, G.F.: Intelligence for an Age of Terror. Cambridge University Press, Cambridge (2009). https://doi.org/10.1017/CBO9780511808708

# Robustness and Cultural Difference on Identifiable Victim Effect

Keisuke Yamamoto[✉]

Development of Social Relations, Faculty of Social Relations, Kyoto Bunkyo University, Kyoto,
Japan
kel.1993m.t.101631@gmail.com

**Abstract.** This study reviews the unresolved identifiable victim effect (IVE) issues. IVE indicates that people are more willing to help an identified victim than to an unidentified number of victims, such as statistical information. Previous studies have demonstrated the robustness of this phenomenon. However, existing studies on IVE have been conducted primarily in Western cultures and their applicability to East Asian cultures has not been sufficiently examined. Recently, some studies have begun to explore cultural differences in IVE and have suggested that IVE may be less likely to occur in East Asian cultures which exhibit a predominantly interdependent construal of self, compared to Western cultures which exhibit a predominantly independent construal of self. This is the first study to confirm the mechanism of IVE and the methods used to suppress it. Furthermore, studies on meta-analyses and cultural differences in IVE are reviewed, and prospects are discussed.

Empathy is widely understood to be a positive affective reaction because it causes altruistic behaviors such as donations and saving lives, as demonstrated by Batson (2010). However, empathy does not always lead to positive outcomes. For example, Batson's et al. (1995) Experiment 2 showed that when participants felt high empathy for a specific individual on a waiting list for medical services, they tended to behave unfairly, changing the order of the waiting list in favor of that individual. This suggests that, while empathy motivates us to provide help by spotlighting specific individuals, it also prevents us from considering others who are not in the spotlight.

Additionally, several studies have reported that people are more willing to provide benefits to specific victims than to nonspecific (statistical) victims (Jenny and Lowenstein 1997). This is called the identifiable victim effect (IVE). In an experiment by Small et al. (2007), when requesting donations for children suffering from food shortages, the donation amount in the example presenting the profile of a seven-year-old girl was approximately doubled compared to the example presenting statistical information about the food crisis in African countries. This finding suggests that people are less likely to feel empathy toward an unspecified number of victims than a specific individual.

The disadvantages of IVE in society are as follows: For example, when a conflict or confrontation occurs, support is needed for victims. If the media only reports on the statistical number of victims, many people will not be able to empathize with them. Conversely, when the media focuses on the plight of specific individuals, many people empathize with them, but their empathy remains with those specific individuals. Thus,

J. Baratgin et al. (Eds.): HAR 2023, LNCS 14522, pp. 152–158, 2024.
https://doi.org/10.1007/978-3-031-55245-8_10

empathy for a particular individual may distract the attention of many other victims (Bloom 2013). As a result, support for victims may not be sufficient.

This study identified unresolved problems related to IVE and summarized the mechanisms, disappearance, robustness, and cultural differences of IVE.

# 1   Mechanism of Identifiable Victim Effect

Why does IVE occur? According to Hsee and Rottenstereich (2004), people are insensitive to numbers, because the impact of the number of victims is weaker than that of their presence or absence. Furthermore, insensitivity to numbers causes the amount of help to not only increase but also decrease toward an unspecified number of victims rather than specific individuals (Slovic 2007; Västfjäll et al. 2014; Small et al. 2007).

A previous study by Butts et al. (2019) have explained this decline from the perspective of the efficacy of help. For example, if a person gives $5 to a specific individual, the recipient receives all the money. However, if they provided $5 to five people, each recipient would only receive $1. Therefore, people may be less motivated to help because they feel that their actions only provide a minor benefit to victims.

However, the decline in help was explained differently depending on the identifiability of the victims in the experimental manipulations. Butts et al. (2019) did not address the manipulation of identifiability. Conversely, Erlandsson et al. (2016) explored a decline in help by manipulating identifiability. The results showed that the decline in help was caused by weakened empathy, rather than by the efficacy of help. IVE can be understood as the interaction between identifiability and the number of victims, rather than as the main effect of the number of victims. Thus, as Erlandsson et al. (2016) claim, it would be more valid for IVE to occur when empathy is weakened.

Why do people find it difficult to empathize with an unspecified number of victims? According to Slovic (2007), the occurrence of empathy that motivates prosocial behaviors requires attention and imagery of the victims. It is easier for people to understand the victims' stories and pay attention to them if their focus is on the suffering of specific individuals. Furthermore, people can more easily imagine the victims' suffering. Conversely, information such as the number of victims and the amount of damage is abstract and does not provide clues for understanding the suffering of others.

# 2   Disappearance and Robustness of Identifiable Victim Effect

How can IVE be suppressed? Small et al. (2007) demonstrated that IVE disappears through analytic thinking based on System 2 in dual process theory (Evans and Stanovich 2013). Affective reactions toward specific individuals lead to increased donations to victims. Accordingly, analytical thinking is predicted to inhibit prodigious help from particular individuals because it keeps individuals calm and reduces empathic affective reactions. In an experiment conducted by Small et al. (2007), thinking style was manipulated by a priming procedure in which an emotional or computational task was conducted before the decision to donate was made. As predicted, participants primed with intuitive thinking donated significantly more money when referring to specific individual victim information than to statistical victim information. In contrast, participants primed with analytical thinking showed no significant difference in donation amounts between statistical and specific individual victims, and the IVE disappeared (Small et al. 2007). Importantly, the disappearance of IVE meant that the donation amount predominantly given to a specific individual was reduced to the same amount given to an unspecified number of victims.

Although it is essential to suppress IVE because it is perceived as an unfair behavior that excessively favors specific individuals, it is also a significant loss to society if overall donations are reduced. Therefore, efforts should be made to enhance the general willingness to help.

How robust is the occurrence of IVE? Lee and Feeley (2016) conducted a meta-analysis of 41 studies on IVE. The effect size of the comparison between identified and unidentified victims was $r = .05$, $z = 2.07$, $p < .05$, 95% CI $= .003-.10$. This effect indicates that helping behavior is more significant in the identified condition than in the unidentified condition. However, the effect size was small.

The experimental conditions differed for each study. Lee and Feeley (2016) explored the effect of moderator variables on IVE. For example, IVE was likely to occur when the victims were children, whereas IVE was unlikely when the victims were adults. Moreover, IVE was more likely when the victim was living in poverty, but less likely in cases of injury or disease. Furthermore, IVE was observed only when the victims were not considered responsible for their plight. Presenting pictures of victims also induced IVE. Surprisingly, differences in group belonging did not affect the development of IVE; even if the victims had the same nationality as the respondents. Finally, they demonstrated the importance of the dependent variable. IVE was likely to be present in studies that considered the actual contribution of money rather than the intention to contribute money or allocate time.

The trolley dilemma is similar to the IVE issue because it focuses on individuals or groups. Decety and Cowell (2015) discuss the connection between trolley problems and empathy. For example, people with low levels of dispositional empathic concern are likely to prefer utilitarian choices in trolley problems (Gleichgerrcht and Young 2013). Therefore, as with IVE, when empathy arises in the trolley problem, it is easier to prioritize individuals over groups. The IVE trends observed in the context of help were similarly demonstrated in the area of moral dilemmas, suggesting that IVE is a general trend.

## 3  Cultural Difference of Identifiable Victim Effect

Are there cultural differences between IVE? Previous studies on IVE have been conducted in Israel (Kogut 2011; Kogut and Ritov 2005a, 2005b; Ritov and Kogut 2011), the United States (Cryder and Loewenstein 2012; Friedrich and McGuire 2010; Small (2007), and Sweden (Erlandsson et al. 2016). Except for Israel, which has a mix of Western and East Asian cultures, most studies have been conducted in Western countries. Few studies have directly examined cultural differences, with the exceptions of Kogut et al. (2015) and Wang et al. (2015). However, data on East Asian cultures are lacking.

If cultural differences exist, how can they be explained? In cultural psychology, cultural differences are often explained based on the framework of cultural construal of the self. From this perspective, we can predict the cultural differences in IVE. Markus and Kitayama (1991) claimed that people in Western cultures, where an independent construal of the self is predominant, are strongly motivated to decide on their behavior and destiny based on their own internal factors (e.g., ability and personality). Conversely, those in East Asian cultures, where an interdependent construal of self is predominant, are motivated to fulfill external expectations. If IVE were generated by differences in the degree of empathy for unspecified and specific individual victims, we would expect the incidence of IVE to vary depending on whether one is more likely to act per the emotional response. According to Markus and Kitayama (1991), Westerners are more likely to make decisions based on their emotional responses compared to Easterners. Thus, in Western cultures, there is expected to be a pronounced tendency to be more willing to help to a specific victim, rather than an unspecified number of victims. Conversely, because Easterners are concerned about whether they will deviate from the group, they empathize with a wider range of others, not just with specific individuals. Therefore, there may not be differences in helping behaviors between a particular victim and an unspecified number of victims among Easterners compared to Westerners. Although few studies have directly examined cultural differences, Wang et al. (2015) and Kogut et al. (2015) provide findings consistent with this prediction.

Wang et al. (2015) explore the differences between Chinese and American IVE. Their experiment manipulated singularity (single victim vs. a group of eight victims) and identification (identified vs. unidentified), as Kogut and Ritov (2005) did. The children who were victims belonged to the same cultural groups as the participants. The dependent variable was helping intention, which was the hypothetical decision without real donations. Resultantly, willingness to contribute was higher in the identified single-victim condition than in the unidentified single-victim condition for American samples. Conversely, the desire to contribute to the Chinese sample did not differ between identified and unidentified single-victim conditions.

Interestingly, a comparison of Chinese participants showed that their willingness to contribute in the group condition was significantly greater than in the single condition for the identified and unidentified groups. Furthermore, the American samples showed a similar tendency only in the unidentified conditions; the willingness to contribute was greater in an unidentified group than in an unidentified single victim. Although these results are ambiguous as to whether they support the hypothesis based on differences in the cultural construal of the self, they nonetheless indicate cultural differences in how IVE manifests.

Kogut et al. (2015) tested cultural differences in IVE from the perspectives of individualism and collectivism. Participants from Western Israel and Bedouin were included in the experiments. Israel's society is a blend of individualistic and collectivist cultures: Western Israel tends toward individualism, while Bedouin tends toward collectivism. Subsequently, these two ethnic groups were compared. Their experiment manipulated the singularity of the victim (a single identified victim vs. a group of eight identified victims) while describing sick children. The dependent variable was substantial donations, and money was transferred to the organization after the experiment. Participants could donate part of the payment for experimental participation and contribute more than the experimental obligation at their own expense, if they wished. The interaction between ethnic group and singularity was significant, indicating that participants in Western Israel donated more to single victims than to groups. Conversely, there were no significant differences in singularities in the Bedouin group. Contributions to the groups did not differ between Western Israel and Bedouin. These results suggest that IVE is likely to occur in Western cultures, where an independent construal of the self is predominant. Simultaneously, IVE is not expected to occur in East Asian Cultures, where the interdependent construal of the self is predominant. However, Kogut et al. (2015) confirmed that Bedouin scored higher on collectivism than Western Israel, whereas no difference was observed in individualism scores. Therefore, it is unclear whether the differences between the above conditions are genuinely based on cultural construal of the self.

Similarly, studies of trolley problems have reported cultural differences. Gold et al. (2014) found that in the trolley problem, Chinese participants were more likely than Russians or Americans to prefer deontological choices that saved a single individual over multiple people. Awad et al. (2020) showed similar results when comparing Americans with Chinese and Japanese populations. Studies dealing with the trolley problem showed a trend opposite to the hypothesis of cultural differences assumed in this study. This discrepancy may be related to the fact that the trolley problem needed judgments to rescue individuals or groups by dichotomy, whereas participants in the experiments focused on IVE made independent decisions about the degree of help for the individual and the group. However, this issue remains unclear, and further studies are required to explain this discrepancy.

## 4  Conclusion

This study summarizes the mechanisms, disappearance, robustness, and cultural differences of IVE, as discussed in previous studies. This study highlights the possibility that IVE is not universally confirmed in culture. The number of existing studies examining cultural differences is small and the robustness of the results regarding cultural differences has not yet been demonstrated. Future studies should examine whether IVE is culturally different and whether it is valid to interpret cultural differences within the framework of independent or interdependent self-construal.

In addition, previous studies (Small 2007) have examined ways to make IVE disappear, but have not explored ways to raise the contribution to an unspecified number of victims to the same level as the contribution to a specific individual. For example, reporting a statistically significant number of victims in the media is unlikely to generate empathy when conflict or confrontation occurs. Focusing on a specific victim can draw attention; however, interest remains limited to empathy for an unfortunate individual. Empathy for an identified individual may divert attention from the growing number of victims of conflict (Bloom 2013). Therefore, it is essential to consider ways to generalize people's attention to the entire group, not just to an identified individual. In the future, it will be necessary to investigate how empathy and the willingness to contribute can be effectively evoked in an unidentified number of victims.

# References

Awad, E., Dsouza, S., Shariff, A., Rahwan, I., Bonneto, J..-F..: Universals and variations in moral decisions made in 42 countries by 70,000 participants. Proc. Nat. Acad. Sci. USA. **117**(5), 2332–2337 (2020). https://doi.org/10.1073/pnas.1911517117

Batson, C.D., Klein, T.R., Highberger, L., Shaw, L.L.: Immorality from empathy-induced altruism: when compassion and justice conflict. J. Pers. Soc. Psychol. **68**(6), 1042–1054 (1995). https://doi.org/10.1037/0022-3514.68.6.1042

Batson, C.D.: Altruism in Humans. Oxford University Press, New York (2010)

Bloom, P.: The baby in the well: The case against empathy. The New Yorker (2013). Retrieved 29 May 2023, from https://www.newyorker.com/magazine/2013/05/20/the-baby-in-the-well

Butts, M.M., Lunt, D.C., Freling, T.L., Gabriel, A.S.: Helping one or helping many? A theoretical integration and meta-analytic review of the compassion fade literature. Organ. Behav. Hum. Decis. Process. **151**, 16–33 (2019). https://doi.org/10.1016/j.obhdp.2018.12.006

Cryder, C.E., Loewenstein, G.: Responsibility: the tie that binds. J. Exp. Soc. Psychol. **48**(1), 441–445 (2012). https://doi.org/10.1016/j.jesp.2011.09.009

Decety, J., Cowell, J.M.: Empathy, justice, and moral behavior. AJOB Neurosci. **6**(3), 3–14 (2015). https://doi.org/10.1080/21507740.2015.1047055

Erlandsson, A., Västfjäll, D., Sundfelt, O., Slovic, P.: Argument-inconsistency in charity appeals: Statistical information about the scope of the problem decrease helping toward a single identified victim but not helping toward many non-identified victims in a refugee crisis context. J. Econ. Psychol. **56**, 126–140 (2016). https://doi.org/10.1016/j.joep.2016.06.007

Evans, J.S.B., Stanovich, K.E.: Dual-process theories of higher cognition: advancing the debate. Perspect. Psychol. Sci. **8**(3), 223–241 (2013). https://doi.org/10.1177/1745691612460685

Friedrich, J., McGuire, A.: Individual differences in reasoning style as a moderator of the identifiable victim effect. Soc. Influ. **5**(3), 182–201 (2010). https://doi.org/10.1080/15534511003707352

Gleichgerrcht, E., Young, L.: Low levels of empathic concern predict utilitarian moral judgment. PloS One. **8**(4), e60418 (2013). https://doi.org/10.1371/journal.pone.0060418

Gold, N., Colman, A.M., Pulford, B.D.: Cultural differences in responses to real-life and hypothetical trolley problems. Judgm. Decis. Mak. **9**(1), 65–76 (2014). https://doi.org/10.1017/S1930297500000499X

Hsee, C.K., Rottenstreich, Y.: Music, pandas, and muggers: on the affective psychology of value. J. Exp. Psychol. Gen. **133**(1), 23–30 (2004). https://doi.org/10.1037/0096-3445.133.1.23

Jenni, K., Loewenstein, G.: Explaining the identifiable victim effect. J. Risk Uncertain. **14**, 235–257 (1997). https://doi.org/10.1023/A:1007740225484

Kogut, T.: Someone to blame: when identifying a victim decreases helping. J. Exp. Soc. Psychol. **47**(4), 748–755 (2011). https://doi.org/10.1016/j.jesp.2011.02.011

Kogut, T., Ritov, I.: The "identified victim" effect: an identified group, or just a single individual? J. Behav. Decis. Mak. **18**(3), 157–167 (2005). https://doi.org/10.1002/bdm.492

Kogut, T., Ritov, I.: The singularity effect of identified victims in separate and joint evaluations. Organ. Behav. Hum. Decis. Process. **97**(2), 106–116 (2005). https://doi.org/10.1016/j.obhdp.2005.02.003

Kogut, T., Slovic, P., Västfjäll, D.: Scope insensitivity in helping decisions: is it a matter of culture and values? J. Exp. Psychol. Gen. **144**(6), 1042–1052 (2015). https://doi.org/10.1037/a0039708

Lee, S., Feeley, T.H.: The identifiable victim effect: a meta-analytic review. Soc. Influ. **11**(3), 199–215 (2016). https://doi.org/10.1080/15534510.2016.1216891

Markus, H.R., Kitayama, S.: Culture and the self: implications for cognition, emotion, and motivation. Psychol. Rev. **98**(2), 224–253 (1991). https://doi.org/10.1037/0033-295X.98.2.224

Ritov, I., Kogut, T.: Ally or adversary: the effect of identifiability in inter-group conflict situations. Organ. Behav. Hum. Decis. Process. **116**(1), 96–103 (2011). https://doi.org/10.1016/j.obhdp.2011.05.005

Slovic, P.: "If I look at the mass I will never act": psychic numbing and genocide. Judgm. Decis. Mak. **2**(2), 79–95 (2007). https://doi.org/10.1017/S1930297500000061

Small, D.A., Loewenstein, G., Slovic, P.: Sympathy and callousness: the impact of deliberative thought on donations to identifiable and statistical victims. Organ. Behav. Hum. Decis. Process. **102**(2), 143–153 (2007). https://doi.org/10.1016/j.obhdp.2006.01.005

Västfjäll, D., Slovic, P., Mayorga, M., Peters, E.: Compassion fade: affect and charity are greatest for a single child in need. PLoS ONE **9**(6), e100115 (2014). https://doi.org/10.1371/journal.pone.0100115

Wang, Y., Tang, Y.-Y., Wang, J.: Cultural differences in donation decision-making. PLoS ONE **10**(9), e0138219 (2015). https://doi.org/10.1371/journal.pone.0138219

# On Independence and Compound and Iterated Conditionals

Angelo Gilio[1] , David Over[2] , Niki Pfeifer[3] , and Giuseppe Sanfilippo[4]([✉])

[1] Department SBAI, University of Rome "La Sapienza", Rome, Italy
`angelo.gilio1948@gmail.com`
[2] Department of Psychology, Durham University, Durham, UK
`david.over@durham.ac.uk`
[3] Department of Philosophy, University of Regensburg, Regensburg, Germany
`niki.pfeifer@ur.de`
[4] Department of Mathematics and Computer Science, University of Palermo, Palermo, Italy
`giuseppe.sanfilippo@unipa.it`

**Abstract.** Understanding the logic of uncertain conditionals is a key problem in the new paradigm psychology of reasoning and related fields. We investigate conjunctions of conditionals, iterated conditionals, and independence within the theory of logical operations on conditionals, where compound conditionals are suitably defined as conditional random quantities. We show how conjunctions of conditionals and conditionals which feature conditionals in the antecedent and the consequent can be rationally interpreted. In particular, we study the behavior of such objects under different logical constraints, by also considering a kind of "independence" property. Unlike alternative approaches, in our framework we avoid counterintuitive consequences, which is necessary for understanding, or improving, human and artificial rationality in general.

**Keywords:** Coherence · Compound and iterated conditionals · Conditional random quantities · Conditional bets · Independence · Probabilistic reasoning

## 1 Introduction

Understanding reasoning about conditionals is key to understanding human rationality and artificial cognition, and knowledge of independence is essential to human rationality and artificial cognition. In this paper, we will consider how compound and iterated conditionals can be used to convey information about independence. We take a probabilistic approach to this topic. Probabilistic analyses have recently become popular in artificial intelligence, philosophy, and in the psychology of reasoning.

In this framework, the basic question is how to evaluate the probability of a conditional *if H then A.* Many philosophers and psychologists interpret the

---

A. Gilio—Retired.

J. Baratgin et al. (Eds.): HAR 2023, LNCS 14522, pp. 159–177, 2024.
https://doi.org/10.1007/978-3-031-55245-8_11

probability of a conditional *if H then A* as the conditional probability $P(A|H)$, known as *Adams's Thesis* ([1]), also called *the Equation* (see, e.g., [13,18,59,60]), or *Conditional Probability Hypothesis (CPH)* ([8,47,57]). Many psychological experiments confirm this hypothesis. People understand the probability of a conditional to be the conditional probability (e.g., [21,24,44,45,48,49,51,54]). For psychological experiments on the Equation and three-valued logics see also [2,3,55] A central problem is reasoning about sentences which include conditionals, like iterated (also called "nested"; see, e.g., [20,37,41]) conditionals or compounds of conditionals (i.e., conjunctions featuring (non-material) conditionals in their conjuncts). As David Lewis has shown with his triviality results ([42]), such objects cannot be represented simply by conditional probability. In our theory of logical operations among conditionals we define compound conditionals as suitable conditional random quantities (see e.g., [30,33]); this allows us to avoid such triviality results ([26,52,57,58]). In the subjective theory of de Finetti, given two events $A$ and $C$, with $A \neq \emptyset$, the probability $P(C|A)$ of the conditional event $C|A$, for an agent, measures their degree of belief that $C$ will be true, by assuming that $A$ is true. The conditional event $C|A$ is looked at as a three-valued object which is *true* when $A$ and $C$ are both true (i.e., $AC$ is true), *false* when $A$ is true and $C$ is false (i.e., $A\overline{C}$ is true), and *void* when $A$ is false (i.e., $\overline{A}$ is true). In numerical terms its indicator, still denoted by $C|A$, takes the value 1, or 0, or $P(C|A)$, according to whether the conditional event is true, or false, or void, respectively. Then, by setting $P(C|A) = x$, the indicator of $C|A$ can be represented as

$$C|A = AC + x\overline{A} = \begin{cases} 1, \text{ if } AC \text{ is true,} \\ 0, \text{ if } A\overline{C} \text{ is true,} \\ x, \text{ if } \overline{A} \text{ is true.} \end{cases} \tag{1}$$

Notice that the prevision of the indicator, denoted by $\mathbb{P}(C|A)$, coincides with $P(C|A)$. Indeed, as $P(AC) = P(C|A)P(A)$, one has

$$\mathbb{P}(C|A) = 1 \cdot P(AC) + 0 \cdot P(\overline{A}C) + P(C|A) \cdot P(\overline{A})$$
$$= P(C|A) \cdot P(A) + P(C|A) \cdot P(\overline{A}) = P(C|A).$$

Of special interest are *conjunctions of conditionals*, like $(A|H) \wedge (B|K)$, and *iterated conditionals*, like $(B|K)|(A|H)$. Such objects, as $A|H$ and $B|K$, may not only be true or false but also void, and must be defined, in order to preserve some basic logical and probabilistic properties, on more than three conditions and hence cannot typically be conditional events any longer ([7,25,35,36,56]). They are defined in the setting of coherence as suitable conditional random quantities.

There are long-standing disagreements about conjunctions of conditionals and how to analyse them in natural language ([6,17,43]): how should conjunctions of the form *if   A   then   C  & if not- A then   C* be interpreted, where "&" denotes a natural language "and". These conjunctions can be used to "sum up" the valid inference of Dilemma: inferring $C$ from a proof of $C$ from $A$ and a proof of $C$ from not-$A$. An acceptable analysis of *if   A    then   C  & if not- A then   C* must have the result that $C$ validly follows from it. We will show below in the formal section of this paper that our account does have this result.

Another important use of these conjunctions is to indicate that two propositions or events are independent. Consider:

(S1) If your children are vaccinated $(H)$, they will not get autism $(A)$.

We could use (S1) in certain contexts to indicate, pragmatically, that developing autism is independent of having a vaccination. To be as clear and explicit as possible about this fact, doctors could assert this conjunction for the benefit of parents whose children did not have autism at birth:

(S2) If your children are vaccinated $(H)$, they will not get autism $(A)$, and if they are not vaccinated (not-$H$), they will not get autism $(A)$.

After all, a use of (S1) could (mistakenly) be interpreted as stating that the vaccination prevents autism, but (S2) makes it perfectly clear that (actually) independence is being conveyed. A formal analysis will be given in Sect. 2.4 and in Sect. 3.1.

One recent view, inferentialism (see, e.g., [16,38,61]), of conditionals like (S1) is that they are somehow "non-standard". A "standard" conditional *if A then C* is supposed to be one supported by a deductive, inductive, abduction, or other relation between $A$ and $C$ ([14,15]). As the term "inferentialism" suggests, this account of conditionals is based on the notion that the conditional is like an inference from $A$ to $C$ in which $A$ supplies a positive reason for believing $C$. If we call a conditional for which such a relation holds between $A$ and $C$ a *dependence conditional*, and a conditional, like (S1), in which $A$ and $C$ are independent an *independence conditional*, then inferentialism states that only dependence conditionals are "standard" conditionals, and that independence conditionals are "non-standard", implying that conjunctions of independence conditionals are "non-standard" as well. We, however, can see nothing non-standard about independence conditionals like (S1) or conjunctions of them like (S2), or non-standard about Dilemma inferences for that matter, and we will focus on conditionals in the form *(if A then C) & (if not-A then C)* in this paper, treating them as standard conditionals that sometimes have the very important pragmatic role of indicating that $A$ and $C$ are independent (for critical comments on inferentialism, see [4,9,10,40,48,51]).

Let us consider an example that we can more easily make probability judgments about. (This example is derived from [17], and developed with Simone Sebben for psychological experiments.) Suppose two people are on the way to an airport in a car. They have enough time to make their flight as long as their car does not break down. There is a probability of 0.5 that the car will break down, and a probability of 0.5 that the passenger will cross their fingers. But the driver trusts the car, and not at all finger crossing, and says to the passenger:

(S3) If you cross your fingers $(F)$, then we will be in time $(T)$, and if you do not cross your fingers $(\overline{F})$, then we will be in time.

Following Lance ([39]), we could attempt to reason about (S3) in the following way. The probability that the car will break down is 0.5. Suppose first the car does not break down. Then (S3) is supposedly "true". Suppose second the car breaks down. Then (S3) is supposedly "false". Thus, (S3) is supposedly "true" half the time and has a probability of 0.5. The problem for Lance with this way of reasoning is that he also wants to hold the Conditional Probability Hypothesis that the probability of a natural language conditional, $P(\text{if } A \text{ then } C)$, is the conditional probability of $C$ *given* $A$. As we have already remarked, there is certainly much to recommend this hypothesis $P(\text{if } A \text{ then } C) = P(C|A)$, which can be traced back to de Finetti ([11,12]). There are strong grounds for it in philosophical logic ([18]) and in experiments in the psychology of reasoning ([46,48,50,51]). By this relation in this example, $P(\text{if } F \text{ then } T) = P(T|F) = 0.5$. Lance thus claims about examples like this that $P((\text{if } F \text{ then } T) \,\&\, (\text{if not-}F \text{ then } T)) = P(\text{if } F \text{ then } T) = P(T|F)$.

However, Lewis ([42]) proved that $P(\text{if } A \text{ then } C) = P(C|A)$ cannot be combined with claiming that these conditionals are simply true or false, as in Lance's attempted reasoning. Cantwell ([6]) has also proven that there is a serious problems with Lance's claims about conjunctions like (S3) and holding that $P(\text{if } F \text{ then } T) = P(T|F)$ and $P((\text{if } F \text{ then } T) \,\&\, (\text{if not-}F \text{ then } T)) = P(\text{if } F \text{ then } T) = P(T|F)$. The underlying problem for Lance can be illustrated by an instance of a conditional that is not a conjunction of conditionals:

(S4) If you cross your fingers ($F$) or we will be in time ($T$), then we will be in time ($T$).

Let $B$ be the car breaks down, and consider the case in which the possibilities below have equal probabilities $\frac{1}{4}$, with $P(BT) = P(\overline{B}\,\overline{T}) = 0$:

$$BF\overline{T}, \; B\overline{F}\,\overline{T}, \; \overline{B}FT, \; \overline{B}\,\overline{F}T.$$

Then by Lance's reasoning, (S4) is "true" 50% of the time, because (S4) is true when $\overline{B}FT$ is true, and when $\overline{B}\,\overline{F}T$ is true, since in both cases it has a true antecedent and a true consequent; moreover (S4) is false when $BF\overline{T}$ is true, because it has a true antecedent and a false consequent, and (S4) is also false when $B\overline{F}\,\overline{T}$ is true, since in this case it is not true for Lance, who only classifies conditionals as true or false. Then, for Lance the conditional (S4), which coincides in our framework with the conditional event $T|(F \vee T)$, "should" have a probability of 0.5. However,

$$P(T|(F \vee T)) = \frac{P(T)}{P(F \vee T)} = \frac{P(\overline{B}FT) + P(\overline{B}\,\overline{F}T)}{P(BF\overline{T}) + P(\overline{B}FT) + P(\overline{B}\,\overline{F}T)} = \frac{2}{3}.$$

Edgington knows that, given Lewis's proof, conditionals cannot be simply true or false in the actual world in general, and she follows Bradley ([5]) in holding that a conditional *if A then C* can only be simply true when $A \,\&\, C$ is true, and otherwise *if A then C* can only be made true by pairs of possible worlds ([19]). For example, suppose not-$F$ holds in the actual world. Then in

Bradley's account, *if F then T* would be made "true" by the actual world and one alternative world where $F \& T$ was true and made "false" by another alternative world where $F$ & not-$T$ was true. We might express this account informally by saying that *if F then T* is true relative to one alternative to the actual world and false relative to another. People may judge that one of these alternatives is more probable than the other, or they may judge them equally probable.

Our approach is different. It goes back to de Finetti ([11,12]) and compares the assertion of an indicative conditional, like *if F then T*, closely to a conditional bet, *if F then I bet that T*. In our analysis, this indicative conditional is true, and the conditional bet is won, when $F \& T$ holds, the indicative conditional is false, and the conditional bet is lost, when $F$ & not-$T$ holds. The indicative conditional and the conditional bet are "void" when not-$F$ holds. We will present our analysis formally below, showing that it implies in general that the expected value, or prevision in de Finetti's terms, of *if A then C* is $P(C|A)$. Our account, which properly manages void cases and where the Import-Export principle is not valid, avoids Lewis' triviality results ([30,57]).

We investigate conjunctions of conditionals *if H then A* & *if K then B* formally below, within an extended account in betting terms. In this extended account, we can say informally that *if H then A* & *if K then B* is fully true when $AHBK$ holds, is partly true when $\overline{H}BK$ holds, or $\overline{K}AH$ holds, is false when $\overline{A}H \vee \overline{B}K$ holds, and is void when $\overline{H}\,\overline{K}$ holds. In our account, the sentence (S3), *if F then T* & *if not-F then T*, has a prevision, i.e., an expected value, of .25 (see Sect. 2.4 for a formal analysis). Our account can be compared to McGee's account ([43]) to a limited extent, but there are, as well, differences between ours and McGee's; for instance, we do not require the general validity of the Import-Export principle.

Finally, we will consider the iterated conditional sentence

(S5) If we will be in time ($T$) when you do not cross your fingers ($\overline{F}$), then we will be in time when you cross your fingers ($F$).

We will show that the "intuitive" assertion "*the probability of the sentence (S5) coincides with $P(T|F)$* " in our formalism is correct. This result is another way of looking at the "independence" of the sentences "*if F then T*" and "*if not-F then T*".

The paper is organized as follows. In Sect. 2 we give an interpretation of the possible values of the conjunction $(A|H) \wedge (B|K)$ of two conditional events $A|H$ and $B|K$. We illustrate a real world application within the context of multiple bets. We also recall the iterated conditional $(B|K)|(A|H)$. Then, we consider the compound conditionals $(A|H) \wedge (A|\overline{H})$ and $(A|H)|(A|\overline{H})$ in order to formalize the intuitions on the sentences (S3) and (S5), by discussing the aspects of "independence". In Sect. 3, we recall a general notion of iterated conditional and we study the particular objects $A|[(A|H) \wedge (A|\overline{H})]$ and $A|(A|H)$, in order to examine the *p-validity* of the inferences from (S2) to $A$, and from (S1) to $A$. Finally, in Sect. 4 we give some conclusions.

## 2  Conjunctions and Iterated Conditionals

In this section we first give an interpretation of the possible values of conjoined conditionals, which is in agreement with [6,43]. Then, we recall the notion of iterated conditionals. In particular, in order to correctly formalize the intuition about the sentences $(S3)$ and $(S5)$, we analyze the conjunction $(A|H) \wedge (A|\overline{H})$ and the iterated conditional $(A|H)|(A|\overline{H})$.

### 2.1  Interpretation of the Possible Values of $(A|H) \wedge (B|K)$

In our approach the conjunction of two conditional events $A|H$ and $B|K$ is a conditional random quantity, defined in the setting of coherence as

$$(A|H) \wedge (B|K) = \begin{cases} 1, & \text{if } AHBK \text{ is true,} \\ 0, & \text{if } \overline{A}H \vee \overline{B}K \text{ is true,} \\ P(A|H), & \text{if } \overline{H}BK \text{ is true,} \\ P(B|K), & \text{if } AH\overline{K} \text{ is true,} \\ \mathbb{P}[(A|H) \wedge (B|K)], & \text{if } \overline{H}\,\overline{K} \text{ is true.} \end{cases} \qquad (2)$$

In more explicit terms, by setting $P(A|H) = x$, $P(B|K) = y$, it holds that

$$(A|H) \wedge (B|K) = (AHBK + x\overline{H}BK + yAH\overline{K})|(H \vee K),$$

with

$$\mathbb{P}[(A|H) \wedge (B|K)] =$$
$$= P(AHBK|(H \vee K)) + xP(\overline{H}BK|(H \vee K)) + yP(AH\overline{K}|(H \vee K)).$$

Notice that by coherence ([30, Theorem 7])

$$\max\{x + y - 1, 0\} \leq \mathbb{P}[(A|H) \wedge (B|K)] \leq \min\{x, y\}. \qquad (3)$$

Intuitively, in agreement with [6,43], we could say that

– when $AHBK$ is true, i.e., both conditional events $A|H$ and $B|K$ are true, then the conjunction is "true", and its numerical value is 1
– when $\overline{A}H \vee \overline{B}K$ is true, that is at least a conditional event is false, the conjunction is "false", and its numerical value is 0
– when $\overline{H}BK$ is true, that is $A|H$ is void and $B|K$ is true, the conjunction is "partly true", and its numerical value is the conditional probability $P(A|H)$,
– when $AH\overline{K}$ is true, that is $A|H$ is true and $B|K$ is void, the conjunction is "partly true", and its numerical value is the conditional probability $P(B|K)$,
– when $\overline{H}\,\overline{K}$ is true, that is both conditional events are void, the conjunction is "void", and its numerical value is its prevision $\mathbb{P}[(A|H) \wedge (B|K)]$.

Then, in a bet on $(A|H) \wedge (B|K)$, the prevision $\mathbb{P}[(A|H) \wedge (B|K)]$ is the amount you agree to pay in order to receive the random win $(A|H) \wedge (B|K)$, that is

– when $AHBK$ is true, you receive 1 (you "win"),

- when $\overline{A}H \vee \overline{B}K$ is true, you receive 0 (you "lose"),
- when $\overline{H}BK$ is true, you receive $x$, (you "partly win"),
- when $AH\overline{K}$ is true, you receive $y$, (you "partly win"),
- when $\overline{H}\,\overline{K}$ is true, you receive back the amount you paid $\mathbb{P}[(A|H) \wedge (B|K)]$.

We observe that $(A|H) \wedge (B|K) = (B|K) \wedge (A|H)$, i.e. conjunction satisfies the commutativity property; moreover, when $H = K$, the conjunction $(A|H)\wedge(B|H)$ reduces to the conditional event $AB|H$.

## 2.2   A Real World Application of the Conjunction $(A|H) \wedge (B|K)$

In this section, based on an example given in [22], we illustrate an instance of a real world interpretation of the conjunction $(A|H)\wedge(B|K)$ in terms of the return of a double-bet. Let us consider two football matches. For each (valid) match the possible outcomes are: *Home win*, *Draw*, and *Away win*. Let us consider two single bets on the two events

$A =$ *The outcome of match 1 is Home win*,
$B =$ *The outcome of match 2 is Away win*.

In a single bet, when a match is not being played or abandoned (i.e., it is not valid) the bet is cancelled and the stake will be refunded (in this case the bet is called off). Let us define the events $H$=*the match 1 is valid*, and $K =$ *the match 2 is valid*. Then, actually, we have to consider two bets on the conditional events:

$A|H =$ *The outcome of match 1 is Home win, given that it is valid*,
$B|K =$ *The outcome of match 2 is Away win, given that it is valid*.

In a bet on $A|H$, with $P(A|H) = x \in [0,1]$, for every $s_1$, you agree to pay $xs_1$ and to receive $s_1(AH + x\overline{H})$, that is you receive $s_1$, or 0, or $xs_1$, according to whether $AH$ is true, or $\overline{A}H$ is true, or $\overline{H}$ is true (bet called off). In equivalent terms, by setting $Q_1 = \frac{1}{x}$ (when $x \neq 0$) and $r_1 = xs_1$, as $s_1 = r_1Q_1$, in a bet on $A|H$ you agree to pay $r_1$ and to receive $Q_1r_1(AH + \frac{1}{Q_1}\overline{H}) = Q_1r_1AH + r_1\overline{H}$. Similarly, in a bet on $B|K$ with $P(B|K) = y$, by setting $Q_2 = \frac{1}{y}$ and $r_2 = ys_2$, you agree to pay $ys_2$ and to receive $s_2(BK + y\overline{K})$, or equivalently as $s_2 = r_2Q_2$, you agree to pay $r_2$ and to receive $Q_2r_2BK + r_2\overline{K}$.

A *double-bet* is a linked series of the two single bets, where the return from one bet is automatically staked on the other bet. Then, if one of the two single bets is void, the double-bet continues on the remaining bet (the double-bet becomes a single bet). In our approach, the return of a double-bet can be related to the conjunction $(A|H)\wedge(B|K)$, when its prevision, $z = \mathbb{P}[(A|H) \wedge (B|K)]$, coincides with the product $xy$ (notice that the product $xy$ satisfies the inequalities in (3)). Based on (2), we observe that

$$(A|H) \wedge (B|K) = AHBK + x\overline{H}BK + y\overline{K}AH + z\overline{H}\,\overline{K}. \qquad (4)$$

Moreover, when $z = xy$ it holds that

$$(A|H) \wedge (B|K) = AHBK + x\overline{H}BK + y\overline{K}AH + xy\overline{H}\,\overline{K}$$
$$= (AH + x\overline{H})(BK + y\overline{K}) = (A|H) \cdot (B|K). \tag{5}$$

In a bet on $(A|H) \wedge (B|K)$, for every $s$, you agree to pay $zs$ and to receive the random win $s(AHBK + x\overline{H}BK + y\overline{K}AH + z\overline{H}\,\overline{K})$. In particular, when $z = xy = \frac{1}{Q_1Q_2}$, by setting $r = xys$, that is $s = Q_1Q_2r$, you agree to pay the amount $r$ and to receive

$$Q_1Q_2r(AHBK + \tfrac{1}{Q_1}\overline{H}BK + \tfrac{1}{Q_2}\overline{K}AH + \tfrac{1}{Q_1Q_2}\overline{H}\,\overline{K})$$
$$= Q_1Q_2rAHBK + Q_2r\overline{H}BK + Q_1rAH\overline{K} + r\overline{H}\,\overline{K}$$
$$= r(Q_1AH + \overline{H})(Q_2BK + \overline{K}),$$

that is to receive

$$\begin{cases} Q_1Q_2r & \text{(win)}, & \text{if } AHBK \text{ is true}, \\ 0 & \text{(lose)}, & \text{if } \overline{A}H \vee \overline{B}K \text{ is true}, \\ Q_2r & \text{(partly win)}, & \text{if } \overline{H}BK \text{ is true}, \\ Q_1r & \text{(partly win)}, & \text{if } AH\overline{K} \text{ is true}, \\ r & \text{(bet called off)}, & \text{if } \overline{H}\,\overline{K} \text{ is true}. \end{cases}$$

When a match is cancelled, the double-bet reverts to a single bet on the remaining match, with the double-bet void when both matches are cancelled, in which case your stake will be refunded.

Based on this normative analysis, new psychological predictions can easily be derived for future experiments: E.g., what exchange price would participants assign in a betting situation according to the described conditions? It is psychologically interesting to investigate whether people assume (as usually being done by bookmakers) or do not assume "independence" among the conditional events in such cases.

## 2.3   The Conjunction $(A|H) \wedge (A|K)$

When $B = A$ formula (2) becomes

$$(A|H) \wedge (A|K) = \begin{cases} 1, & \text{if } AHK \text{ is true}, \\ 0, & \text{if } \overline{A}H \vee \overline{A}K \text{ is true}, \\ P(A|H), & \text{if } A\overline{H}K \text{ is true}, \\ P(A|K), & \text{if } AH\overline{K} \text{ is true}, \\ \mathbb{P}[(A|H) \wedge (A|K)], & \text{if } \overline{H}\,\overline{K} \text{ is true}, \end{cases} \tag{6}$$

with $\mathbb{P}[(A|H) \wedge (A|K)]$ satisfying the inequalities ([33, Theorem 9])

$$P(A|H)P(A|K) \le \mathbb{P}[(A|H) \wedge (A|K)] \le \min\{P(A|H), P(A|K)\}. \tag{7}$$

Moreover, when $HK = \emptyset$, it holds that $\mathbb{P}[(A|H) \wedge (B|K) = P(A|H)P(B|K)]$ and, in particular, $\mathbb{P}[(A|H) \wedge (A|K) = P(A|H)P(A|K)]$, as shown in the next section.

## 2.4   The Conjunction $(A|H) \wedge (A|\overline{H})$

The conjunction $(A|H) \wedge (B|K)$, when $B = A$ and $K = \overline{H}$, reduces to the conjunction $(A|H) \wedge (A|\overline{H})$. By setting $P(A|H) = x$ and $P(A|\overline{H}) = y$, from (2) one has

$$(A|H) \wedge (A|\overline{H}) = (A|H) \cdot (A|\overline{H}) = \begin{cases} 0, & \text{if } \overline{A}H \vee \overline{A}\,\overline{H} \text{ is true,} \\ x, & \text{if } A\overline{H} \text{ is true,} \\ y, & \text{if } AH \text{ is true.} \end{cases}$$

Then $(A|H) \wedge (A|\overline{H}) = xA\overline{H} + yAH$. Moreover,

$$\begin{aligned} \mathbb{P}[(A|H) \wedge (A|\overline{H})] &= x\,P(A\overline{H}) + y\,P(AH) \\ &= x\,P(A|\overline{H})P(\overline{H}) + y\,P(A|H)P(H) = xy\,P(\overline{H}) + xy\,P(H) = xy\,. \end{aligned} \tag{8}$$

In particular, $\mathbb{P}[(A|H) \wedge (A|\overline{H})] = 1$ if and only if $P(A|H) = P(A|\overline{H}) = 1$. Moreover, in this particular case it holds that $(A|H) \wedge (A|\overline{H}) = A\overline{H} + AH = A$, with $P(A) = P(A|H)P(H) + P(A|\overline{H})P(\overline{H}) = P(H) + P(\overline{H}) = 1$.

The previous analysis, applied with $A = T$ and $H = F$, allows to correctly formalize our intuition about the conjunction (S3), that is, from $P(T|F) = P(T|\overline{F}) = 0.5$ it follows that $\mathbb{P}[(T|F) \wedge (T|\overline{F})] = P(T|F)P(T|\overline{F}) = 0.25$.

*Remark 1.* Let us consider two conditional events $A|H$ and $B|K$, with $P(A|H) = P(B|K) = 0.5$. To the pair $(A|H, B|K)$ we can associate the four conditional constituents ([32]):

$$(A|H) \wedge (B|K), \ (A|H) \wedge (\overline{B}|K), \ (\overline{A}|H) \wedge (B|K), \ (\overline{A}|H) \wedge (\overline{B}|K),$$

which are such that

$$(A|H)\wedge(B|K)+(A|H)\wedge(\overline{B}|K)+(\overline{A}|H)\wedge(B|K)+(\overline{A}|H)\wedge(\overline{B}|K) = A|H+\overline{A}|H = 1.$$

Then,

$$\mathbb{P}[(A|H)\wedge(B|K)]+\mathbb{P}[(A|H)\wedge(\overline{B}|K)]+\mathbb{P}[(\overline{A}|H)\wedge(B|K)]+\mathbb{P}[(\overline{A}|H)\wedge(\overline{B}|K)] = 1.$$

In particular

$$P(A|H) = \mathbb{P}[(A|H) \wedge (B|K)] + \mathbb{P}[(A|H) \wedge (\overline{B}|K)], \tag{9}$$

and hence, if $\mathbb{P}[(A|H) \wedge (B|K)] = P(A|H)$, then $\mathbb{P}[(A|H) \wedge (\overline{B}|K)] = 0$. By applying the previous reasoning to the sentence (S3), with any (natural language) conjunction & satisfying (9), that is

$$P(T|F) = P[(T|F)\&(T|\overline{F})] + P[(T|F)\&(\overline{T}|\overline{F})],$$

from $P[(T|F)\&(T|\overline{F})] = P(T|F) = 0.5$ it would follow the unlikely result that $P[(T|F)\&(\overline{T}|\overline{F})] = 0$.

In our approach, when $HK = \emptyset$ it holds that

$$\mathbb{P}[(A|H) \wedge (B|K)] = \mathbb{P}[(A|H) \cdot (B|K)] = P(A|H)P(B|K);$$

see [28, Section 5], where this result is interpreted as a case of uncorrelation between two random quantities. Then, concerning (S3), we have that $\mathbb{P}[(T|F) \wedge (T|\overline{F})] = \mathbb{P}[(T|F) \wedge (\overline{T}|\overline{F})] = \mathbb{P}[(\overline{T}|F) \wedge (T|\overline{F})] = \mathbb{P}[(\overline{T}|F) \wedge (\overline{T}|\overline{F})] = 0.25$.

We note that the object $(A|H) \wedge (A|\overline{H})$ is relevant to connexive logic [62], as its negated version corresponds to the well-known connexive principle *Aristotle's Second Thesis* $(\sim((H \to A) \wedge (\sim H \to A)))$. A systematic study on connexive principles within our framework of logical operations on conditionals has been done in [52], where it is shown that Aristotle's Second Thesis is not valid. This result follows by observing that the conjunction $(A|H) \wedge (A|\overline{H})$ is not constant and equal to 0. However, in a framework where the Aristotle's Second Thesis were valid it would follow, for instance, that the sentence (S3) should be always false. Moreover, experimental psychological data suggests that most people interpret Aristotle's Second Thesis as not valid [53].

## 2.5    The Iterated Conditional $(B|K)|(A|H)$

The notion of an iterated conditional $(B|K)|(A|H)$ is based on a structure like (1), that is $\square|\bigcirc = \square \wedge \bigcirc + \mathbb{P}(\square|\bigcirc)\overline{\bigcirc}$, where $\square$ denotes $B|K$, $\bigcirc$ denotes $A|H$, $\square \wedge \bigcirc$ is the conjunction $(B|K) \wedge (A|H)$, and where we set $\mathbb{P}(\square|\bigcirc) = \mu$. In the framework of subjective probability $\mu = \mathbb{P}(\square|\bigcirc)$ is the amount that you agree to pay, by knowing that you will receive the random quantity $\square \wedge \bigcirc + \mathbb{P}(\square|\bigcirc)\overline{\bigcirc}$. Then, given two conditional events $A|H$ and $B|K$, with $AH \neq \emptyset$, and a coherent assessment $(x, y, z)$ on $\{A|H, B|K, (A|H) \wedge (B|K)\}$, the iterated conditional $(B|K)|(A|H)$ is defined as (see, e.g., [27,28,30]):

$$(B|K)|(A|H) = (B|K) \wedge (A|H) + \mu \overline{A}|H = \begin{cases} 1, & \text{if } AHBK \text{ is true,} \\ 0, & \text{if } AH\overline{B}K \text{ is true,} \\ y, & \text{if } AH\overline{K} \text{ is true,} \\ x + \mu(1-x), & \text{if } \overline{H}BK \text{ is true,} \\ \mu(1-x), & \text{if } \overline{H}\overline{B}K \text{ is true,} \\ z + \mu(1-x), & \text{if } \overline{H}\overline{K} \text{ is true,} \\ \mu, & \text{if } \overline{A}H \text{ is true,} \end{cases}$$

(10)

where

$$\mu = \mathbb{P}[(B|K)|(A|H)] = \mathbb{P}[(B|K) \wedge (A|H) + \mu \overline{A}|H] = z + \mu(1-x).$$

Then

$$z = \mathbb{P}[(B|K) \wedge (A|H)] = \mathbb{P}[(B|K)|(A|H)]P(A|H) = \mu x \, ;$$

in particular, $z = \mu$ when $x = 1$. Notice that, when $x > 0$, a bet on $(B|K)|(A|H)$ is called off when $\overline{A}H \vee \overline{H}\,\overline{K}$ is true. When $x = 0$, the bet is called off when $\overline{A} \vee \overline{H}$ is true. Moreover, in the particular case where $H = K = \Omega$, it holds that $\mu = P(B|A)$ and $(B|\Omega)|(A|\Omega) = AB + \mu\overline{A} = AB + P(B|A)\overline{A} = B|A$.

## 2.6   The Iterated Conditional $(A|H)|(A|\overline{H})$

We now examine the formal aspects related with the sentence $(S5)$. We observe that

$$(A|H)|(A|\overline{H}) = (A|H) \wedge (A|\overline{H}) + \mu(\overline{A}|\overline{H}) = \begin{cases} x, & \text{if } A\overline{H} \text{ is true,} \\ y + \mu(1-y), & \text{if } AH \text{ is true,} \\ \mu(1-y), & \text{if } \overline{A}H \text{ is true,} \\ \mu, & \text{if } \overline{A}\,\overline{H} \text{ is true.} \end{cases}$$

Then, by the linearity of prevision

$$\mu = xy + \mu(1-y),$$

that is $\mu y = xy$. Then $\mu = x$, when $y > 0$. Moreover, when $y = 0$, it can be verified that

$$(A|H)|(A|\overline{H}) = \begin{cases} x, & \text{if } A\overline{H} \text{ is true} \\ \mu, & \text{if } A\overline{H} \text{ is false} \end{cases} = xA\overline{H} + \mu(1 - A\overline{H}) \in \{x, \mu\}.$$

Hence, in a bet on $(A|H)|(A|\overline{H})$, we agree to pay $\mu$, by receiving $x$, when $A\overline{H}$ is true, or by receiving back the paid amount $\mu$, when $A\overline{H}$ is false (in this case the bet is called off). Then, by coherence, $\mu = x$. Therefore in all cases $\mathbb{P}[(A|H)|(A|\overline{H})] = P(A|H)$. By setting $A = T$ and $H = F$, we can apply the previous analysis to the sentence $(S5)$, by showing that our account is in agreement with the "intuitive" assertion that "*the probability of the sentence $(S5)$ coincides with $P(T|F)$*". Likewise, $\mathbb{P}[(A|\overline{H})|(A|H)] = P(A|\overline{H})$. In the context of connexive principles and compounds of conditionals ([52]), the objects $(A|\overline{H})|(A|H)$ and $(A|H)|(A|\overline{H})$ can be interpreted as the unnegated versions (i.e., the consequent of the main connective is not negated) of the *variations of Boethius' theses* $(H \to A) \to \neg(\neg H \to A)$ and $(\neg H \to A) \to \neg(H \to A)$, respectively (called (B3) and (B4) in [23]). In [52] these principles are not valid, which is consistent with the present context.

*Remark 2.* Notice that in the case of simple conditionals the relation $P(E|H) = P(E)$, when $P(E) \in ]0,1[$, does not follow from some logical relation between $E$ and $H$, but from probabilistic evaluations of the involved events (stochastic independence is a subjective relation). Indeed, in this case coherence allows any value in $[0,1]$ for $P(E|H)$. However, differently from the case of simple conditionals, concerning the particular object $(A|H)|(A|\overline{H})$ it holds that coherence always requires that $\mathbb{P}[(A|H)|(A|\overline{H})]$ must be equal to $P(A|H)$. Then, in this case $A|H$ is "independent" from $A|\overline{H}$ for any coherent evaluation.

Moreover in general, when $HK = \emptyset$, it holds that $\mathbb{P}[(A|H)|(B|K)] = P(A|H)$ and $\mathbb{P}[(B|K)|(A|H)] = P(B|K)$, with in particular $\mathbb{P}[(A|H)|(A|K)] = P(A|H)$ and $\mathbb{P}[(A|K)|(A|H)] = P(A|K)$. Therefore, when $HK = \emptyset$, $A|H$ and $B|K$ (in particular, $A|H$ and $A|K$) are "independent". Indeed, by setting

$$P(A|H) = x, \quad P(B|K) = y, \quad \mathbb{P}[(A|H) \wedge (B|K)] = z, \quad \mathbb{P}[(B|K)|(A|H)] = \mu,$$

and by recalling formula (10), it holds that $z = \mu x$. Moreover, under the hypothesis $HK = \emptyset$, it holds that $z = xy$ and hence, when $x > 0$, one has $\mu = y$. If $x = 0$ (and hence $z = 0$), by observing that $AHBK = AH\overline{B}K = \emptyset$, formula (10) becomes

$$(B|K)|(A|H) = (A|H) \wedge (B|K) + \mu \overline{A}|H = \begin{cases} y, & \text{if } AH\overline{K} \text{ is true,} \\ \mu, & \text{if } AH\overline{K} \text{ is false.} \end{cases}$$

Then, by coherence, (when $x = 0$) it holds that $\mu = y$ and the iterated conditional $(B|K)|(A|H)$ is constant and coincides with $y = P(B|K)$.

Therefore, when $HK = \emptyset$, in all cases it holds that $\mathbb{P}[(B|K)|(A|H)] = P(B|K)$. Likewise, $\mathbb{P}[(A|H)|(B|K)] = P(A|H)$.

## 3   On General Iterated Conditionals

We recall that a family of conditional events $\mathcal{F} = \{E_i|H_i, \ i = 1, \ldots, n\}$ is p-consistent if and only if the assessment $(1, 1, \ldots, 1)$ on $\mathcal{F}$ is coherent. Moreover, a p-consistent family $\mathcal{F}$ p-entails a conditional event $E|H$ if and only if the unique coherent extension on $E|H$ of the assessment $(1, 1, \ldots, 1)$ on $\mathcal{F}$ is $P(E|H) = 1$ (see, e.g., [29]). We say that the inference from $\mathcal{F}$ to $E|H$ is p-valid if and only if $\mathcal{F}$ p-entails $E|H$. The characterization of p-entailment using the notion of conjunction has been given in [31]. In particular, given a p-consistent family $\mathcal{F}$ of $n$ conditional events it holds that. $\mathcal{F}$ p-entails $E|H$ if and only if $\mathcal{C}(\mathcal{F}) \leq E|H$, where $\mathcal{C}(\mathcal{F}) = \bigwedge_{E_i|H_i \in \mathcal{F}} E_i|H_i$. Let two logically independent events $A$ and $H$ be given. Then, the assessment $P(A|H) = P(A|\overline{H}) = 1$ is coherent and the family $\{A|H, A|\overline{H}\}$ is p-consistent. As $P(A) = P(A|H)P(H) + P(A|\overline{H})P(\overline{H})$, from $P(A|H) = P(A|\overline{H}) = 1$ it follows that $P(A) = 1$; thus the inference from $\{A|H, A|\overline{H}\}$ to $A$ is p-valid ([1]). In this section we show that this result is in agreement with the formalism of iterated conditionals. We first recall a general notion of iterated conditional given in ([34]). Let us consider two finite families of conditional events $\mathcal{F}_1$ and $\mathcal{F}_2$. We set

$$\mathcal{C}(\mathcal{F}_1) = \bigwedge_{E|H \in \mathcal{F}_1} E|H, \ \ \mathcal{C}(\mathcal{F}_2) = \bigwedge_{E|H \in \mathcal{F}_2} E|H,$$

and by definition ([31])

$$\mathcal{C}(\mathcal{F}_1) \wedge \mathcal{C}(\mathcal{F}_2) = \mathcal{C}(\mathcal{F}_1 \cup \mathcal{F}_2) = \bigwedge_{E|H \in \mathcal{F}_1 \cup \mathcal{F}_2} E|H.$$

Then, under the assumption that $\mathcal{C}(\mathcal{F}_1)$ eq0, the iterated conditional $\mathcal{C}(\mathcal{F}_2)|\mathcal{C}(\mathcal{F}_1)$ is defined as

$$\mathcal{C}(\mathcal{F}_2)|\mathcal{C}(\mathcal{F}_1) = \mathcal{C}(\mathcal{F}_1) \wedge \mathcal{C}(\mathcal{F}_2) + \mu(1 - \mathcal{C}(\mathcal{F}_1))$$

where $\mu = \mathbb{P}[\mathcal{C}(\mathcal{F}_2)|\mathcal{C}(\mathcal{F}_1)]$. In a bet on $\mathcal{C}(\mathcal{F}_2)|\mathcal{C}(\mathcal{F}_1)$ the quantity $\mu$ is the amount to be paid in order to receive the random amount $\mathcal{C}(\mathcal{F}_2)|\mathcal{C}(\mathcal{F}_1)$. The characterization of the p-entailment in terms of iterated conditionals has been given in [34] (see also [26]). More precisely, given a p-consistent family $\mathcal{F}$ of $n$ conditional events it holds that $\mathcal{F}$ p-entails $E|H$ if and only if the iterated conditional $(E|H)|\mathcal{C}(\mathcal{F})$ is constant and equal to 1.

## 3.1    On the Iterated Conditional $A|[(A|H) \wedge (A|\overline{H})]$

By setting $\mathcal{F}_1 = \{A|H, A|\overline{H}\}$ and $\mathcal{F}_2 = \{A\}$, we will show that the p-entailment from $\{A|H, A|\overline{H}\}$ to $A$ can be characterized by the property that the iterated conditional $\mathcal{C}(\mathcal{F}_2)|\mathcal{C}(\mathcal{F}_1) = A|((A|H) \wedge (A|\overline{H}))$ is constant and coincides with 1 ([26,34]).

Let us consider the following complex conditional sentence:

(S6) If $((A$ when $H)$ and $(A$ when not-$H))$, then $A$,

which we represent by the iterated conditional $A|((A|H) \wedge (A|\overline{H}))$. We will show the p-validity of the inference (S6) from the conjunction "$A$ when $H$ and $A$ when not- $H$" to the conclusion $A$.

We set $P(A|H) = x, P(A|\overline{H}) = y, \mathbb{P}[A|((A|H) \wedge (A|\overline{H}))] = \mu$. Then

$$A \wedge (A|H) \wedge (A|\overline{H}) = (A|H) \wedge (A|\overline{H}) = \begin{cases} 0, & \text{if } \overline{A} \text{ is true,} \\ y, & \text{if } AH \text{ is true,} \\ x, & \text{if } A\overline{H} \text{ is true,} \end{cases}$$

and

$$\mu[1 - (A|H) \wedge (A|\overline{H})] = \begin{cases} \mu, & \text{if } \overline{A} \text{ is true,} \\ \mu(1-y), & \text{if } AH \text{ is true,} \\ \mu(1-x), & \text{if } A\overline{H} \text{ is true.} \end{cases}$$

Therefore

$$A|((A|H) \wedge (A|\overline{H})) = (A|H) \wedge (A|\overline{H}) + \mu[1 - (A|H) \wedge (A|\overline{H})] =$$
$$= \begin{cases} \mu, & \text{if } \overline{A} \text{ is true,} \\ y + \mu(1-y), & \text{if } AH \text{ is true,} \\ x + \mu(1-x), & \text{if } A\overline{H} \text{ is true,} \end{cases}$$

where we assume that $((A|H) \wedge (A|\overline{H}))$ does not coincide with the constant 0, that is $(x, y) \neq (0, 0)$. By the linearity of prevision and by (8), it holds that

$$\mu = \mathbb{P}[A|((A|H) \wedge (A|\overline{H}))] = xy + \mu(1 - xy)$$

or equivalently $\mu xy = xy$. Then, $\mu = 1$ when $x > 0$ and $y > 0$. When $x = 0$ and $y > 0$, it holds that

$$A|((A|H) \wedge (A|\overline{H})) = \begin{cases} \mu, & \text{if } \overline{A} \vee A\overline{H} \text{ is true,} \\ y + \mu(1-y), & \text{if } AH \text{ is true.} \end{cases}$$

By coherence, $\mu = y + \mu(1 - y)$, that is $\mu y = y$, and hence $\mu = 1$. Likewise, when $y = 0$ and $x > 0$, it follows that $\mu = 1$. Thus, $\mu = 1$ in all cases, and hence the iterated conditional $A|((A|H) \wedge (A|\overline{H}))$ is constant and coincides with 1.

By applying the previous analysis to the sentence (S2), we can infer the p-validity of the inference from "*if your children will not get autism (A), when they are vaccinated (H), and your children will not get autism (A), when they are not vaccinated ($\overline{H}$)*" to the conclusion "*your children will not get autism (A)*".

## 3.2    On the Two Iterated Conditionals $A|(A|H)$ and $A|(A|\overline{H})$

Notice that, differently from the iterated conditional $A|((A|H) \wedge (A|\overline{H}))$ which coincides with the constant 1, we can verify that both the iterated conditionals $A|(A|H)$ and $A|(A|\overline{H})$ do not coincide with the constant 1. In other words, both the inferences from $A|H$ to $A$ and from $A|\overline{H}$ to $A$ are not p-valid. Indeed, by setting $P(A|H) = x$ and $\mathbb{P}[A|(A|H)] = \eta$, it holds that

$$A|(A|H) = A \wedge (A|H) + \eta\,(1 - A|H) = \begin{cases} 1, & \text{if } AH \text{ is true,} \\ x + \eta\,(1 - x), & \text{if } A\overline{H} \text{ is true,} \\ \eta\,(1 - x), & \text{if } \overline{A}\,\overline{H} \text{ is true,} \\ \eta, & \text{if } \overline{A}H \text{ is true,} \end{cases}$$

which, in general, does not coincide with the constant 1. For instance, if we evaluate $P(A|H) = P(A|\overline{H}) = P(H) = \frac{1}{2}$, it holds that

$$A|(A|H) = \begin{cases} 1, & \text{if } AH \text{ is true,} \\ \frac{1}{2} + \frac{1}{2}\,\eta, & \text{if } A\overline{H} \text{ is true,} \\ \frac{1}{2}\,\eta, & \text{if } \overline{A}\,\overline{H} \text{ is true,} \\ \eta, & \text{if } \overline{A}H \text{ is true,} \end{cases}$$

with

$$\eta = P(AH) + (\frac{1}{2} + \frac{1}{2}\,\eta)P(A\overline{H}) + \frac{1}{2}\,\eta P(\overline{A}\,\overline{H}) + \eta P(\overline{A}H) = \frac{1}{4} + \frac{1}{8} + \frac{1}{8}\,\eta + \frac{1}{8}\,\eta + \frac{1}{4}\,\eta,$$

that is $\eta = \frac{3}{4}$. Therefore, the inference from the premise "if $H$ then $A$" to the conclusion $A$ is not p-valid. By recalling the sentence (S1), if we evaluate that the probability is high of getting autism (event $A$), given vaccination (event $H$), it does not follow that the probability of getting autism (event $A$) is high.

By a similar analysis, by setting $P(A|\overline{H}) = y$ and $\mathbb{P}[A|(A|\overline{H})] = \nu$, it holds that

$$A|(A|\overline{H}) = A \wedge (A|\overline{H}) + \nu\,(1 - A|\overline{H}) = \begin{cases} 1, & \text{if } A\overline{H} \text{ is true,} \\ y + \nu\,(1 - y), & \text{if } AH \text{ is true,} \\ \nu\,(1 - y), & \text{if } \overline{A}H \text{ is true,} \\ \nu, & \text{if } \overline{A}\overline{H} \text{ is true,} \end{cases}$$

which, in general, does not coincide with the constant 1. Therefore, the inference from the premise "if $\overline{H}$ then $A$" to the conclusion $A$ is not p-valid. Recalling the sentence (S1), if we evaluate that the probability is high of getting autism (event $A$), given no vaccination (event $\overline{H}$), it does not follow that the probability of getting autism (event $A$) is high.

On the contrary, as shown in Sect. 3.1, if we evaluate that the probability of getting autism (event $A$), given vaccination (event $H$), and the probability of getting autism (event $A$), given no vaccination (event $\overline{H}$), are both high, then the probability of getting autism (event $A$) is high.

*Remark 3.* What about the iterated conditional $((A|H) \wedge (A|\overline{H}))|A$ ? It can be verified that the inference from the premise $A$ to the conjunction $(A|H) \wedge (A|\overline{H})$

is not p-valid, that is from $P(A) = 1$ it does not follow that $\mathbb{P}[(A|H) \wedge (A|\overline{H})] = 1$. Indeed, the assessment $(1, y, 1)$ on $\{A|H, A|\overline{H}, H\}$, with $y < 1$, is coherent. Then $P(A) = P(A|H)P(H) + P(A|\overline{H})P(\overline{H}) = 1$, while $\mathbb{P}[(A|H) \wedge (A|\overline{H})] = y < 1$.

We also observe that, by setting $P(A|H) = x$, $P(A|\overline{H}) = y$, $\mathbb{P}[((A|H) \wedge (A|\overline{H}))|A] = \eta$,

$$((A|H) \wedge (A|\overline{H}))|A = (A|H) \wedge (A|\overline{H}) + \eta\,(1 - A)$$
$$= \begin{cases} \eta, & \text{if } \overline{A} \text{ is true,} \\ y, & \text{if } AH \text{ is true,} \\ x, & \text{if } A\overline{H} \text{ is true,} \end{cases}$$

and hence, by coherence, $\eta$ must belong to the convex hull of the set of values $\{x, y\}$, that is $\eta \in [\min\{x, y\}, \max\{x, y\}]$. Moreover, by the linearity of prevision, $\eta = xy + \eta(1 - P(A))$, that is $\eta P(A) = xy$. In particular, when $P(A) > 0$, by setting $P(H) = t$ it holds that

$$\eta = \mathbb{P}[((A|H) \wedge (A|\overline{H}))|A] = \frac{P((A|H) \wedge (A|\overline{H}))}{P(A)} = \frac{xy}{xt + y(1 - t)},$$

with $\eta \in [x, y]$ when $x < y$, and $\eta \in [y, x]$ when $x > y$.

## 4   Conclusions

In this paper, we investigated conjunctions of conditionals, iterated conditionals, and (stochastic) independence within the theory of logical operations on conditionals (i.e., interpreted by suitable conditional random quantities). We presented a formal and a more intuitive approach compared to alternative approaches in the literature and also avoided Lewis' triviality results. Our probabilistic analysis also lays the groundwork for expanding the current domain of new paradigm psychology of reasoning from basic conditionals to complex conditional structures, like nested and compound conditionals.

We gave an interpretation of the possible values of the conjunction $(A|H) \wedge (B|K)$ in agreement with [6,43]. This interpretation shows why more than three values are needed. We also illustrated this situation through a real world application within the context of multiple bets from which psychological predictions for future experiments can easily be derived.

Then, we formalized intuitions about the sentences (S3) and (S5) by using the compound conditionals $(A|H) \wedge (A|\overline{H})$ and $(A|H)|(A|\overline{H})$. These two objects can be related to the non-validity of Aristotle's Second Thesis and of the variation (B4) of Boethius' Thesis, respectively. Connexivity is a desired property of inferentialist accounts of conditionals (since *if A then not-A* also violates basic inferentialist intuitions), but more work is needed to fully understand formally the relations of inferentialism and our account.

We also discussed "independence" in sentences (S3) and (S5), based on the coherence conditions $\mathbb{P}[(A|H) \wedge (A|\overline{H})] = P(A|H)P(A|\overline{H})$ and $\mathbb{P}[(A|H)|(A|\overline{H})] = P(A|H)$.

We recalled a general notion of iterated conditional $\mathcal{C}(\mathcal{F}_2)|\mathcal{C}(\mathcal{F}_1)$. In particular we showed that the object $A|[(A|H) \wedge (A|\overline{H})]$ is constant and equal to 1, by characterizing the p-validity of the inference from the premise $(A|H) \wedge (A|\overline{H})$ to the conclusion $A$. In our framework, this inference amounts to the inference from $\{A|H, A|\overline{H}\}$ to $A$.

Finally, concerning the sentence (S1), we discussed the p-invalidity of the inferences from $A|H$ to $A$, and from $A|\overline{H}$ to $A$, by studying the iterated conditionals $A|(A|H)$ and $A|(A|\overline{H})$, respectively.

The present paper contributes to the understanding of judgments about independence and human and artificial rationality. It also provides a rationality framework for future experimental studies, which are needed to investigate its psychological plausibility.

**Acknowledgements.** We thank three anonymous reviewers, John Cantwell, and the chair for useful comments and suggestions. G. Sanfilippo is member of the INdAM-GNAMPA research group and is supported by the FFR 2024 project of University of Palermo, Italy.

# References

1. Adams, E.W.: The Logic of Conditionals. Reidel, Dordrecht (1975)
2. Baratgin, J., Over, D.E., Politzer, G.: Uncertainty and the de Finetti tables. Think. Reason. **19**(3–4), 308–328 (2013)
3. Baratgin, J., Politzer, G., Over, D.E., Takahashi, T.: The psychology of uncertainty and three-valued truth tables. Front. Psychol. **9**, 1479 (2018)
4. Bourlier, M., Jacquet, B., Lassiter, D., Baratgin, J.: Coherence, not conditional meaning, accounts for the relevance effect. Front. Psychol. 14 (2023)
5. Bradley, R.: Multidimensional possible-world semantics for conditionals. Philos. Rev. **121**(4), 539–571 (2012)
6. Cantwell, J.: Revisiting McGee's probabilistic analysis of conditionals. J. Philos. Log. **51**(5), 973–1017 (2022)
7. Castronovo, L., Sanfilippo, G.: Iterated conditionals, trivalent logics, and conditional random quantities. In: Dupin de Saint-Cyr, F., Ozturk-Escoffier, M., Potyka, N. (eds.) Scalable Uncertainty Management. SUM 2022. LNCS, vol. 13562, pp. 47–63. Springer, Cham (2022). https://doi.org/10.1007/978-3-031-18843-5_4
8. Cruz, N.: Deduction from uncertain premises? In: Elqayam, S., Douven, I., Evans, J.S.B.T., Cruz, N. (eds.), Logic and Uncertainty in the Human Mind: A Tribute to David E. Over, pp. 27–41. Routledge, Oxon (2020)
9. Cruz, N., Baratgin, J., Oaksford, M., Over, D.E.: Centering and the meaning of conditionals. In: Papafragou, A., Grodner, D., Mirman, D., Trueswell, J. (eds.), Proceedings of the 38 th Annual Meeting of the Cognitive Science Society, pp. 1104–1109, Austin, TX, 2016. The Cognitive Science Society (2016)
10. Cruz, N., Over, D.E.: Independence conditionals. In: Kaufmann, S., Over, D.E., Sharma, G. (eds.) Conditionals – Logic. Linguistics. and Psychology, pp. 223–233. Palgrave Macmillan, London (2023)
11. de Finetti, B.: La logique de la probabilité. In: Actes du Congrès International de Philosophie Scientifique, Paris, 1935 , pp. IV 1–IV 9 (1936)

12. de Finetti, B.: Foresight: its logical laws, its subjective sources. In: Studies in subjective probability, pp. 55–118. Krieger, Huntington (1980). English translation of La Prévision: Ses Lois Logiques, Ses Sources Subjectives, Annales de l'Institut Henri Poincaré, 17:1–68 (1937)
13. Douven, I., Dietz, R.: A puzzle about Stalnaker's hypothesis. Topoi, pp. 31–37 (2011)
14. Douven, I., Elqayam, S., Krzyanowska, K.: Inferentialism: A manifesto. In: Kaufmann, S., Over, D.E., Sharma, G. (eds.) Conditionals-Logic. Linguistics and Psychology, pp. 175–221. Palgrave Macmillan, London (2023)
15. Douven, I., Elqayam, S., Singmann, H., van Wijnbergen-Huitink, J.: Conditionals and inferential connections: toward a new semantics. Think. Reason. **26**(3), 311–351 (2020)
16. Douven, I.: The Epistemology of Indicative Conditionals: Formal and Empirical Approaches. Cambridge University Press, Cambridge (2015)
17. Edgington, D.: Matter-of-fact conditionals. Proc. Aristot. Soc. **65**, 185–209 (1991)
18. Edgington, D.: On conditionals. Mind **104**, 235–329 (1995)
19. Edgington, D.: Compounds of conditionals, uncertainty, and indeterminacy. In: Logic and Uncertainty in the Human Mind, pp. 43–56. Routledge, Abingdon (2020)
20. Egré, P., Rossi, L., Sprenger, J.: De Finettian logics of indicative conditionals part I: trivalent semantics and validity. J. Philos. Log. (2020)
21. Evans, J.S.B.T., Handley, S.J., Over, D.E.: Conditionals and conditional probability. J. Exp. Psychol. Learn. Memory Cogn. **29**(2), 321–355 (2003)
22. Flaminio, T., Gilio, A., Godo, L., Sanfilippo, G.: Compound conditionals as random quantities and Boolean algebras. In: Proceedings of the 19th International Conference on Principles of Knowledge Representation and Reasoning, KR 2022, pp. 141–151 (2022)
23. Francez, N.: Natural deduction for two connexive logics. IfCoLog J. Log. Appl. **3**, 479–504 (2016)
24. Fugard, A.J.B., Pfeifer, N., Mayerhofer, B., Kleiter, G.D.: How people interpret conditionals: shifts towards the conditional event. J. Exp. Psychol. Learn. Memory Cogn. **37**(3), 635–648 (2011)
25. Gilio, A., Over, D.E., Pfeifer, N., Sanfilippo, G.: On trivalent logics, compound conditionals, and probabilistic deduction theorems (submitted). Preprint: http://arxiv.org/abs/2303.10268
26. Gilio, A., Pfeifer, N., Sanfilippo, G.: Probabilistic entailment and iterated conditionals. In: Elqayam, S., Douven, I., Evans, J.S.B.T., Cruz, N. (eds.), Logic and Uncertainty in the Human Mind: A Tribute to David E. Over, pp. 71–101. Routledge, Oxon (2020)
27. Gilio, A., Sanfilippo, G.: Conditional random quantities and iterated conditioning in the setting of coherence. In: van der Gaag, L.C. (ed.) Symbolic and Quantitative Approaches to Reasoning with Uncertainty. ECSQARU 2013. LNCS, vol. 7958, pp. 218–229. Springer, Berlin, Heidelberg (2013). https://doi.org/10.1007/978-3-642-39091-3_19
28. Gilio, A., Sanfilippo, G.: Conjunction, disjunction and iterated conditioning of conditional events. In: Kruse, R., Berthold, M., Moewes, C., Gil, M., Grzegorzewski, P., Hryniewicz, O. (eds.) Synergies of Soft Computing and Statistics for Intelligent Data Analysis. AISC, vol. 190, pp. 399–407. Springer, Berlin, Heidelberg (2013). https://doi.org/10.1007/978-3-642-33042-1_43
29. Gilio, A., Sanfilippo, G.: Probabilistic entailment in the setting of coherence: the role of quasi conjunction and inclusion relation. Int. J. Approx. Reason. **54**(4), 513–525 (2013)

30. Gilio, A., Sanfilippo, G.: Conditional random quantities and compounds of conditionals. Stud. Log. **102**(4), 709–729 (2014)
31. Gilio, A., Sanfilippo, G.: Generalized logical operations among conditional events. Appl. Intell. **49**(1), 79–102 (2019)
32. Gilio, A., Sanfilippo, G.: Algebraic aspects and coherence conditions for conjoined and disjoined conditionals. Int. J. Approx. Reason. **126**, 98–123 (2020)
33. Gilio, A., Sanfilippo, G.: Compound conditionals, Fréchet-Hoeffding bounds, and Frank t-norms. Int. J. Approx. Reason. **136**, 168–200 (2021)
34. Gilio, A., Sanfilippo, G.: Iterated conditionals and characterization of p-entailment. In: Vejnarova, J., Wilson, N. (eds.) Symbolic and Quantitative Approaches to Reasoning with Uncertainty. ECSQARU 2021. LNCS, vol. 12897, pp. 629–643. Springer, Cham (2021). https://doi.org/10.1007/978-3-030-86772-0_45
35. Gilio, A., Sanfilippo, G.: On compound and iterated conditionals. Argumenta **6**(2), 241–266 (2021)
36. Gilio, A., Sanfilippo, G.: Subjective probability, trivalent logics and compound conditionals. In: Égré, P., Rossi, L. (eds.), Handbook of Three-Valued Logics. MIT Press, Cambridge (forthcoming)
37. Kaufmann, S.: Conditionals right and left: probabilities for the whole family. J. Philos. Log. **38**, 1–53 (2009)
38. Krzyżanowska, K., Wenmackers, S., Douven, I.: Inferential conditionals and evidentiality. J. Log. Lang. Inform. **22**(3), 315–334 (2013)
39. Lance, M.: Probabilistic dependence among conditionals. Philos. Rev. **100**(2), 269–276 (1991)
40. Lassiter, D.: Decomposing relevance in conditionals. Mind & Language (2022)
41. Lassiter, D., Baratgin, J.: Nested conditionals and genericity in the de Finetti semantics. J. Philos. **10**(1), 42–52 (2021)
42. Lewis, D.: Probabilities of conditionals and conditional probabilities. Philos. Rev. **85**(3), 297–315 (1976)
43. McGee, V.: Conditional probabilities and compounds of conditionals. Philos. Rev. **98**(4), 485–541 (1989)
44. Oaksford, M., Chater, N.: Bayesian Rationality: The Probabilistic Approach to Human Reasoning. Oxford University Press, Oxford (2007)
45. Oberauer, K., Wilhelm, O.: The meaning(s) of conditionals: conditional probabilities, mental models and personal utilities. J. Exp. Psychol. Learn. Memory Cogn. **29**(4), 680–693 (2003)
46. Over, D.E.: The development of the new paradigm in the psychology of reasoning. In: Logic and Uncertainty in the Human Mind, pp. 243–263. Routledge, Abingdon (2020)
47. Over, D.E., Cruz, N.: Suppositional theory of conditionals and rationality. In: Knauff, M., Spohn, W. (eds.), Handbook of Rationality, pp. 395–404. MIT Press (2021)
48. Over, D.E., Cruz, N.: Indicative and counterfactual conditionals in the psychology of reasoning. In: Kaufmann, S., Over, D.E., Sharma, G. (eds.) Conditionals – logic. linguistics, and psychology, pp. 139–174. Palgrave Macmillan, London (2023)
49. Pfeifer, N.: Experiments on Aristotle's thesis: towards an experimental philosophy of conditionals. Monist **95**(2), 223–240 (2012)
50. Pfeifer, N.: The new psychology of reasoning: a mental probability logical perspective. Think. Reason. **19**(3–4), 329–345 (2013)
51. Pfeifer, N.: The logic and pragmatics of conditionals under uncertainty: a mental probability logic perspective. In: Kaufmann, S., Over, D.E., Sharma, G. (eds.)

Conditionals – Logic. Linguistics, and Psychology, pp. 73–102. Palgrave Macmillan, London (2023)

52. Pfeifer, N., Sanfilippo, G.: Connexive logic, probabilistic default reasoning, and compound conditionals. Studia Logica, in press. https://doi.org/10.1007/s11225-023-10054-5

53. Pfeifer, N., Schöppl, L.: Experimental philosophy of connexivity. In: Wansing H., Omori, H. (eds.) 60 Years of Connexive Logic. Trends in Logic. Springer, Cham (to appear). Preprint: https://philpapers.org/rec/PFEEPO

54. Pfeifer, N., Tulkki, L.: Conditionals, counterfactuals, and rational reasoning. An experimental study on basic principles. Minds Mach. 27(1), 119–165 (2017)

55. Politzer, G., Over, D.E., Baratgin, J.: Betting on conditionals. Think. Reason. 16(3), 172–197 (2010)

56. Sanfilippo, G.: Lower and upper probability bounds for some conjunctions of two conditional events. In: Ciucci, D., Pasi, G., Vantaggi, B. (eds.) Scalable Uncertainty Management. SUM 2018. LNCS, vol. 11142, pp. 260–275. Springer, Cham (2018). https://doi.org/10.1007/978-3-030-00461-3_18

57. Sanfilippo, G., Gilio, A., Over, D.E., Pfeifer, N.: Probabilities of conditionals and previsions of iterated conditionals. Int. J. Approx. Reason. 121, 150–173 (2020)

58. Sanfilippo, G., Pfeifer, N., Over, D.E., Gilio, A.: Probabilistic inferences from conjoined to iterated conditionals. Int. J. Approx. Reason. 93(Supplement C), 103–118 (2018)

59. Stalnaker, R.C.: Probability and conditionals. Philos. Sci. 37(1), 64–80 (1970)

60. Van Fraassen, B.C.: Probabilities of conditionals. In: Harper, W.L., Hooker, C.A. (eds.) Foundations of Probability Theory, Statistical Inference, and Statistical Theories of Science. The University of Western Ontario Series in Philosophy of Science, vol. 6a, pp. 261–308. Springer, Dordrecht (1976). https://doi.org/10.1007/978-94-010-1853-1_10

61. Vidal, M., Baratgin, J.: A psychological study of unconnected conditionals. J. Cogn. Psychol. 29(6), 769–781 (2017)

62. Wansing, H.: Connexive logic. In: Zalta, E.N. (ed.), The Stanford Encyclopedia of Philosophy. Summer 2022 edition (2022)

# Interplay of Conditional Reasoning and Politeness: The Role of Speaker Relationships in the Japanese Context

Hiroko Nakamura[1,2]([✉]) [iD], Nao Ogura[1], Kazunori Matsumoto[1],
and Tatsuji Takahashi[1]

[1] Tokyo Denki University, Ishizaka, Hatoyama-Machi, Hiki-Gun, Saitama 350-0394, Japan
`21hz003@ms.dendai.ac.jp`
[2] Japan Society for the Promotion of Science, Kojimachi Business Center Building, 5-3-1
Kojimachi, Chiyoda-Ku, Tokyo 102-0083, Japan

**Abstract.** This study investigates how inter-speaker relations influence the interpretation of conditional statements "*if p1 then q*" and conditional reasoning, focusing on Japanese participants. The suppression paradigm of conditional reasoning reveals that adding additional conditional "*if p2 then q*" can suppress MP reasoning, "*p1, therefore q*". Based on politeness theory, the study examines whether the interpretation of ambiguous additional conditionals varies according to the relationship between the speakers, and how this variation affects conditional reasoning. Results indicated that introducing an additional conditional decreased the perceived plausibility of conditional and inverse sentences, but the relationship between the speakers had no significant impact on the interpretation of additional conditionals and probability assessments. The presence of additional conditionals influenced MP, MT, and DA reasoning, but AC reasoning was not impacted. The study underscores the importance of context and cultural nuances in the interpretation of conditional statements and highlights the need for further research to understand the factors that influence the interpretation of conditional sentences in different cultural settings.

**Keywords:** conditional reasoning · politeness · pragmatics

## 1 Introduction

### 1.1 The Suppression of Conditional Reasoning

Conditional reasoning is reasoning based on the conditional statement "*if p then q*". The four forms of conditional reasoning are: Modus Ponens (MP) is "*p, therefore q*", Modus Tollens (MT) is "*not-q, therefore not-p*", Affirming the Consequence (AC) is "*q, therefore p*", Denying the Antecedent (DA) is "*not-p, therefore not-q*". MP and MT are logically valid reasonings, while AC and DA are not. Research has shown that human reasoning isn't always logically consistent. Even logically valid MP reasoning can be influenced and suppressed by the information and context in which it is presented.

This work was supported by Grant-in-Aid for JSPS Fellows 22KJ2787 and Fund for the Promotion of Joint International Research 21KK0042.

J. Baratgin et al. (Eds.): HAR 2023, LNCS 14522, pp. 178–189, 2024.
https://doi.org/10.1007/978-3-031-55245-8_12

In the standard suppression paradigm for conditional reasoning, three premises like the following are presented:

(1a) *If p1, then q*;
(1b) *If p2, then q*:
(1c) *p1*.

Where the antecedent *p* is different but the consequent *q* is the same, the task is to reason about the conditional statement (1a). For instance, Byrne [1] presented the following:

(2a) *If she has an essay to write, she will study late in the library.*
(2b) *If the library stays open, she will study late in the library.*

Byrne showed that the MP inference "*She has an essay to write; therefore, she will study late in the library*" is suppressed by the introduction of the additional conditional (2b). The additional conditionals like (2b) lead to the inference "*Even if she has an essay to write, if the library does not stay open, she cannot study late.*" Thus, the additional statement is seen as correcting (2a), reducing its certainty and suppressing MP.

However, not all additional conditional statements suppress MP. For example, the following additional conditional statement:

(2c) *If she has some textbooks to read, she will study late in the library.*

This conditional is more likely to be interpreted as an alternative rather than as a correction. It could be interpreted as "*Whether she has an essay to write or she has some textbooks to read, in either case, she will study late in the library.*" This additional conditional (2c) doesn't reduce the certainty of (2a) or cause suppression of MP.

### 1.2 Politeness and Interpretation of Ambiguous Conditional

In some cases, it is not clear whether an additional conditional is a correction or an alternative to the conditional statement. The interpretation of ambiguous additional conditional varies depending on the context [2, 3]. Consider the following example:

(3a) Alan says, "*If we use Lupin marshmallow, then the candy will be soft*".
(3b) Ben says, "*If we use Maujy marshmallow, then the candy will be mellow*".

Research indicates that the additional conditional (3b) can be interpreted either as a correction, meaning "*If we use Maujy marshmallow, then the candy will be mellow, but not if we use Lupin marshmallow,*" or as an alternative, meaning "*Whether we use Lupin marshmallow or Maujy marshmallow, in either case, the candy will be mellow*". The suppression of the MP inference depends on the relationship between Alan and Ben [3]. A negative relationship leads to interpretation as a correction, while a positive relationship leads to interpretation as an alternative. Moreover, if the additional conditional is seen as a correction, the certainty of the conditional (3a) is reduced, resulting in the suppression of MP.

Demeure et al. applied politeness theory [4], which theorizes strategic language use in interpersonal communication, to explain how relationship variations between speakers influence the interpretation of additional conditionals. Politeness theory posits that individuals have two faces: a positive face, associated with a need for approval, and a negative face, associated with a desire to protect personal freedoms. Individuals employ face-saving strategies to avoid threatening situations and to maintain their own and others' public image. Correction is perceived as a face-threatening action, and to mitigate this, individuals often employ ambiguous expressions as a face-saving strategy. These ambiguous expressions allow room for the statement to be interpreted either as a correction or in its literal sense.

Brown & Levinson [4] proposed that the use of politeness strategies is influenced by the distance between speakers, the power of the interlocutor, and the extent to which a request is perceived as an imposition within a particular culture. It is widely accepted that individuals use politeness strategies to communicate corrections, and this depends on their relationship with the interlocutor. In addition, the dynamics between speakers play a role in how individuals interpret ambiguous expressions. Demeure [3] explained that as emotional distance increases, there is a higher propensity to use ambiguous expressions as a politeness strategy to convey corrections. Consequently, when there is a significant emotional distance between speakers, ambiguous additional conditionals are more likely to be interpreted as corrections.

## 1.3  Politeness Strategies and Suppression of Reasoning in Japanese

This study investigates how inter-speaker relationships impact the interpretation of additional conditionals and suppression of conditional reasoning in Japanese participants. The use of indirect expressions in situations such as requests and refusals as a means of conveying politeness is widespread in various cultures, although there are differences between them. For example, Chinese individuals have a stronger preference for indirect expressions compared to Americans [5]. Studies have found that Japanese people prefer and are more likely to use indirect expressions when making requests [6]. Furthermore, the theory of politeness has been shown to be applicable to the Japanese context [7]. However, when Japanese expressed their requests, psychological distance and power relations were affected more by the use of honorific expressions than by ambiguous expressions [8]. Previous research suggests that the use of honorifics is indicative of facework or politeness behaviors in Japan [9]. Therefore, in situations where face may be threatened, Japanese speakers may prefer to use honorifics instead of indirect expressions. As a result, psychological distance may not have a significant effect on Japanese people's use of indirect or ambiguous expressions, even when corrections are needed. Therefore, we hypothesize that the speaker relationship will not affect the interpretation of ambiguous additional conditionals, nor will it affect probability judgments or performance in conditional reasoning.

The previous study [3] focused on how additional conditionals affect MP and the plausibility of conditional statements. This study examines the effect of additional conditionals on all forms of conditional reasoning and the probability of both conditional (*if p then q*) and inverse (*if not-p then not-q*) statements. Conditional statements are often interpreted as (defective) biconditional, with the inverse "*if not-p then not-q*" inferred from "*if p then q*". In causal reasoning (e.g., "*if cause then effect*"), models acknowledging the bidirectional nature of cause and effect tend to have a better fit compared to models that don't [10]. Additionally, it is observed that (defective) biconditional interpretations in causal conditional reasoning are more prevalent among individuals with lower cognitive abilities [11]. In the realm of biconditional statements, not only MP and MT, but also AC and DA are logically valid reasonings. Viewing conditional statements as biconditional can affect the plausibility of the conditional statement "*if p then q*" as well as the inverse "*if not-p then not-q*", and likewise the AC and DA inferences.

Regarding the suppression of conditional reasoning, MP and MT reasoning are known to be suppressed by disabling conditions, while AC and DA reasoning are suppressed by alternative causes [12–14]. The disabling condition prevents '*p*' from leading to '*q*', thereby questioning the sufficiency of '*p*' causing '*q*'. Consequently, MP and MT reasoning are suppressed. When an additional conditional is interpreted as a corrective (e.g., "*If we use Maujy marshmallow, the candy will be mellow, but not with Lupin marshmallow*"), it serves as a disabling condition, undermining the sufficiency of the conditional "*If p, then always q*" and thus suppressing MP and MT reasoning. Conversely, an alternative cause questions the necessity of '*p*' causing '*q*', leading to the suppression of AC and DA inferences. When the additional conditional statement is interpreted as an alternative (e.g., "*Whether we use Lupin marshmallow or Maujy marshmallow, the candy will be mellow in both cases*"), it serves as an alternative cause, undermining the necessity of "*If not-p, then always not-q*", and likely leading to the suppression of AC and DA reasoning.

In this study, we test the following hypotheses about how interspeaker relationships impact the plausibility of conditional sentences and conditional reasoning through the interpretation of ambiguous additional conditional among Japanese participants:

1. Japanese participants are less likely to use ambiguous and indirect expressions for politeness [8, 9]. The differences in the interpretations of ambiguous additional conditional, as shown by French participants [3], may not be observed in Japanese.
2. When an ambiguous additional conditional is interpreted as a correction, it acts as a disabling condition and reduces the sufficiency of '*p*' causing '*q*' [12–14]. Consequently, additional conditional may reduce the plausibility of the conditional and may suppress MP and MT reasonings.
3. When an ambiguous additional conditional is interpreted as an alternative, it reduces the necessity of '*p*' causing '*q*' [12–14]. Consequently, the additional conditional may reduce the plausibility of the inverse and suppress AC and DA reasonings.

## 2  Experiment

### 2.1  Method

**Participants.** The study recruited 180 Japanese participants through the crowdsourcing service CrowdWorks (www.crowdworks.jp). Data from 157 participants who provided complete responses (76 females and 81 males; mean age 42.12 years, SD = 9.3) were included in the analysis.

**Experimental Design.** The experiment was structured with a one factor (relationship to the speaker making the additional conditional statement) that had three levels (good, bad, none), using a between-participants design.

**Materials.** The experiment used a fictional scenario involving two characters, A and B, who are employees of a candy company. They make conditional statements about candy. In the none condition, character B is absent and no additional conditional statement is made.

A. "*If we use ingredient X, then the candy will taste good*" (conditional statement).
B. "*If we use ingredient Y, then the candy will taste good*" (additional conditional statement).

Participants were asked to complete three tasks: probability judgment, interpretation of an additional conditional statement and conditional reasoning.

In the probability judgment task, participants rated the probability of the conditional statement "P (*if p then q*)" being true and its inverse "P (*if ¬p then ¬q*)" – "*If we do not use ingredient X, then the candy will not taste good*" on a 7-point scale ranging from 1 (very low likelihood) to 7 (very high likelihood).

The additional condition interpretation task was given only to participants in the 'good' and 'bad' conditions. In this task, participants responded to the interpretation of the additional conditional "*If we use ingredient Y, then the candy will taste good.*" They were asked to indicate whether they interpreted this as a correction ("*If we use ingredient Y, then yes, the candy will taste good, but not if we use ingredient X*") or an alternative ("*Whether we use ingredient X or ingredient Y, in either case, the candy will taste good*").

In the conditional reasoning task, participants were shown the conditional statement, "*If we use ingredient X, then the candy will taste good*," along with the minor premises for the MP, MT, AC, and DA reasonings, and were asked to rate the validity of the conclusion on a 5-point scale, with 1 indicating strong disagreement and 5 indicating strong agreement with the conclusion.

**Procedures.** The experiment was conducted online Qualtrics software. Participants were initially presented with an informed consent form, and their agreement to participate was recorded. They were then randomly assigned to the reasoning task and the probability judgment task. Participants in the 'good' and 'bad' conditions also completed an additional conditional interpretation task. Additionally, participants underwent an instructional manipulation check and demographic information questions.

## 2.2 Results

The data were analyzed using R software [15].

**Interpretation of the Additional Conditional.** Participants who interpreted the additional condition as a correction were 38% in the good interspeaker relationship condition and 57% in the bad interspeaker relationship condition. Chi-squared test results indicated that there was no significant difference in the interpretation of the additional conditional statement based on whether the relationship was good or bad, $\chi^2$ (1) = 3.09, $p = .08$.

**Probability Judgment Task.** A two-factor mixed analysis of variance (ANOVA) with three levels for relationship (good, bad, none) and two levels for form (conditional, inverse) was conducted to analyze the scores on the probability judgment task (Fig. 1). The main effects of relationship $f$ (2, 154) = 10.9, $_g\eta^2 = .07$, $p < .001$ and form $f$ (1, 154) = 84.8, $_g\eta^2 = .20$, $p < .001$ were significant. Follow-up testing revealed that the bad and good relationship conditions received lower probability ratings than the none condition for both the conditional and inverse forms. The inverse form was rated as less likely than the conditional form.

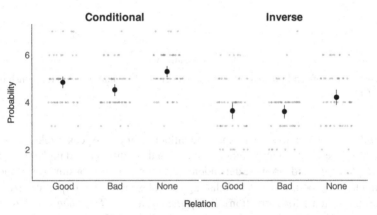

**Fig. 1.** Probability of conditional and inverse as a function of the inter-speaker relationship, Mean and 95% CI.

**Conditional Reasoning Task.** We conducted a two-factor mixed ANOVA with three levels for relationship (good, bad, none) and four levels for reasoning form (MP, MT, DA, AC) on the acceptance of conditional reasoning (Fig. 2). The Greenhouse-Geisser correction was applied to adjust for violations of sphericity. There was a significant interaction between relationship and reasoning form, $f$ (5.71, 440.05) = 2.9, $_g\eta^2 = .21$, $p = .009$. MP reasoning was more likely to be judged incorrect when the relationship was bad than when no additional conditional was present $f$ (2, 154) = 3.1, $_g\eta^2 = .04$, $p = .048$. MT reasoning was more likely to be judged incorrect when the relationship was bad or good than when no additional conditional was present $f$ (2, 154) = 6.6, $_g\eta^2 = .08$, $p = .002$. In DA, participants were more likely to rate the reasoning as incorrect when the relationship was good than when the relationship was bad or no additional conditional $f$ (2, 154) = 9.7 $_g\eta^2$ = .11, $p$ < .001. There was no effect of relationship on AC $f$ (2, 154) = 3.0 $_g\eta^2$ = .04, $p$ = .053.

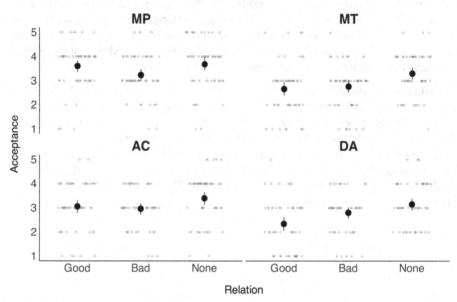

**Fig. 2.** Acceptance of conditional reasonings as a function of the inter-speaker relationship, Mean and 95% *CI*.

**Path Analysis.** Path analysis examined the influence of inter-speaker relations on conditional inference via the interpretation of additional conditional and probability assessments. MP, MT, AC, and DA were dependent variables, and relationship was an independent variable (0 = good, 1 = bad). Mediating variables included interpretation (coded as 0 for alternative and 1 for correction), P (*if p then q*) and P (*if ¬p then ¬q*). Confidence intervals were obtained using the bootstrap method with 1000 iterations.

There was a significant effect of the relationship on interpretation $\beta$ = .19, $p$ = .05, where the 'correction' interpretation was more frequent in the 'bad' relation. However, neither the relationship nor the interpretation had significant effects on P (*if p then q*) $\beta$ = −.17, $p$ = .08 or P (*if ¬p then ¬q*) $\beta$ = −.01, $p$ = .88.

For MP (Fig. 3), there was a significant direct effect of P (*if p then q*) on MP reasoning $\beta$ = .38, $p$ < .001), where higher values of P (*if p then q*) were associated with MP being more likely judged as correct. The direct effects of relationship $\beta$ = −.12, $p$ = .26, interpretation $\beta$ = −.10, $p$ = .27, P (*if ¬p then ¬q*) $\beta$ = .11, $p$ = .15, and the total effect of relationship $\beta$ = −.18, $p$ = .06 were not significant.

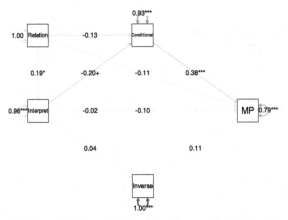

**Fig. 3.** Path analysis for the effect ($\beta$ coefficients) of inter-speaker relationship on the acceptance of modus ponens (*** $p < .001$, ** $p < .01$, * $p < .05$, + $p < .10$).

For MT (Fig. 4), none of the variables showed a significant effect on MT reasoning. This includes relationship $\beta = .07, p = .46$, interpretation $\beta = -.01, p = .93$, P (*if p then q*) $\beta = .11, p = .41$, P (*if ¬p then ¬q*) $\beta = .04, p = .77$, and total effect of relationship $\beta = .06, p = .56$.

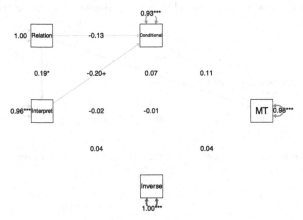

**Fig. 4.** Path analysis for the effect ($\beta$ coefficients) of inter-speaker relationship on the acceptance of modus tollens (*** $p < .001$, ** $p < .01$, * $p < .05$, + $p < .10$).

For AC (Fig. 5), none of the variables showed a significant effect on AC reasoning. This includes relationship $\beta = -.25, p = .79$, interpretation $\beta = -.25, p = .78$, P (*if p then q*) $\beta = .12, p = .19$, P (*if ¬p then ¬q*) $\beta = .13 \, p = .24$, and the total effect of relationship $\beta = -.05, p = .62$.

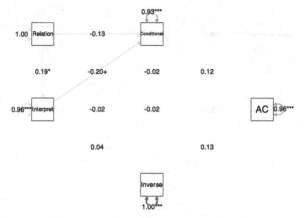

**Fig. 5.** Path analysis for the effect ($\beta$ coefficients) of inter-speaker relationship on the acceptance of affirming the consequence $(^{***} p < .001, \, ^{**} p < .01, \, ^* p < .05, \, ^+ p < .10)$.

For DA (Fig. 6), the direct effects of the relationship $\beta = .20, p = .03$, P (*if* ¬*p then* ¬*q*) $\beta = .36, p < .001$, and the total effect of relationship $\beta = .23, p = .02$ on DA reasoning were significant. When relationship was bad, DA was more likely to be judged as correct. However, the effects of interpretation $\beta = .12, p = .20$ and P (*if p then q*) $\beta = -.13, p = .18$ were not significant. Participants were more likely to judge DA as correct when there was a negative relationship and when the inverse probability was high.

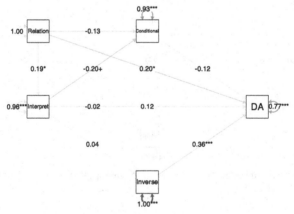

**Fig. 6.** Path analysis for the effect ($\beta$ coefficients) of inter-speaker relationship on the acceptance of denying the antecedent $(^{***} p < .001, \, ^{**} p < .01, \, ^* p < .05, \, ^+ p < .10)$.

## 3  Discussion

In this study, we examined the effect of inter-speaker relations on the interpretation of an ambiguous additional conditional "*if p2 then q*", and whether the interpretation of the additional conditional affects the perceived plausibility of conditional statements "*if p1 then q*" and reasoning among Japanese participants.

Regarding our first hypothesis, which posited that Japanese participants would be unlikely to be influenced by inter-speaker relations when interpreting additional conditionals due to their use of honorifics instead of ambiguity for politeness, the results somewhat supported this view. There was no significant statistical difference in the interpretation of additional conditional based on inter-speaker relationships. Also, there was no evidence that additional sentences were more often interpreted as corrections when the relationship was bad. Although path analysis suggested that bad inter-speaker relations led to a slightly higher likelihood of interpreting additional conditionals as corrections, the effect size was small ($\beta = .19$) compared to a medium or larger effect ($\beta = .32$) in a previous study with French participants [3].

Given the Japanese tendency to use honorifics rather than indirect expressions for politeness [9], it's reasonable to think that interpersonal relationships might influence the use of honorifics rather than ambiguous terms. As a result, the influence of speaker relationships on interpreting ambiguous additional conditionals may be more subtler for Japanese participants than for Western participants in earlier studies [3].

Regarding the second and third hypotheses, which predicted that P (*if p2 then q*) would be lower (and MP and MT suppressed) if the additional conditional sentence was seen as a correction, and P (*if not-p1 then not-q*) would be lower (and AC and DA suppressed) when it was interpreted as an alternative, the results were not consistent with these predictions. Neither the relationship between speakers nor the interpretation of the additional conditional had any effect on probability judgments. Rather, the mere presence of an additional conditional reduced the perceived plausibility of the conditional, regardless of how it was interpreted. The information extracted from the additional conditional may differ depending on the focus of the evaluation. For instance, the disabling cause might be emphasized when assessing the probability of a conditional, while alternatives might be emphasized when evaluating an inverse probability.

An additional conditional affected MP, MT, and DA reasonings. MP was reduced in situations of bad relationships and low plausibility of the conditional statement, somewhat in line with previous findings. However, the relationship between the speakers did not significantly affect the interpretation of the additional conditional or the evaluation of the probability of the conditional sentence'. The bad relationship itself may have led to negative evaluations of the conditional sentences.

MT reasoning was suppressed when additional conditional statements were present, regardless of the relationship between the speakers. The interpretation of additional conditionals or probability assessments didn't affect MT either. Endorsing MT reasoning, which follows the logic "*not-q, therefore not-p1*", depends on the sufficiency of the statements "*if p1 then always q*" and "*if not-q then always not-p1*". The reasoning "*not-q, therefore always not-p1*" can be suppressed whether the additional conditional is interpreted as a correction ("*if not-q, then not-p2, not always not-p1*") or as an alternative ("*if not-q, then not-p1 or not-p2, not always not-p1*").

For DA reasoning, suppression was more pronounced when the relationship between the speakers was good than when it was bad, and the DA was more readily accepted when the estimated probability of the inverse was higher. However, the interpretation of the additional condition had no effect on DA. The additional conditional statement mentions the possibility of *"p2 (not-p1) therefore q"*, which could have suppressed the DA *"not-p1, therefore always not-q"* regardless of the interpretation.

Additional conditional statements had no observed effect on AC reasoning. The necessity that *"p1 must occur for q to occur"* affects the endorsement of AC *"q, therefore p1"*, Regardless of how the additional conditional is interpreted, as a correction (*"If p1, it is not always q, but it's not necessarily the case that if q then not-p1"*) or as an alternative (*"If q, it could be either p1 or p2, so it cannot be concluded that if q then not-p1"*), AC reasoning may not be suppressed.

The purpose of this study was to test whether speaker relations can influence how additional conditional sentences inhibit conditional inference in Japanese participants. However, we did not find a significant effect of inter-speaker relations on the interpretation of the additional conditional, and no significant effect of the interpretation on probability judgments and reasonings. These results differ from previous studies with French participants [3]. A study of cultural differences in conditional reasoning [16] found that reasoning performance on abstract conditional tasks was similar among Japanese, Chinese, and French participants. French and Chinese participants showed similar reasoning for the sufficiency conditional, *"If Sophie is in the living room, then Mary is in the kitchen,"* and the necessity conditional statement, *"Only if Sophie is in the living room, then Mary is in the kitchen"* [18]. However, in a betting context, such as *"I bet $100 for if A then C,"* research has shown that the French evaluate the plausibility of a conditional sentence by focusing only on the case *A*, while the Japanese and Chinese evaluated the plausibility by considering both *A* and *not-A* cases [17]. Cultural differences in reasoning may be more pronounced in concrete, contextually informative cases, such as the one presented here, than in abstract scenarios.

The use and interpretation of appropriate linguistic expressions, such as politeness strategies, differ across cultures, and such cultural differences in pragmatic aspects are thought to lead to cultural differences in conditional reasoning. Conditional sentences often serve to communicate predictions, permissions, and warnings in conversations [17]. Future research should aim to elucidate the factors and cultural differences that influence the interpretation of conditional sentences and reasoning within conversational contexts.

# References

1. Byrne, R.M.: Suppressing valid inferences with conditionals. Cognition **31**, 61–83 (1989). https://doi.org/10.1016/0010-0277(89)90018-8
2. Stevenson, R.J., Over, D.E.: Reasoning from uncertain premises: effects of expertise and conversational context. Think. Reason. **7**, 367–390 (2001). https://doi.org/10.1080/135467 80143000080
3. Demeure, V., Bonnefon, J.F., Raufaste, E.: Politeness and conditional reasoning: interpersonal cues to the indirect suppression of deductive inferences. J. Exp. Psychol. Learn. Mem. Cogn. **35**, 260–266 (2009). https://doi.org/10.1037/a0013799

4. Brown, P., Levinson, S.C., Levinson, S.C.: Politeness: Some Universals in Language Usage. Cambridge University Press (1987)
5. Guo, Y.: Chinese and American refusal strategy: a cross-cultural approach. Theory Pr. Lang. Stud. **2**, 247–256 (2012). https://doi.org/10.4304/tpls.2.2.247-256
6. Okamoto, S.: Linguistic styles of requests. Japanese J. Exp. Soc. Psychol. **26**, 47–56 (1986). https://doi.org/10.2130/jjesp.26.47
7. Kiyama, S., Tamaoka, K., Takiura, M.: Applicability of Brown and Levinson's politeness theory to a non-Western culture. SAGE Open **2**, 215824401247011 (2012). https://doi.org/10.1177/2158244012470116
8. Hirakawa, M., Fukada, H., Higuchi, M.: A test of the politeness theory regarding factors influencing the use of request expressions. Japanese J. Exp. Soc. Psychol. **52**, 15–24 (2012). https://doi.org/10.2130/jjesp.52.15
9. Ide, S.: Formal forms and discernment: two neglected aspects of universals of linguistic politeness. Multilingua. **8**, 223–248 (1989). https://doi.org/10.1515/mult.1989.8.2-3.223
10. Takahashi, T., Oyo, K., Tamatsukuri, A., Higuchi, K.: Correlation detection with and without the theories of conditionals. Logic and Uncertainty in the Human Mind: A Tribute to David E. Over, pp. 207–222 (2020)
11. Evans, J.S.B.T., Handley, S.J., Neilens, H., Over, D.: Understanding causal conditionals: a study of individual differences. Q. J. Exp. Psychol. **61**, 1291–1297 (2008). https://doi.org/10.1080/17470210802027961
12. Thompson, V.A.: Conditional reasoning: the necessary and sufficient conditions. Can. J. Exp. Psychol./Revue canadienne de psychologie expérimentale. **49**, 1–60 (2007). https://doi.org/10.1037/1196-1961.49.1.1
13. Cummins, D.D., Lubart, T., Alksnis, O., Rist, R.: Conditional reasoning and causation. Mem. Cognit. **19**, 274–282 (1991). https://doi.org/10.3758/bf03211151
14. Thompson, V.A.: Interpretational factors in conditional reasoning. Mem. Cognit. **22**, 742–758 (1994). https://doi.org/10.3758/BF03209259
15. R Core Team: R: A Language and Environment for Statistical Computing. R Foundation for Statistical Computing, Vienna, Austria (2022)
16. Evans, J.S.B.T.: The social and communicative function of conditional statements. Mind Soc. **4**, 97–113 (2005). https://doi.org/10.1007/s11299-005-0003-x
17. Nakamura, H., Shao, J., Baratgin, J., Over, D.E., Takahashi, T., Yama, H.: Understanding conditionals in the east: a replication study of with easterners. Front. Psychol. **9**, 505 (2018). https://doi.org/10.3389/fpsyg.2018.00505
18. Shao, J., Tikiri Banda, D., Baratgin, J.: A study on the sufficient conditional and the necessary conditional with chinese and french participants. Front. Psychol. **13**, 787588 (2022). https://doi.org/10.3389/fpsyg.2022.787588

# From Classical Rationality to Quantum Cognition

Pierre Uzan[⊠] [iD]

Habilitation in Philosophy of Science, Catholic University of Paris, CHArt Laboratory, Paris, France
uzanpier@gmail.com

**Abstract.** This article highlights the difficulties of "classical" rationality, grounded on classical set-based logic and classical probability calculus, and explains how they can be overcome. Classical rationality may have gone some way towards realizing the age-old project of mechanizing thought, thus making possible the dazzling development of artificial intelligence we are witnessing. However, as shown from experimental data, it suffers from many biases that make it incapable of reliably modeling mental processes. This article shows that mental processes can be more reliably modelled within a generalized probability theory grounded on the vector-space formalism of quantum theory. The reason is that such a quantum-like approach to cognition is capable of accounting for the contextual, order and interference effects inherent to most of mental processes. An important consequence of this necessary shift, from classical to quantum cognition, is that Bayes' rule, which plays a fundamental role in categorization tasks, must be replaced by a generalized probabilistic rule capable of accounting for order effects. It is shown that implementing this new probabilistic rule could significantly improve the current deep learning algorithms of artificial emotional intelligence.

**Keywords:** Classical rationality · Quantum cognition · Bayes' rule · Emotional Artificial Intelligence

## 1 Introduction: From the Language of Human Thought to Classical Logic

Leibniz's project was to create a system of ideographic signs, the *lingua characteristica universalis* in direct contact with the ideas they express and in which human reasoning could be translated into the form of a calculation [1]. In this language of human thoughts, inspired from Chinese ideography, ideas could be broken down into simple ones, forming the alphabet of human thought, while the combination of its "letters" would enable invention, not just deduction. Following this idea, Boole proposed in his *Laws of Thought* [2] to study "the fundamental laws of the operations of the mind, expressing them in the symbolic language of calculus". His "algebra of human thoughts" clearly reduces the language of human thought to *classical propositional logic*. A reduction which

J. Baratgin et al. (Eds.): HAR 2023, LNCS 14522, pp. 190–209, 2024.
https://doi.org/10.1007/978-3-031-55245-8_13

echoes with more recent proposals, like Fodor's hypothesis of a mental language, the "Mentalese", whose terms would be structured as logical statements, using the classically defined logical connectors [3]. In the same vein, let us finally mention Frege's seminal work on classical predicate logic. If in his *Begriffsschrift* [4], he set out to take up Leibniz's unfinished project of developing a universal ideographic language of human thoughts, his "formal language of pure thought constructed on the model of arithmetic" is nothing but the *classical predicate calculus*.

The connection between this symbolic approach to thought in terms of classical logic and the project of its mechanization was finalized by the work of Turing and Church. According to Church-Turing's thesis, "computable" functions (in intuitive sense) are those whose values can be calculated by a Turing machine [5, 6]. Moreover, Turing showed that there exists a universal Turing Machine which can have as input (written on the tape) the description of the functioning of any Turing Machine and the input of the latter, which is the theoretical model of modern computers. Mental processes, understood as calculation governed by classical logic, could thus be mechanized. However, the project of mechanisation of thought has been realized thanks to a clear shift in the search for the "universal language of human thought" : Leibniz' initial idea of a universal *ideographic language*, as a direct and reliable expression of human thoughts, has clearly been reduced to the idea that mental processes can be modeled by *classical logic*.

It could be argued that, more recently, this project of mechanization of thought has been realized within an alternative approach to mental processes which does not rely on the symbolic approach to thought. According to this alternative, explicitly materialist approach, thought would *emerge* from the brain, as its biological product. This emergentist or "connectionist" approach inspired by the functioning of the brain overcomes several limitations of the symbolic approach, like the fragility and the rigidity of its programs, and made it possible to mechanize important cognitive tasks, such as pattern recognition, learning or visual perception. However, *realizing these tasks appeals to Bayes' inference rule, whose validity still relies on the relevance of classical logic and classical probability calculus for modeling mental processes.*

This article emphasizes several difficulties of the modelling of human cognition by the classical rationality, grounded on classical logic and classical probability calculus (Sect. 1). It then briefly presents the quantum-like approach to cognition where all these issues can find elegant and satisfactory solutions (Sect. 2). Section 3 deals with the key ingredient which is currently used for integrating uncertainty in cognition and decision-making, namely Bayes' rule. It is recalled that the latter is valid only if the order in which the events considered is indifferent. A new, probabilistic rule capable of accounting for order effects is then shown within quantum cognition (Sect. 3.2) and the way it can be implemented in the field of artificial emotional intelligence is explained in Sect. 4.

## 2 The Failure of Classical Rationality in the Current Models of Mental Processes

As emphasized by many authors, including Cruz, Baratgin, Bruza, Busemeyer, Wang and Aerts [7–10], several "fallacies" in human reasoning seem to occur when cognition and decision-making are modeled according to classical rationality. These apparent fallacies of human reasoning mainly regard conjunctive, disjunctive and conditional inferences, as much as decision making under uncertainty. Several apparent "fallacies" of human reasoning have been analyzed in detail and attempts to explain or overcome them have been provided in the previously mentioned references [7–10]. However, for a question of length of the article, and as suggested by an anonymous referee, we will here essentially focus on a few of them, those that involve *typically non-classical features of human rationality*, namely *context effects*, referring to works including those of Hampton and Uzan [15, 16], *order effects*, in agreement with Wang and Busemeyer's more formal presentation [10] *and interference effects*, referring to the works of Aerts, Busemeyer and Uzan [8, 9, 16]. Note that a particular focus on the study of order effects will introduce to the *non-classical Bayesian modelling* developed in Sect. 3.3 and its application in the field of emotional intelligence.

## 2.1 The Conjunction Fallacy

The conjunction fallacy refers to the fact that, *in contradiction with the classical probability calculus*, human subjects often assign a probability of occurrence of the conjunction of two events greater than the probability of the occurrence of each of them: $P(A \wedge B)$ could be greater than $P(A)$ and $P(B)$ while, as a direct consequence of Kolmogorov's axioms of classical probability calculus, the probability $P(A \wedge B)$ cannot exceed either $P(A)$ or $P(B)$[1]. This apparent "fallacy" of human reasoning about conjunction emphasizes *the role of the context* where these events occur: the belief that B is true when A is known to be true can be greater than the belief that B is true without knowing that A is true because A can play the role of a context in which the conjunction $A \wedge B$ becomes more probable than A and B.

This emphasis on contextuality questions the thesis of compositionality of meaning whose roots lie in the principle of verifunctionality of classical logic. According to the thesis of compositionality of meaning, which was for example supported by Fodor [11], the meaning of an utterance would be determined exclusively by those of its components and the way in which they combine. However, everyday experience shows us that the meaning that we attribute to a linguistic expression depends on the whole sentence or even the whole text in which it is used, and this meaning is thus defined by its contribution to this global linguistic context. This property of *contextuality* of natural language has been clarified and generalized by several authors, including Stalnacker [12], following the founding works of Austin [13] and Wittgenstein [14].

The failure of the thesis of compositionality of meaning, and then that of the principle of verifunctionality of classical logic, has been illustrated by an experimental study conducted by Hampton [15]. This experiment quantifies the role played by the context is in a very simple case, by studying the way we assign meaning to a concept obtained by composing the two concepts "food" and "plant" and gives very significant results. A study of this experiment and of its possible quantum-like interpretation has been developed by Uzan in reference [16].

---

[1] The reason of this constraint is that the classical probability calculus is isomorphic to the Boolean algebra constituted by the set of parts of a set ordered by the inclusion relation and provided with the operations of complementation, intersection and union.

## 2.2 Order Effects in Decision-Making

In a survey realized in 1997 (September 6–7) and involving 1002 respondents, half of the participants were asked the two questions 'is Clinton honest and trustworthy?', noted as A hereafter, *and then* 'is Gore honest and trustworthy?', noted as B hereafter, while the other half were asked the same pair of questions in the opposite order. As reported by Moore [17], the list of answers for the two groups shows that Clinton received *50% agreement when asked first* (which defines the "non-comparative" context) but *57% when asked second* (which defines the "comparative" context because this answer can be influenced by the first one). It also shows that Gore received *68% when asked first and 60% when asked second*. This difference in the frequencies of the respondents' answers shows that *the order in which the questions are asked is significant* since the frequency of the positive answers to the same question depends on whether this question is asked first or second. Focusing for example on positive answers for both questions A and B, respectively noted as Ay and By, this order effect can be expressed by the following difference: P (Ay By) $\neq$ P (By Ay), where P (Ay By) is the probability of responding "yes" to question A followed by "yes" to question B, and P (By Ay) is the probability of obtaining the same answer to these questions asked in the inverse order.

Moore calls this type of question order effect "consistency effect" to denote the fact that the difference between the probabilities of positive answers for questions A and B decreases from the non-comparative context to the comparative context, which is here the case since in the non-comparative context P (By) − P (Ay) = 18% while in the comparative context P (Ay/n By) − P (By/n Ay) = 3%. Note that other types of order effects in decision making have been observed in other similar survey experiments, for example a "contrast" order effect showing that, unlike the previous consistency effect, the difference between these probabilities is amplified in the comparative context.

## 2.3 Order Effects in the Domain of Emotions

Order effects do not only occur in decision making but in *all* mental processes where subjective experience is involved. In particular, significant order effects can be observed and quantified in the domain of emotions, and should then be taken into account in the cutting-edge research in artificial intelligence (see Sect. 4). As can be observed, and in contrast with the possible assignation of intrinsic properties to physical objects in classical physics, experienced emotions cannot be regarded as intrinsic features of a person since they are continuously changing according to our life experience. Their nature and their intensity are highly *contextual* since they strongly depend on our personal past and present experience of life, on our social environment and even on what we felt just a moment before. For example, asking a subject about her degree of happiness and asking the same question after reminding her of a sad event in her life generally provides different results. As was the case for the previous example of surveys with two successive questions, the order effects relative to emotions can be evaluated from data on successive measurements of the intensities of emotions experienced by subjects. These intensities can be collected by asking them to report discrete values on a graduate scale or to report them continuously, using a continuous response digital interface on which the subject moves a stylus or finger [18]. A precise study of the non-commutativity of emotional observables has been provided by Uzan in reference [19] and is briefly reported in Sect. 2.1.1, after introduction of the appropriate quantum-like formalism. In this reference, a method for computing their degree of non-commutativity is provided. This computation is explicitly developed for the couple of observables Anger and Disgust.

Moreover, emotional observables do not generally commute neither with their physiological correlates nor with their behavioral correlates. In a similar manner, this can be established from data about the joint measurement of emotional observables and their physiological or behavioral correlates reported in the literature [20–23]. For example, Kassam's and Mendes's article [22] shows on experimental basis, that the very act of reporting one's own emotional state can drastically change one's physiological and behavioral "responses", this effect being particularly significant for subjects conditioned in angry state. In this experiment, the subjects are conditioned in such an angry state by delivering them a negative feedback to a difficult task they have done - for example by telling them that they are incompetent. The observed physiological responses are here evaluated by the values of cardiovascular observables, like heart rate and pre-ejection period, which can be measured almost continuously. The behavioral responses are evaluated by external experimenters through videos showing the participants performing the required tasks, by noting for example their facial expression and their body movements. Kassam's and Mendes' study clearly shows that for these subjects, conditioned in angry state, the changes in the values of the cardiovascular and behavioral observables are significantly different depending on whether or not they report their emotional state (of anger, in this example). This tells us that the successive measurement of emotional and physiological or behavioral observables gives rise to order effects because if these observables were commuting their values would not be inter-dependent, as is actually the case.

## 2.4 Ellsberg's Paradox of the Classical Decision-Making Theory

Tversky and Shafir [24] have demonstrated on experimental grounds a significant violation of the principle of consistency of possible choices in classical decision theory, *which questions the validity of the law of total probabilities of the classical probability calculus.* This violation gives rise to Ellsberg's paradox regarding the behavior of economic agents [25].

The principle of consistency of possible choices, known in the literature as the "sure thing principle", formulated as follows by Savage [26]:

"If A is preferred to B when X is realized and if A is preferred to B when X is not realized, then A is preferred to B even if it is not known whether X is or is not realized."

This principle was tested by Tversky and Shafir in the following experiment: participants are (actually) offered to play a game where the chance of winning $200 is the same as the chance of losing $100, that is, 1/2. Participants are offered the game a first time and asked if they would like to play a second time in the following three cases (the population being divided into three groups of equal size): (a) they know they won the first time, (b) they know they lost the first time and (c) they don't know whether they won or lost the first time. The results are as follows: 69% of participants in group (a) decide to play again, 59% of participants in group (b) decide to play again, while only 36% of participants in group (c), whose participants don't know whether they won or lost, decide to play again. This result is at odds with the "sure thing" principle mentioned above, since if the latter applied, the probability to play again for group (c) would be 64%. More precisely, according to the law of total probability of the classical probability calculus, this probability would be equal to half the sum of the probabilities measured for groups (a) and (b) respectively, which define exclusive sequences of events:

$$P \text{ (to play again)} = P \text{ (to have won the first time and to play again)}$$
$$+ \, P \text{ (to have lost the first time and to play again)}$$
$$= (1/2) \, [P \text{ (to play again knowing that you won the first time)}$$
$$+ \, P \text{ (to play again knowing you lost the first time)].}$$

This *violation of classical probability calculus*, which at first glance seems to indicate that human beings would be prone to "fallacious" reasoning, is in fact a typical example of how they "reason" and make decisions.

## 3  How Overcoming the Failure of Classical Rationality?

The previous developments show some important limitations in the classical approach to mental processes and thus corresponding limitations in the development of artificial intelligence-whose first aim is to simulate these mental processes. Many proposals have been done to overcome these limitations according to two different ways. The first one involves pragmatic or heuristic considerations on real human cognition processes, but they do not question the validity of classical rationality.

For example, Grice [27] has proposed a purely pragmatic, socially-based, explanations of the disjunction fallacy, according to which disjunctive statement is estimated to be less probable than one of its component statements. This author has suggested that it would be inappropriate, even misleading, for a person to endorse in a conversation the disjunction of two statements when one of its components is regarded as more informative about the situation considered. Gigerenzer and Todd [28] have suggested that, in contrast to what is generally supposed in logical modelling of human reasoning, the human is not endowed with supernatural powers of reasoning, limitless knowledge, and endless time but that "decisions in the real world requires a more psychologically plausible notion of bounded rationality", namely the use of *"tool boxes" of heuristics* that shorten and greatly simplify and accelerate the decision making process and preserve the validity of classical rationality. In order to illustrate such considerations, these authors mention the simple decision tree often used by physicians for quickly classifying incoming heart attack patients into high and low risk patients. In the same order of ideas, Tversky and Kahneman [29] mention three mental operations generally used in decision making under uncertainty, which lead to systematic and predictable errors of reasoning and should then better understood in order to improve judgments and decisions in situations of uncertainty while preserving classical rationality.

A very different approach to deal with these apparent "fallacies" of human cognition consists in questioning the logical basis of classical rationality. For example, in the field of three-valued logics, initially proposed by Łukasiewicz and further developed by Reichenbach, Frege and other famous logicians, de Finetti [30] has proposed a three-valued treatment of conditional sentences in order to characterize uncertainty in judgements. In contrast with the classical truth-value assignation of conditional, Finetti assigns the truth-value "undefined" to a conditional inference when the antecedent is false. More accurately, fuzzy logic [31], a form of many-valued logic in which the truth value of variables may be any real number between 0 and 1, can be used to quantify the indetermination of the uncertainty of a judgment, since it captures the idea that our reasoning can only be "approximate" and introduces a notion of "degree of truth" of a statement –which can be defined as a probability. Let us also mention the development of non-monotonic logics [32] capable of dealing with the process of restriction of knowledge due to the reception of a new piece of information –a property characteristic of our daily reasoning, which cannot be accounted by classical logic.

We do not here question the fecundity of these non-classical logics, neither the utility of pragmatic and heuristic approaches to cognition, which are or could be used successfully to model very *specific* decision-making tasks. However, we will present here a possible, elegant solution, *capable of comprehensively dealing with all these supposed "fallacies" of human reasoning and decision-making processes.* Pursuing the project of developing a logical approach to thought and cognitive processes, we believe that these require, rather than a panoply of pragmatic-heuristic "recipes" specific to each situation studied, a unifying modeling, which supposes *the development of a single appropriate logic.* This synthetic solution will be explained by first setting its theoretical framework, which is different from the classical, set theoretical framework, and by then proposing solutions for the previously presented difficulties of the classical approach to cognition and decision-making. This new approach to human cognition, called "quantum cognition" in the literature, can deal with the important properties of cognitive processes that cannot be tackled by the classical models of cognition, namely contextuality, order effects and interferences effects.

## 3.1 A Survey of Quantum Cognition

"Quantum cognition" deals with mental processes within the same *mathematical framework* than that of quantum theory but it has *a priori* nothing to do with physics. It does not at all refer to the "physicality" of the world, through its parts, like elementary particles or black holes, or to physical concepts, like those of energy, mass or velocity. Quantum cognition focus on the properties of mental states and the rules that govern mental processes, independent of any physical reference. It has been developed for a few decades by several authors, including Busemeyer and Bruza [8], Aerts *et al.* [9], Aerts and Sozzo [9]. Its basic idea is to represent geometrically the cognitive state of a subject (which can also be understood as her "mental state" or her "belief state") by a vector of an appropriate Hilbert space spanned by all her possible cognitive states and to represent its transition into a new cognitive state by its projection onto the vector subspace associated to this new cognitive state. Like for physical observables in quantum theory, mental observables, which are the properties of any cognitive state that can be evaluated or "measured", are represented by Hermitian operators forming a *non-commutative* algebra and whose (real) eigenvalues are the possible results of their measurement. In Dirac notations, the transition of the cognitive state $|\psi\rangle$ to the new one obtained by measuring[2] the observable **A** with the outcome $A_i$ is represented by the projection $\mathbf{P}_{Ai}|\psi\rangle$, where $\mathbf{P}_{Ai}$ is the projector $|A_i\rangle\langle A_i|$ onto the eigenspace associated with $A_i$, which, *for sake of simplicity of presentation*[3], has been assumed to be one-dimensional, meaning that it is spanned by the vector $|A_i\rangle$ of H). This transition can be geometrically illustrated as follows (Fig. 1):

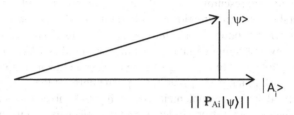

**Fig. 1.** A geometrical representation of the measurement of the observable **A** with result Ai

Now, using Born's rule, which gives the only probability measure that can be assigned to this Hilbert space, the probability for a subject in state $|\psi\rangle$ to transit into the new state $\mathbf{P}_{Ai}|\psi\rangle$, is computed as the square of the modulus of its projection onto the subspace associated with the result $A_i$:

$$P\,(A) = \|\,\mathbf{P}_A\,|\psi\rangle\|^{\,2} = \langle\psi|\,\mathbf{P}_A|\psi\rangle.$$

As will be shown hereafter, the previously mentioned difficulties of classical cognition and classical decision-making can find fruitful representations and satisfactory solutions within this quantum-like approach. Only some of them are briefly presented hereafter.

### 3.1.1  Evaluation of the Degree of Complementarity for Emotions

A fruitful representation of the order effect in the domain of emotions and its behavioural and physiological correlates can be provided within the theoretical framework of quantum cognition. Such an evaluation of the degree of non-commutativity (or "complementarity") of emotions is required for improving the capacity of machines (robots) to simulate emotional intelligence (see Sect. 4).

---

[2] "Measuring" in a very general sense, which can be understood as observing or even experiencing in the case of emotional observables (see Sect. 2.1.2).

[3] Of course, the eigenspace associated with a particular eigenvalue of **A** is generally multi-dimensional. This point is important since, as recalled by Boyer-Kassem [34], only multi-dimensional state-spaces (relative to what he calls "degenerate" situations) can rigorously explain the conjunction fallacy, thus showing the superiority of quantum cognition on classical rationality in this case.

Uzan [19] has provided an estimation of the degree of non-commutativity of emotions from experimental data reported in Prkachin *et al.*'s article [20], which report the average intensity of five emotions experienced by subjects conditioned in target emotional by Lang's method [33]. Conditional probabilities of experiencing an emotion A for a subject conditioned in an emotional state B can be evaluated by the ratio of the average intensity of experienced emotion A to the sum of the average intensities for the five basic emotions. In accordance with the presentation of the previous section, a geometrical representation of the corresponding vector-states |A> and |B> is provided. In this representation, each considered emotion E is associated with the two-dimensional, orthonormal E-basis { |E>, ⌐|E> }, the vector |E> representing a state of extreme emotion E (with a rating of 7) and the vector ⌐|E> a state where no emotion E is felt (rating 0). The reference [19] provides a detailed evaluation of the commutator of the observables Anger and disgust, respectively noted as **A** and **D**, by first computing the probability amplitude <D|A> of the transition from the state of pure anger to the state of pure disgust and then computing the outer product **D** = |D><D| in the basis { |A>, ⌐|A> }. The norm of the commutator [**A**, **D**], which evaluates the degree on complementarity of the observables Anger and Disgust is equal to 0.449, which shows a strong order effect in the evaluation of the intensity of the successive experiences of Anger and Disgust. More generally, it can be shown that all the other couples of emotions considered in Prkachin *et al.*'s data [20] do not commute and that many other examples of couples of non-commutative emotional observables can be computed from the relevant experimental data reported in the literature [21–23].

### 3.1.2  Solving Ellsberg's Paradox

This generalized probabilistic framework can also be applied to deal with decision-making in uncertain situation. For example, Ellsberg's paradox [26] bearing on the behavior of economic agents in a situation of uncertain knowledge (see Sect. 1.4) can be solved within this mathematical framework. As mentioned in Sect. 1.4, Ellsberg's paradox puts into question the law of total probabilities of classical probability calculus. According to the quantum-like approach to cognition and decision making presented here, the paradoxical difference between the probability of the agent's choice computed by classical probability calculus and the experimental result comes from the *interference term* between the possible belief states of a participant in the uncertain situation c) defined in Sect. 1.4 as the uncertainty about the outcome of the first game –if she won or she lost. This violation of the classical calculus of probability, which at first glance seems to indicate that human beings are prone to "fallacious" reasoning, is, in fact, a typical example of how we "reason" and make decisions in uncertain situations.

Several authors, including Busemeyer and Bruza [8], Aerts and Sozzo [36] and al-Nowaihi *et al.* [37] provide clear presentations of the quantum decision theory where Ellsberg's paradox can be solved. However, for a question of place, we will here present a simple solution that highlights the key-ingredient of the quantum-like approach to decision-making under uncertainty, namely the interference effects responsible for this human "fallacy". Consider the two possible sequences of mental states h1 and h2 (or possible "stories") leading a participant to the decision to play again. History h1 is defined by the sequence of states (I, G, J) and history h2 is defined by the sequence of states (I, ⌐G, J), where I is the "initial" cognitive state of the subject, G is her mental state when she learned that she won the first time, ⌐G is her mental state when she learned that she lost the first time and J is her mental state when she makes the decision to play again (see Fig. 2 below).

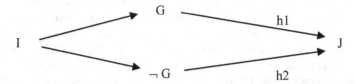

**Fig. 2.** The two possible stories leading to the decision to play again.

The law of total probabilities of the classical probability calculus tells us that:

$$P(J) = (1/2)[P(J/G) + P(J/\neg G)],$$

which is, as we have just seen, in contradiction with the experimental data. On the other hand, this paradox can be resolved if we assume here that, contrary to what would be required by a presupposed "realism of mental states", the participant who does not know whether or not she won the first time *is himself in a state of superposition of states* G and ⌐G. Let |g| Let |g> this state of superposition, which is represented by a unit vector of the configuration space of mental states (Hilbert space), and can be written as:

$$|g> = (1/\sqrt{2})(|G> + |\neg G>),$$

where |G> and |⌐G> are the representative unit vectors of the states G and ⌐G, while the normalization coefficient $(1/\sqrt{2})$ is calculated considering the equality of the probabilities of winning and losing the first time. Applying now Born's rule to the transition from |g> to the state |J> describing the decision to play again, we can calculate the value of the probability of replaying according to this model (see Appendix 1). This probability contains an *interference term* expressing the fact that the initial sate |g> is a state of superposition. The presence of this term can explain with accuracy the difference between the experimental results reported above (Sect. 1.4) and what we should find by applying the law of total probabilities.

## 4  Bayesian Models of Cognition Revisited

### 4.1  Bayesian Rationality

Bayesian rationality is the idea that cognitive processes must be modeled in probabilistic terms, by updating the prior distribution when a new event occurs - instead of applying rigid rules of deduction [38, 39]. The current Bayesian models of cognition, which are developed in order to model high-level cognitive processes along the connectionist approach, appeal to Bayes' inference in order to update the prior probabilities. This can be checked in the range of perception, categorization, language processing and emotion recognition [40–46].

Bayes' inference is thus regarded as the key ingredient used for integrating uncertainty in cognition and decision-making. It is used to model many areas of human activities, like finance (for modelling risk), medicine (for diagnostic and decision making) or meteorology (for weather forecasting), and the success of Bayesian models of cognition seems certain, as well in reasoning, learning or making decision. Accordingly, Bayesian networks, are used in machine learning whose applications have been developed in image processing, neuroscience and medical diagnostics, among other domains. Bayes' inference extends the framework of binary classical logic by taking into account the uncertainty in the knowledge of premises and the acquisition of information, which is evaluated in terms of probabilities. And even when probabilities are interpreted subjectively, as degrees of belief [47], Bayesian reasoning still satisfies the rules of the classical probability calculus, as shown by Cox-Jaynes theorem [48], which strengthens the idea that Bayesian inference would be totally appropriate for modeling mental processes. It is thus supposed to correctly represent the way we reason, we learn and make decision in uncertain situation, and therefore to be a key-ingredient for developing artificial intelligence. However, as will be recalled hereafter, Bayes' rule is valid under the assumption that the order in which are evaluated the considered observables does not matter *while, as shown in* Sect. 1, *most of mental processes do not satisfy this condition of commutativity.*

## 4.2  Bayes' Rule

Bayes' rule allows to calculate how a priori probabilities are updated when new information is gathered: the posterior probability P (A/E) of the event A given the evidence E, which can denote in particular the observation of some feature of the situation under consideration, is calculated from the prior probability P (A), estimated before the occurrence (or the knowledge of the occurrence) of E, and the "likelihood" P (E/A), which is the conditional probability of observing E when A is realized:

$$P(A/E) = [P (A).P(E/A)]/P(E), \tag{1}$$

where P(E) is the probability of occurrence of the event E, which is a priori estimated independently of A.

Bayes' rule is derived from the definition of conditional probability in the classical probability calculus:

$$P(A/E) = P(A \text{ and } E)/P(E),$$

where *"A and E" has no temporal connotation*, meaning that A and E can occur in any temporal order or be simultaneous. This order-independent definition of P (A and B), the joint probability of the conjunction of A and B, gives rise to the "rule of multiplication" of the classical probability calculus:

$$P(A \text{ and } E) = P(A/E). P (E) = P(E/A).P(A), \tag{2}$$

and Bayes' rule (1) is then straightforwardly obtained by dividing the two terms of the second equality of (2) by P(E). The relation (2) is thus valid on the condition that the value of the joint probability P (A and E) is independent from the order of occurrence of A and E. But in the case P (A and E) depends on the order of occurrence of A and E, one must clearly distinguish the calculation of P (A and then E) = P (E/A). P (A) from that of P (E and then A) = P (A/E). P (E). In this case Bayes' rule (1) is not verified since (2) is no longer valid. This means that *Bayes' rule is valid only on condition that the order in which the events under consideration occur does not matter.*

A careful study of the paradigmatic applications of Bayes' rule given in academic textbooks and in the relevant literature shows that the events considered in the presented experimental situations are regarded as purely intrinsic features of reality, independent of each other and independent of our means of observation. For example, in the paradigmatic application of the drawing of balls in two urns, noted as I and II, Bayes' rule is used to evaluate the conditional probability that a ball of a certain color was drawn from urn I or urn II. However, in this situation the colors of the balls are intrinsically and once and for all defined, and consequently the observables measuring the proportions of balls of a given color in each urn can be regarded as intrinsic properties of the physical world. As a consequence, the order in which they are evaluated does not matter and Bayes' rule can be successfully applied. Similarly, in the paradigmatic case of diagnostic testing [48], the involved observables, which respectively measure the state of health of a patient and its contamination rate can be regarded as independent, intrinsic properties of reality and are then assumed to commute. However, as shown previously (Sect. 1), it is not the case for mental processes: *the condition of commutativity of the relevant observables is generally not fulfilled – which seriously questions the reliability of the current Bayesian models of cognition.*

### 4.3 A New Probabilistic Rule of Inference

Fortunately, Bayes' rule can be generalized in order to account for the non-commutativity of mental observables which, as explained in Sect. 1, is involved in most of mental processes. This generalization, which can be done within the quantum-like framework presented in Sect. 2, is justified by the will to continue working in the paradigm of Bayesian rationality, while making it capable to deal with order effects. Moreover, such a generalization of Bayes' rule seems to be quite necessary in the field of artificial emotional intelligence (see Sect. 4).

To compute the conditional probability P (E/F) for the cognitive events E and F, let us first compute within the quantum-like framework presented here the difference of the probabilities of the two opposite sequences of events E and F:

$$P (E F) - P (F E) = <\psi|P_E P_F P_E |\psi>-<\psi|P_F P_E P_F |\psi>$$

$$= <\psi|P_E P_F P_E - P_F P_E P_F |\psi> \qquad (3)$$

$$= <\psi|[P_E, P_F] (P_E + P_F - I) |\psi>,$$

where the last equality has been obtained by factorizing the expression between the bra $<\psi|$ and the ket $|\psi>$, and by using the definition of the commutator $[P_E, P_F] = P_E P_F - P_F P_E$. Defining the operator. Defining the operator $Q$ as:

$$Q =_{df} [P_E, P_F] (P_E + P_F - I),$$

the difference $P(E\ F) - P(F\ E)$ can be written as the expectation value of $Q$ in the mental state $|\psi>$:

$$P(E\ F) - P(F\ E) = <\psi|\ Q\ |\psi> \equiv <Q>_\psi. \qquad (4)$$

Using relation (4) and the definition of conditional probabilities $P(F/E)$ and $P(F/E)$, one can show the following new rule of probabilistic inference:

$$P(F/E) = [P(E/F) \times P(F) + <Q>_\psi] / P(E). \qquad (5)$$

This new generalized probabilistic rule of inference computes the conditional probability of occurrence of the event F given the occurrence (or the knowledge of the occurrence) of the event E for a subject in the mental state $|\psi>$. Its *classical limit*, when the projectors associated to the event E and F are commuting, and then when $<Q>_\psi = 0$, , is of course nothing but Bayes' rule (1): $P(F/E) = [P(E/F) \times P(F)]/P(E)$.

## 5   The Quantum-Like Approach to Cognition Implemented in Emotional AI

The most impressive tasks of artificial intelligence, like learning, categorizing or perceiving, which are presently developed according to the connectionist approach (artificial neural networks), utilize probabilistic algorithms and thus appeal to Bayes' rule in order to compute the update the probabilities that can be assigned to new data. In order to take into account the order effects inherent to most of mental processes, these algorithms should then be revisited. Still keeping the probabilistic, Bayesian approach to cognition, such a change only regards the inference rule that must be used to update the probabilities of realization of some assumptions when new information is gathered. These models should use the general probabilistic rule (5) instead of Bayes' rule (1) insofar as the latter cannot account for order effects. Let us focus in the following on the important case of *emotional* artificial intelligence[4], which play a crucial role in any aspect of life [49, 50].

Emotional intelligence involves verbal and non-verbal communication, like reading persons' face expression, observing body movements and postures, and physiological manifestations. Simulating emotional intelligence algorithmically is now an important subject of research in the field of artificial intelligence, namely for improving human-machine interaction. It first requires *emotion recognition* on which we will focus hereafter.

Emotion "recognition" is a classification task from multimodal sensory, behavioral or physiological data [51]. It uses several types of sensors that detect speech signal, voice tone, facial expressions and body language, and appeal to data on previously observed correlations (called above "the common-view correspondence") between, on the one hand, the nature and intensities of emotions, and, on the other hand, the values

---

[4] The cases of categorization and visual perception have been addressed by Uzan in reference [] and will not be further detailed here for a question of place.

of physiological and behavioral observables. These data are analyzed by deep learning algorithms that compute the most probable emotional state that can be assigned to a subject by analyzing her behavior, like her facial expression or her body language and posture [52, 53], which seems the most informative observations even realized from afar and from any angle of view.

This task requires computing the conditional probabilities, noted as $P(E_k/\{B_j\})$, that the emotional observable experienced by a subject takes the value $E_k$ given the observation of a set of behavioral observables $\{B_j\}$, like her facial expression or her body language. However, as explained in Sect. 1.3, emotional observables do not commute with each other and do neither commute with their behavioral and physiological correlates. Consequently, classifying emotions experienced by a subject from the knowledge of her behavioral features thus requires considering these order effects, *which means that the generalized probabilistic rule (5) must be used instead of Bayes' classical inference rule*. This requires to first compute the commutators $[\mathbf{E_k}, \mathbf{B_j}]$ of the relevant observables $\mathbf{E_k}, \mathbf{B_j}$, which, as mentioned above (Sect. 2.3), can be computed from data about their joint measurement. These data can be found, for example, in the very comprehensive study published in the articles by Duran and Fernandez-Dols [54] and by Barrett *et al.* [21]. Such a computation, which allows asserting the relevance of the non-classical Bayesian approach introduced in Sect. 3, is detailed in Appendix 2.

Revisiting the emotion recognition task thus consists in changing the current deep learning algorithms by grounding the computation of the most probable emotion given a facial expression on the generalized probabilistic rule (5) instead on Bayes' rule, which requires a preliminary computation of the commutators of type $[E_i, F_k]$, for all considered couples of emotional and behavioral observables, and then the computation of the term $<\mathbf{Q}>_\psi$ in order to correctly evaluate the conditional probabilities $P(E_i/F_k)$ by rule (5).

Moreover, let us notice that emotion recognition is only the first step to achieve in order to "understand and to reason with emotions". In particular, in order for a subject to respond appropriately to a situation, she must be able to make *predictions* about the emotional state of her social environment given her present emotional state and new data she can gather, like information about the change of their face expression, their voice tone or even their physiological changes [55], [56]. *This requires again the use of the generalized probabilistic rule (5) capable of accounting for the non-commutativity of the relevant emotional/behavioral and emotional/physiological couples of observables.*

As a final remark, one can ask whether the aforementioned changes of the algorithms that are currently used for simulating emotional intelligence would apply for the simulation of *all* aspects of mental activity insofar as emotions are involved in them to varying degrees. For example, as is well known, "negative" emotions, like anger or sadness, can disturb our concentration and our capacity of memorization, while "positive" emotions, like a feeling of happiness, can improve our ability to perform these same tasks[5]. Following this idea, quantum cognition would be the fundamental ingredient for modeling mental processes and thus for improving the algorithms of artificial intelligence.

---

[5] This well-known point has been analyzed by Wang and Ross [58] and reported in the field of psychoanalysis [59].

## Appendix 1: Interference of Cognitive States in Tversky's and Shaffir's Experiment [25]

Referring to Fig. 2 of Sect. 2.1, the probability to play again for a subject in the state of superposition |g> can be computed within quantum cognition as follows, $\Pi(J)$ being the projector onto the state |J> corresponding to her decision to play again:

$P_Q(J) = <g|\Pi(J)|g>$

$= (1/2) [<G|\Pi(J)|G> + <\rceil G|\Pi(J)|\rceil G> + <G|\Pi(J)|\rceil G> + <\rceil G|\Pi(J)|G>]$

$= (1/2)[P (J/G) + P(J/\rceil G)] + ||<\rceil G|\Pi(J)|G>|| \cos w,$

where to the first term of this sum (which is the result $P_{class} (J)$ according to classical probability calculus) is added the interference term:

$$Int = ||<\rceil G|\Pi(J)|G>|| \cos w,$$

where w is the argument of the complex number $<\rceil G|\Pi(J)|G>$.

The presence of this term can thus explain the difference between the experimental result reported above (see Sect. 1.4) and what we should find by applying the law of total probabilities, that is, $P_{class} (J)$. This interference term can be evaluated from the conditional probabilities $P(J/G)$ and $P(J/\rceil G)$:

$$Int = (P(J/G).(P(J/\rceil G))^{1/2}. \cos w.$$

For w = 116°, we compute that Int = −0.28 and therefore:

$$P_Q(J) = P_{class}(J) - 0.2 = 0.36,$$

which is exactly Tversky's and Shaffir's experimental result.

The angle w, which determines the magnitude of the interference term, therefore evaluates the psychological factor characterizing the (average) sensitivity of the participants to the information that tells them (or not) that they have won or lost in the first game. In other words, w measures their ability to modify their decision to play again upon receipt of this information.

## Appendix 2: Dealing with the Emotion Recognition Task Within Quantum Cognition.

To show that the generalized probabilistic rule (5) of Sect. 3.2, which accounts for order effects, must be applied (instead of Bayes' rule) in emotion recognition tasks, one has to demonstrate that the difference $P (E F_E) - P (F_E E)$ is effectively equal to the average value of the operator $\mathbf{Q}$ for an initial, neutral state of the subject. Following the method presented in reference [19], one can compute the commutators $[\mathbf{E_i}, \mathbf{F_k}]$ involved in this task and show the expected equality. The data that are used in this computation are drawn from Barrett *et al.*'s article [21] (see Sect. 4.1).

The first required experimental conditional probability P (E/F$_E$) can be found on page 36 in Barrett *et al.*'s article [21]. For example, on the first line of Fig. 11, we can read that a subject presented with the "common view" facial expression of Anger labels it as an expression of Anger with the rate 39.92/71.92 = 0. 555. In this rate, 39.92 is the average number of subjects (average over the entire population of subjects tested) who reported an emotion of Anger given the presentation of its common view facial expression, while 71.92 is the sum of the values corresponding to the subject's possible answers. That is, P (A/F$_A$) = 0.555.

Since the same common view facial expression of Anger is presented to all the tested subjects, we can write, *I n this specific experiment*, that its prior probability is P (A) = 1, and that, consequently:

$$P(F_A A) = P(A/F_A) = 0.555.$$

The second conditional probability P (F$_E$/E) can be found on page 19, figure 6, of Barrett *et al.*'s article [21]. This experiment provides, for each of the six basic emotions, the average proportion of subjects *conditioned in this emotional state* who move their face according to the corresponding common view facial expression of this emotion. This experiment thus evaluates the average correlation degree between each of the six experienced emotions they are conditioned and their respective "common view" facial expression. Let us again focus on Anger, whose common view facial expression is characterized by brows furrowed, eyes wide, lips tightened and pressed together –see figure 2A of Barrett *et al.*'s article [21]. Figure 6 shows that for Anger this conditional experimental probability is P (F$_A$/A) = 0.22.

Again, since in this experiment the subjects are all conditioned in a same specified emotion (of anger), one can say that the event A is certain, that is, its prior probability of occurrence is P (A) = 1. The conditional probability P (F$_A$/A) is then equal, in this specific experiment, to the sequential probability P (A F$_A$) of being first conditioned in an emotional state A *and then* showing the facial expression state F$_A$. That is, P (A F$_A$) = P (F$_A$/A) = 0.22. According to Bayes' rule, P (F$_A$/A) and P (A/F$_A$) would here be identical, *which is wrong. Bayes' rule cannot thus explain the difference between the two sequential probabilities* P (F$_A$ A) and P (A F$_A$). By contrast, this difference can be computed by using the new, probabilistic rule derived in Sect. 3.2, which generalizes Bayes' rule in order to account for the complementarity of the observables considered. To show this, one has to first compute the commutator [**A, F$_A$**] of these two observables in order to evaluate the term <**Q** >$_\psi$ that appears in this rule, for an initial "neutral" cognitive state |ψ> of the subject, and to check that:

$$P (A F_A) - P (F_A A) = < \mathbf{Q} >_\psi \tag{6}$$

For computing the expectation value < **Q** >$_\psi$ of the operator **Q** in neutral state, one have to first decompose the state |F$_A$> on the two-dimensional basis {|A>, |⌐A>}:

$$|F_A> = (0.555)^{1/2}|A> + (1 - 0.555)^{1/2}|⌐A>.$$

The matrix of the projector $\mathbf{F_A}$ in this two-dimensional state space representation can then be computed, as the outer product $|F_A><F_A|$:

$$\mathbf{F_A} = \begin{bmatrix} 0.555 & 0.497 \\ 0.497 & 0.445 \end{bmatrix}$$

while he matrix of the observable $\mathbf{A} = |A> <A|$ in this basis is $A = \begin{bmatrix} 1 & 0 \\ 0 & 0 \end{bmatrix}$.

The commutator $[\mathbf{A}, \mathbf{F_A}] = \mathbf{A}\,\mathbf{F_A} - \mathbf{F_A}\mathbf{A}$ can then be computed:

$$[\mathbf{A}, \mathbf{F_A}] = 0.497 \begin{bmatrix} 0 & 1 \\ -1 & 0 \end{bmatrix}.$$

and the operator $\mathbf{Q}$ is then:

$$\mathbf{Q} = [\mathbf{A}, \mathbf{F_A}]\,(\mathbf{A} + \mathbf{F_A} - \mathbf{I}) = \begin{bmatrix} 0.247 & -0.276 \\ -0.276 & -0,247 \end{bmatrix}$$

The subject's "neutral" cognitive state can be reasonably defined by the equiprobable superposition of the six basic emotions (which are those considered in Barrett's article) and can thus be written as:

$$|\psi> = (1/6)^{1/2}|A> + (5/6)^{1/2}|\neg A>.$$

Consequently, one can compute $<\mathbf{Q}>_\psi$:

$$<\mathbf{Q}>_\psi = <\psi\,|\,\mathbf{Q}\,|\,\psi> \approx 0.370.$$

Within the quantum-like approach presented above, one can thus find that 1) the difference $P\,(A\,F_A) - P\,(F_A A)$ is not null (contrary to what would be derived from Bayes' rule) and that 2) a theoretical evaluation of this difference from the generalized Bayes' rule of Sect. 3.2, which is derived within the framework of quantum cognition, gives the value 0.370. This result is not so far from the *experimental* difference:

$$P_{exp}(A\,F_A) - P_{exp}(F_A A) = 0.555 - 0.22 = 0.335,$$

since the value of $P_Q\,(A/F_A) = 0.555\ 0.370 = 0.185$ computed with relation (4) is situated *within the confidence interval (error bar) of the average experimental value* (0.22) reported in Barrett et al.'s article [21], on figure 6.

Of course, the adequacy of the probabilistic rule (5) can be checked for the other basic emotions considered in Barrett et al.'s article [21]. For example, regarding Disgust, one can find on figure 11, p. 36, of Barrett *et al.*'s article (2019) that $P\,(F_D\,D) = 0.406$, while from Fig. 6, p. 19 of this article, one has $P\,(D\,F_D) = 0.24$. According to a quite similar method to that used above, one can compute, for a neutral subject's state, that $<\mathbf{Q}>_\psi = <\psi|\mathbf{Q}|\psi> \approx 0.165$, which is *very close* from the experimental difference $P\,(F_D\,D) - P\,(D\,F_D) = 0.166$.

# References

1. Leibniz, G.W.: Dissertatio de Arte Combinatoria. Leipzig (1966). Original version in Latin at Biblioteca Nazionale Centrale di Firenze. https://archive.org/details/ita-bnc-mag-000008 44-001
2. Boole, G.: 1854–1992: Les lois de la pensée. Vrin, Paris (1992)
3. Fodor, J.: LOT 2: The Language of Thought Revisited. Oxford University Press, Oxford (2008)
4. Frege, G.: Begriffschrift. English translation in Jean Van Heijenoort (ed.): From Frege to Gödel, pp. 1–83. Cambridge: Harvard University Press (1967) (1879)
5. Turing, A.: On Computable Numbers, with an Application to the Entscheidungsproblem. Proc. Lond. Math. Soc. **42**, 230–265 (1936)
6. Church, A.: An unsolvable problem of elementary number theory. Am. J. Math. **58**, 245–263 (1939)
7. Cruz, N., Baratgin, J., Oaksford, M., Over, D.P.: Bayesian reasoning with ifs and ands and ors. Front. Psychol. **6**(192), 10.3389 (2015)
8. Busemeyer, J.R., Bruza, P.: Quantum models of cognition and decision. Cambridge University Press, Cambridge (2012)
9. Aerts, D., Sozzo, S., Gabora, L., Veloz, T.: Quantum structure in cognition: Fundamentals and applications. In: ICQNM 2011: The fifth international conference on quantum, nano and micro technologies (2011)
10. Wang, Z., Busemeyer, J.R.: A quantum question order model supported by empirical tests of an a priori and precise prediction. Top. Cogn. Sci. **5**(4), 689–710 (2013)
11. Fodor, J.: The Elm and the Expert, Cambridge. MIT Press, MA (1994)
12. Stalnaker, R.: Context and content. Oxford University Press, Oxford (1999)
13. Austin, J.L.: Philosophical papers. In: Urmson and Warnock (eds.) Oxford University Press (1961)
14. Wittgenstein, L.: Investigations philosophiques. In: Anscombe et Rush Rhees, G.E.M. (ed.). French translation, Gallimard, coll. « Tel », 1986 (1953)
15. Hampton, J.A.: Overextension of conjunctive concepts: evidence for a unitary model for concept typicality and class inclusion. J. Exp. Psychol. Learn. Mem. Cogn. **14**, 12–32 (1988)
16. Uzan, P.: Psychologie cognitive et calcul quantique. Implications Philosophiques, « Episté-mologie » section (2014)
17. Moore, D.W.: Measuring new types of question order effects. Public Opin. Q. **66**(1), 80–91 (2002)
18. Geringer, J.M., Madsen, K.M., Gregory, D.: A fifteen-year history of the continuous response digital interface: Issues relating to validity and reliability. Bulletin of the Council for Research in MusicEducation **160**, 1–15 (2004)
19. Uzan, P.: Complementarity in psychophysics. In: Atmanspacher, H., Filk, T., Pothos, E. (eds.) Quantum Interaction 2015. Lecture Notes in Computer Science, Vol. 9535, pp. 168–178. Springer (2016)
20. Prkachin, K.M., Williams, R.M., Zwaal, C., Mills, D.E.: Cardiovascular changes during induced emotion: an application of Lang's theory of emotional imagery. J Psychos Res **47**(3), 255–267 (1999)
21. Barrett, L.F., Adolphs, R., Marsella, S., Martinez, A.M., Pollak, S.D.: Emotional expressions reconsidered: challenges to inferring emotion from human facial movements. Psychol. Sci. Pub. Inter. **20**(1), 1–68 (2019)
22. Kassam, K.S., Mendes, W.B.: The effects of measuring emotion: physiological reactions to emotional situations depend on whether someone is asking. PLoS ONE **8**(6), e64959 (2013)

23. Kreibig, S.D., Wilhelm, F.H., Roth, W., Gross, J.J.: Cardiovascular, electrodermal, and respiratory response patterns to fear- and sadness-induced films. Psychophysiology **44**, 787–806 (2007)
24. Tversky, A., Shafir, E.: Choice under conflict: the dynamics of deferred decision. Psychol. Sci. **3**(6), 358–361 (1992)
25. Ellsberg, D.: Risk, Ambiguity and the Savage Axioms. Quaterly Journal of Economics **75**(4), 643–669 (1961)
26. Savage, L.J.: The Foundations of Statistics. Wiley, New York (1954)
27. Grice, H.P.: Studies in the Way of Words. Harvard University Press, Cambridge, MA (1989)
28. Gigerenzer, G., Todd, P.M.: Simple heuristics that make us smart. Oxford Univ. Press, New York (1999)
29. Tversky, A., Kahneman, D.: Judgment under uncertainty: Heuristics and biases. Science **185**, 1124–1131 (1974)
30. de Finetti, B.: La logique de la probabilité. Actes du congrès international de philosophie scientifique **4**, 31–39 (1936). English translation: The logic of probability. Philosophical Studies **77**(1), 181–190 (1995)
31. Dernoncourt, F.: La Logique Floue: entre raisonnement humain et intelligence artificielle. www.academia.edu/1053137/La_Logique_Floue_entre_raisonnement_humain_et_intelligence_artificielle (2011)
32. Ginsberg, M.L.: Readings in nonmonotonic reasoning. Morgan Kaufmann, San Francisco (1987)
33. Pothos, E.M., Busemeyer, J.R.: A quantum probability explanation for violations of "rational" decision theory. Proc. R. Soc. B **276**, 2171–2178 (2009)
34. Aerts, D., Sozzo S.: Quantum Structure in Economics: The Ellsberg Paradox. arXiv:1301. 0751 v1 [physics.soc-ph] (2013)
35. al-Nowaihi, A., Dhami, S., Wei, M.: Quantum Decision Theory and the Ellsberg Paradox (2018). CESifo Working Paper Series No. 7158
36. Lang, P.J.: A bio-informational theory of emotional imagery. Psychophysiology **16**, 495–512 (1979)
37. Boyer-Kassem, T., Duchêne, S., Guerci, E.: Quantum-like models cannot account for the conjunction fallacy. Theory Decision **81**, 479–510 (2016)
38. Griffiths, L., Kemp, T., Tenenbaum, J.B.: Bayesian models of cognition. Cambridge University Press, The Cambridge Handbook of Computational Psychology (2008)
39. Oaksford, M., Chater, N.: Bayesian rationality: the probabilistic approach to human reasoning. Oxford University Press (2007)
40. Mamassian, P., Landy, M., Maloney, L.T.: Bayesian modelling of visual perception. In: Rao, R.P.N. (ed.) Probabilistic Models of the Brain: Perception and Neural Function, pp. 13–36. MIT Press (2002)
41. Perfors, A., Tenenbaum, J.B.: Learning to learn categories. In: Taatgen, N., van Rijn, H., Schomaker, L., Nerbonne, J. (eds.) Proceedings of the 31st annual conference of the cognitive science society, pp. 136–141. Cognitive Science Society, Austin, TX (2009)
42. Xu, F., Tenenbaum J.B.: Word Learning as Bayesian Inference: Evidence from Preschoolers. Psychol Rev Apr **114**(2), 245–72 (2007)
43. Tarnowski, P., Kołodziej, M., Majkowski, A., Rak, R.J.: Procedia *Emotion recognition* using facial expressions. Computer Science **108**, 1175–1184 (2017)
44. Estevam, R., Hruschka,, do Carmo Nicoletti, M.: Roles played by bayesian networks in machnie learning: an empirical investigation. In: Ramanna, Jain and Howlett eds: Emerging Paradigms In Machine Learning, Chap. 5 (2013)
45. Kubota, T.: Artificial intelligence used to identify skin cancer (2017). http://news.stanford.edu/2017/01/25/artificial-intelligence-used-identify-skin-cancer/

46. De Finetti, B.: Logical foundations and measurement of subjective probability. Acta Physiol. (Oxf) **34**, 129–145 (1970)
47. Cox, R.T.: Probability, Frequency, and Reasonable Expectation. Am. Jour. Phys. **14**, 1–13 (1946)
48. Broemeling, L.D.: Bayesian methods for medical test accuracy. Diagnostics (Basel, Switzerland) **1**(1), 1–35 (2011)
49. Wang, Q., Ross, M.: Culture and memory. In: Kitayama, S., Cohen, D. (eds.) Handbook of cultural psychology, pp. 645–667. Guilford, New York, NY (2007)
50. Salovey, P., Mayer, J.D.: Emotional intelligence. Imagin. Cogn. Pers. **9**, 185–211 (1990)
51. Li, S., Deng, W.: Deep facial expression recognition: A Survey. Computer science, Psycxhology (2018)
52. Santhoshkumar, R., Geetha, M.: Deep learning approach for emotion recognition from human body movements with feedforward deep convolution neural networks. Procedia Computer Science **152**, 158–165 (2019)
53. Singh, M., Majumder A., Behera L.: Facial expressions recognition system using Bayesian inference. In: 2014 International Joint Conference on Neural Networks (IJCNN), pp. 1502–1509 (2014)
54. Duran, J.I., Fernandez-Dols, J-M.: Do emotions result in their predicted facial expressions? A meta-analysis of studies on the link between expression and emotion. PsyArXiv: https://psyarxiv.com/65qp7 (2018)
55. Wang, Q., Ross, M.: Culture and memory. In: Kitayama, S., Cohen, D. (eds.), Handbook of cultural psychology, pp. 645–667. Guilford, New York, NY (2007)

# Are Humans Moral Creatures? A Dual-Process Approach for Natural Experiments of History

Hiroshi Yama[(✉)]

Osaka Metropolitan University, Osaka 558-8585, Japan
yama.hiroshi1204@gmail.com

**Abstract.** The decline in violence and growing awareness of human rights can be viewed as natural experiments in history answering the question if humans are moral creatures. A dual-process approach, which assumes the intuitive and reflective systems, is adopted to examine how the reflective system controls the intuitive system, which possibly produces our cruelty. Reading a story may enhance our reflective system to suppress our cruelty. However, the power of the reflective system is weak in arousing people's action against violence. Hence, theory of mind (ToM) accompanied by emotional empathy is necessary for the suppression of cruelty. However, ToM is modular in nature and one of the subsystems of the intuitive system; hence, empathy is narrow in its focus, which possibly causes social fragmentation and political polarisation.

**Keywords:** Morality · Dual-process model · Empathy · Story

## 1 Introduction

Are humans moral creatures? This is the question that has been asked by philosophers and psychologists throughout the ages. Although many researchers have attempted to answer it, the ultimate answer has not yet been found. However, humans are becoming more moral not only by evolutionary selection but also by the accumulation of cultural products such as moral sense, laws and awareness of human rights.

Non-violence is an index of human morality. For example, Daly and Wilson (1988) demonstrated a decrease in the number of murders from the time of hunter-gatherers to the contemporary world. Especially, the typical murders in which a man loses his temper and kills someone have decreased worldwide. A consequence of this trend in industrial countries is that about 10 murders per million people in a year are committed. However, about 300 occur among the San of the Kalahari, known as the least aggressive of the hunter-gatherers. This suggests that murders were more prevalent in the time of hunter-gatherers than in this contemporary world.

There are many possible factors for the decrease in violence. Reading is an important factor in this study. Pinker (2011) stated that the prevalence of novels was a possible reason for the decline in war, torture, cruel punishment and religious persecution (including witch trials) in the 17th century (the Enlightenment in Europe). For instance, the Frenchman Jean de La Fontaine, wrote many fables inspired by Aesop's Fables. La Fontaine's

J. Baratgin et al. (Eds.): HAR 2023, LNCS 14522, pp. 210–220, 2024.
https://doi.org/10.1007/978-3-031-55245-8_14

fables were popular and provided a model for subsequent fabulists across Europe and numerous alternative versions in French. Daniel Defoe wrote *Robinson Crusoe* and Jonathan Swift wrote *Gulliver's Travels* in the eighteenth century England.

However, he did not explain how the reading of novels enhanced human morality. Yama (2020) adopted a dual-process approach. Dual-process theorists supported the intuitive and reflective systems. In this study, I focus on empathy controlled by the reflective system.

## 2 Dual-Process Approach

**Table 1.** Features attributed to the two systems of cognition.

| System 1 | System 2 |
|---|---|
| Evolutionarily old | Evolutionarily current |
| Unconscious, preconscious | Conscious |
| Implicit knowledge | Explicit knowledge |
| Automatic | Controlled |
| Fast | Slow |
| Parallel | Sequential |
| Low-capacity demand | High-capacity demand |
| Intuitive | Reflective |
| Contextualised | Abstract |
| Domain-specific | Domain-general |
| Associative | Rule based |
| Connected to simple emotion | Connected to complex emotion |
| Independent of general intelligence | Linked to general intelligence |

Several dual-process theories have been proposed over the past three decades (Evans, 2010; Stanovich, 2009). The hot debates about this approach will not be discussed in this chapter. However, some examples are as follows. First, the question is whether two different systems should underlie two types of processing, respectively. Second, if systems are supposed, then what should they be called? Stanovich (1999) used the generic terms System 1 and System 2 to label the two different sets of properties. System 1 is generally reliable but can lead to fallacies and biases, whereas System 2 allows human reasoning to follow normative rules. The former is an evolutionarily old heuristic system, and the latter is an evolutionarily recent analytic system. As shown in Table 1, System 1 supports processing that is implicit, automatic, fast, intuitive, contextual and associative, whereas System 2 supports processing that is explicit, controlled, slow, reflective, abstract and rule based.

Instead of distinguishing between Systems 1 and 2, this study distinguishes between the intuitive system, which needs fewer cognitive resources and is a fast and automatic process, and the reflective system, which is supported by cognitive capacity and supervise the former process. Therefore, our research question is how the fast and automatic process, which sometimes triggers irrational and/or non-adaptive behaviours, is controlled and revised by a supervising (reflective) system (e.g. Strack and Deutsch, 2004). The reflective system pursuits human normative rationality in nature.

Can the reflective system control the process of or revise the output of the intuitive system? Evans (2007) proposed three possible models for the relation between the two systems. The first is the pre-emptive conflict resolution model, which assumes that a decision is made at the outset as to whether an intuitive system or reflective system process will control the response. The second is the parallel-competitive model, which assumes that each system works in parallel to produce a putative and provisional response, sometimes resulting in conflict that needs to be resolved. The third is the default-interventionist model, which involves the cueing of default responses by the intuitive system that may or may not be altered by subsequent interventions of the reflective system.

The second and third models will be discussed in this chapter. The third default-interventionist model assumes that, although the intuitive system produces an irrational response, the reflective system revises the output. In case of cognitive tasks, if the reflective system works with sufficient cognitive capacity, cognitive biases produced by the intuitive system are likely to be suppressed by it. Its evidence was provided by Stanovich and West (1998) who posited that the higher the IQ, the less biased a person's thinking. The IQ score is an index of the cognitive capacity that supports the functions of the reflective system. However, apart from cognitive tasks, revisions are not always positive. As the intuitive system (an evolutionarily old heuristic system) serves genetic interests, revision by the reflective system (the evolutionarily recent analytic system) can be a rebellion in this sense.

Although the reflective system tries to suppress irrational responses, it is plausible that outputs from the two systems coexist. This is comparable with the parallel-competitive model. For instance, some couples in a Christian culture avoid a wedding on Friday the 13th. Although their reflective system judges it as superstitious to regard Friday the 13th as inauspicious, their intuitive system brings fear when they think of their wedding on that day. Their fear is not completely suppressed by the reflective system. As shown in Table 1, the intuitive system is more strongly associated with a single (and generally strong) emotion than the reflective system. When an output of the intuitive system is associated with strong emotion such as fear, it is less likely to be suppressed.

In this sense, human morality reasoning, which is very likely associated with strong emotion such as disguise, is less likely to be suppressed. Haidt (2012) demonstrated that the reflective system does not suppress intuitive moral judgement, such as an incest taboo, but rationalises the judgement. One of the origins of this taboo is to avoid the increase in frequency of birth defects by recessive heredity. However, some participants of the study of Haidt (2012) pointed out that this possibility, even with the probability of pregnancy, was almost naught.

## 3  The Enhancement of Compassion, Decrease in Violence and Growing Awareness of Human Rights in History

Has the reflective system controlled the intuitive system so that it suppresses and/or revises the immoral outputs? As mentioned in the Introduction, humans have enjoyed the decline in violence and the growing awareness of human rights in history. Is this the consequences of the control of the reflective system? Humans have domesticated themselves in the history of evolution (Wrangham, 2019). Animals have two kinds of aggressions: Reactive and proactive aggression. The former is characterised by emotional lability, whereas the latter is driven by low emotionality. According to Wrangham (2019), the former has been reduced in the process of evolution.

Since the beginning of history, most of the enhancement of morality and the decline in violence are caused by the control of the reflective system. The first epoch was roughly between 500 and 300 BCE when similar religious traditions with an unprecedented emphasis on self-discipline and asceticism, including Buddhism, Jainism, Brahmanism, Daoism, Second Temple Judaism, and Stoicism emerged. Karl Jaspers named the time the 'Axial Age'. According to Baumard et al. (2015), increased affluence caused the historical changes. The affluence impacted human motivation and reward systems, nudging people away from short-term strategies and promoting long-term strategies (self-control techniques and cooperative interactions). The long-term strategies are the products of the reflective system, which made people think of themselves.

The second period focused on in this study is the 17th century, when war, torture, cruel punishment and religious persecution in Europe declined. According to Pinker (2011), two possible factors influenced these changes. The first was the establishment of absolute monarchies, which worked as a 'Leviathan' to suppress civil wars and conflicts between feudal lords in France and Germany (Holy Roman Empire). The second factor was the prevalence of novels. Johannes Gutenberg's invention, the printing press, began publishing novels in the sixteenth century. Jean de La Fontaine wrote many fables inspired by *Aesop's Fables*, which were the cultural products of the 'Axical Age' in the 17th century France. Daniel Defoe wrote *Robinson Crusoe* and Jonathan Swift wrote *Gulliver's Travels* in the eighteenth century England. *The Sorrows of Young Werther* (*Die Leiden des jungen Werthers*) by Johann Wolfgang von Goethe was also published in Germany around this time. Furthermore, although Pinker (2011) did not refer to it, very similar trends occurred in Japan. The Japanese saw a great decline in violence in the 17th century. This is called Pax Tokugawana. Similar to the trends in Europe, many novels and plays by Monzaemon Chikamatsu and Saikaku Ihara were published and many plays were performed in theatres at this time.

The third era of interest is the time after World War II. Although Pinker (2011) stated that this period was characterised as no great war after 1945 and no war and no genocide, the Russia-Ukraine war broke out in 2022. The third era was also characterised as a time of a decline in murder and violence in many countries. Furthermore, a growing awareness of human rights over the past 70 years has been prevalent. This includes a decline in the prejudice and discrimination against people of coloured skin, females, ethnic minorities and sexual minorities such as LGBTQ. After the establishment of the Civil Rights Act in the Unites States, the decline in racism has been unprecedented, along with a decline in sexism owing to feminism. For example, a social movement called 'Me

Too' (an awareness campaign against sexual abuse, sexual harassment and rape culture) encouraged people (not only females) to publicise their experiences of sexual abuse or sexual harassment. The phrase 'Me Too' was initially used on social media in 2006 by an activist Tarana Burke. The awareness of human rights was also for children. Various forms of child maltreatment and victimisation declined as much as 40%–70% from 1993 until 2004, including sexual abuse, physical abuse, sexual assault, homicide, aggravated assault, robbery and larceny (Jones et al., 2006).

The most important social factor underlying the trends after World War II is the spread of higher education with the university-going rate over 50% in many countries. Therefore, the decline in illiteracy by the spread of primary education have steadily enhanced awareness of human rights in history. However, the concepts of human rights, democracy and equality are mostly taught and discussed at universities. Therefore, we see the rapid rise of human awareness in this century. Furthermore, with the great surge in publication culture around the world have become increasingly popular. Very likely, the cultural practice of reading novels has been enhanced because of these publications, which has caused the very similar effect as the one shown in the 17th century. Furthermore, the spread of education may have caused the Flynn effect (Flynn, 2012), which is the substantial and long-sustained increase in both fluid and crystallised intelligence test scores that were measured in many areas of the world. Noteworthy, was a negative Flynn effect that has also been reported in Europe (e.g. Dutton and Lynn, 2014). Not only did the increase in test scores reach a ceiling but a slight decrease in these scores was observed in some countries. The reasons for this decrease are the increase in the number of non-European immigrants who have not received higher education and the negative association between the educational level and number of children in a family. As shown in Table 1, the intelligence test score reflects the cognitive capacity that supports the functions of the reflective system. Then, does the reflective system control the intuitive system? This is discussed in the next section.

## 4   Control by the Reflective System

To the question if humans are moral creatures, we are not able to give a completely definitive answer. However, they are certainly becoming less violent and more aware of human rights. These historical changes can be viewed as natural experiments. The next step is then to find their independent variables and propose possible hypotheses in the dual-process approach.

One of the biggest psychological factors for the changes after the dawn of human history may be the improvement of the function of the reflective system. The hardware of the human brain, such as the proportion of the neocortex, has not drastically improved since the beginning of Homo sapiens. However, the increased availability of this capacity in the reflective system is plausible. Specifically, these changes can be attributed to the increase in software efficiency.

Some research results show the relationship between the index of the efficiency of the reflective system (e.g. IQ) and some indexes of morality. The positive correlation indicates that, despite the imperfect control by the reflective system for human morality, we can expect to see a high-level of morality if the system fully functions. For example,

Stanovich and West (1998) posited that the higher the IQ, the less biased is a person's thinking. This means that the reflective system corrects or suppresses biases when working sufficiently. Biases, such as illusive thinking can be used to evoke hatred. Hence, the reflective system can contribute to the decline in hatred based on bias. For example, Swami et al. (2014) demonstrated that a stronger belief in conspiracy theories was significantly associated with lower analytic thinking and open-mindedness. Thus, the analytic thinking and open-mindedness are the indexes of the capacity of the reflective system. Reviewing the papers on fake news, Pennycook and Rand (2021) claimed that analytic thinking can weaken the belief in fake news. The belief in conspiracy theories and fake news can be irrational tools for blaming others, the target of hatred.

Another methodology to investigate the power of the reflective system is to restrict the usage of the reflective system such as giving time pressure to participants. For example, Verkuyten et al. (2022) investigated people's political intolerance against immigrant-origin individuals with dual citizenship. They demonstrated that tolerance was more likely to be greater when people engaged in deliberative (reflective) thinking in which they recognised and considered the equal rights of all citizens. This shows that reflective thinking can increase political tolerance. Furthermore, Verkuyten et al. (2023) found that the deliberate (reflective) thinking increased tolerance of minority beliefs and practices.

Some researchers agree that the capacity of the reflective system is almost identical to working memory capacity focus on the self-regulatory process of working memory. For example, Hofmann et al. (2008) identified the individual differences in working memory capacity to control for self-regulatory behaviour. They found that automatic attitudes toward the temptation of interest such as sexual interest behaviour and the consumption of tempting food had a stronger influence on behaviour for individuals who scored low in working memory capacity. These results demonstrate the importance of working memory capacity for everyday self-regulation, which is one of the aspects of human morality.

Can these results be evidence of the claim that the growing awareness of human rights and the decline in violence after the World War II be attributed to the Flynn effect? As Pinker pointed out the prevalence of novels to be a factor of the reduction in violence in the 17th century Europe, this possibility is considered in the next section. Although Pinker (2011) did not refer to the concept of mindreading, it is the key term if the practice of reading was responsible for people developing an aversion to cruelty. Mindreading is based on the theory of mind (ToM) subsystem of the intuitive system (Baron-Cohen, 1995; Leslie, 1992). ToM enables the construct of a theory to explain others' behaviours. As a subsystem of the intuitive system, ToM processing is automatic, domain-specific and connected to simple emotion. The ToM subsystem works automatically when information of human action is put in and is accompanied by emotional responses. This evolved to help humans, as a social mammal, to understand each other.

However, the ToM subsystem works with the cognitive capacity of the reflective system in the real world (e.g., Carlson et al., 2004). Although the subsystem is modular, the reflective system probably monitors how it works and adapts it to an actual social environment. He et al. (2019) explored the relation between ToM and the working memory capacity. They focused on social working memory as identified by Meyer et al. (2012), who found evidence to support the distinction between social and non-social working memory using a functional MRI technic. The former is for social cognitive information, including people's mental states, traits and relationships. According to them, the medial frontoparietal regions are implicated in social cognition, whereas the lateral frontoparietal system is implicated in non-social forms of working memory. He et al.'s (2019) study revealed that the social working memory capacity increased between 3- and 6-year-olds and positively predicted preschoolers' ToM scores. However, non-social working memory capacity did not. Their study indicates that the capacity of working memory in the reflective system controls the ToM subsystem so that it works in the contemporary social environment.

## 5   Do Stories Enhance ToM and Empathy?

If the ToM subsystem works for mindreading of a miserable victim, people feel pity for the victim. Otherwise, if it works for mindreading of a brutal tyrant, people feel strong anger towards the target person. This power of emotion can be a driving force for action to improve society. According to Pinker (2011), it is plausible that this emotional power influenced the changes in the 17th century Europe. It is also plausible that this power can be the drive for the growing awareness of human rights in the contemporary society, where people watch unprecedented amounts of publication and broadcasting.

Before proceeding, I need to explain some terminology. Empathy concerns our ability to share affective states with others, whereas ToM represents our ability to interpret their mental state, their intentions and beliefs. Furthermore, there are two kinds of empathy: cognitive and emotional empathy. Cognitive empathy involves knowing how other people think and feel, while emotional empathy involves feeling another person's emotions. This is supported by ToM. These are in the intuitive system, but ToM and cognitive empathy are more likely controlled by the reflective system than emotional empathy. Sympathy differs from empathy in that it does not require people to fully engage in the experience of a target person but involves their concern. Because of this narrowness, it is the product of the intuitive system.

Do reading novels and watching dramas enhance the work of the ToM subsystem? Mar and Oatley (2008) focused on narrative fiction. They did not discuss the modularity of ToM much but propose a general mechanism of reading. The literary narratives provide mental models or simulations of the social world via abstraction and create a deep simulative experience of social interactions for readers. This simulation facilitates the communication and understanding of social information. Engaging in the simulative experiences of fiction, literature can facilitate the understanding of others. Djikic et al. (2013) investigated the potential of literature to increase empathy. They used a questionnaire that measures lifelong exposure to fiction and nonfiction. Their participants read either an essay or a short story and were tested for cognitive and affective empathy. Their results suggest a role of fictional literature in facilitating development of cognitive empathy. Therefore, although Djikic et al. (2013) demonstrated the enhancing effect of the experience of reading fiction and nonfiction, their focus was not on emotional empathy but on cognitive empathy. Instead of the distinction between cognitive and emotional empathy, Kidd and Castano (2013) adopt the distinction between cognitive and affective ToM. The former works to infer what others think, whereas the latter infers what others feel. They found that reading literary fiction led to better performance on tests of both affective ToM and cognitive ToM compared with reading nonfiction and popular fiction. Their results show that reading literary fiction temporarily enhances ToM.

These studies do not perfectly support the inference that the high awareness of human rights in contemporary society has been achieved by people's emotional power via the prevalence of novels, because reading fiction and nonfiction do not directly influence emotional empathy. However, because of the modularity of ToM, emotional ToM or emotional empathy sometimes causes negative consequences. These are discussed in the next section. The framework to explain how stories have nourished human compassion and morality is shown in Fig. 1.

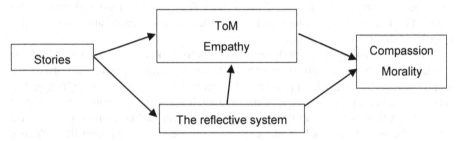

**Fig. 1.** Stories have nourished human compassion and morality

## 6   The Power and the Limitation of Story

Although Bloom (2016) agree that social improvement of human moral judgement and behaviour is attributed to human empathy, he points out the narrowness of empathy. His argument is the following syllogism: ToM is modular and emotional empathy is a consequence of ToM processing; hence emotional empathy is modular. Despite the assumption that the reflective system controls the work of ToM, its control is not perfect.

As stated repeatedly in this paper, the power of emotional empathy is strong for the growing awareness of human rights in contemporary society. However, the emotional empathy does not always lead to people's rational judgement and behaviours. For example, news about a disaster causes many people to have emotional empathy and to donate some money. This shows people's compassion in contemporary society. However, people's compassion is biased by emotional empathy. First, it is because TV stations broadcast disasters which are expected to be interesting for people. Eisensee and Strömberg (2007) analysed the coverage of natural disasters from 1968 to 2002, and found that although the casualties are many, coverage of epidemics, droughts and food shortages are less likely to be broadcasted than those of volcano disasters, earthquakes and storms because they are less spectacular. Another example is the case when Jessica McClure, who was 18 months old, fell into a well in 1987. Rescuers tried to free her, but her rescue was much more difficult than initially anticipated. This rescue effort was given live media coverage and, consequently, viewers donated over 700,000 dollars to the rescue effort. This sum could have helped save many malnourished infants in poor countries. However, it is difficult even for UNICEF to gather such an amount from donations. This psychological mechanism can be applied to the 'identifiable victim effect' (Kogut and Ritov, 2005).

Gottschall (2021) reveals a dark side to storytelling. Stories have certainly contributed to the development of human morality. However, storytelling can also be the main force separating people. Since it can more easily manipulate people's minds than logical arguments or scientific thinking, once a nationalistic story is made and told by an autocratical dictator, it is likely that the nation will wage a war. Gottschall argues that societies succeed or fail depending on how they manage these tensions. Buffone and Poulin (2014) provided empirical evidence to show that empathy for others motivate aggression on their behalf. Their study examined potential predictors of empathy-linked aggression and an empathy target's distress state. They found that induced empathy for a person combined with physiological changes led to an increased amount of trouble assigned to that person's competitor.

It is very plausible that empathy is a key term to resolve the paradox that both the growth of the awareness of human rights, social fragmentation and political polarisation are proceeding. It is difficult for people with different values to exist peacefully together. Furthermore, because of the spread of the Internet, it is easier for people to express their own opinion than before, and there are political battles and no compromise of each other on the web. Pennycook et al. (2023) demonstrated that, although the reflective system (cognitive sophistication) enhances pro-scientific judgement generally, it is used to confirm their political view about some issues such as global warming.

## 7  Conclusion

Are humans moral creatures? The natural experiment of historical changes in the time of Axial Age, 17th century Europe, and the time after World War II reveals that humans have been becoming more moral in the sense of decline in violence and enhancement of compassion, and the growing awareness of human rights. This has been achieved by the accumulation of moral cultures such as legal systems and ethical customs. These moral cultures are based on human psychological mechanism of the reflective system and ToM, which is a subsystem of the intuitive system. However, this current speculation has not yet been supported by empirical data. Therefore, empirical psychological studies are needed in the near future.

## References

Bloom, P.: Against empathy: The case for rational compassion. Ecco (2016)

Baron-Cohen, S.: Mindblindness: An essay on autism and theory of mind. MIT Press/Bradford, Cambridge, MA (1995)

Baumard, N., Hyafil, A., Morris, I., Boyer, P.: Increased affluence explains the emergence of ascetic wisdoms and moralizing religions. Current Biology 25(1), 10–15 (2015). https://doi.org/10.1016/j.cub.2014.10.063

Buffone, A.E., Poulin, M.J.: Empathy, target distress, and neurohormone genes interact to predict aggression for others-even without provocation. Personality and Social Psychology Bulletin 40(11), 1406–1422 (2014). https://doi.org/10.1177/0146167214549320

Carlson, S.M., Moses, L.J., Claxton, L.J.: Individual differences in executive functioning and theory of mind: An investigation of inhibitory control and planning ability. J. Experim. Child Psychol. **87**(4), 299–319 (2004). https://doi.org/10.1016/j.jecp.2004.01.002

Daly, M., Wilson, M.: Homicide. Aldine de Gruyter, New York (1988)

Djikic, M., Oatley, K., Moldoveanu, M.C.: Reading other minds: effects of literature on empathy. Scientific Study of Literature **3**(1), 28–47 (2013). https://doi.org/10.1075/ssol.3.1.06dji

Dutton, E., Lynn, R.: A negative Flynn effect in Finland, 1997–2009. Intelligence **41**(6), 817–820 (2014). https://doi.org/10.1016/j.intell.2013.05.008

Eisensee, T., Strömberg, D.: News droughts, news floods, and U. S. disaster relief. The Quarterly Journal of Economics **122**(2), 693–728 (2007). https://doi.org/10.1162/qjec.122.2.693

Evans, J.S.T.B.T.: On the resolution of conflict in dual process theories of reasoning. Thinking & Reasoning **13**, 321–339 (2007)

Evans, J.S.T.B.T.: Thinking twice: Two minds in one brain. Oxford University Press, Oxford, UK (2010)

Flynn, J.R.: Are we getting smarter? Rising IQ in the twenty-first century. Cambridge University Press, New York (2012)

Gottschall, J.: The story paradox: How our love of storytelling builds societies and tears them down. Basic Books (2021)

Haidt, J.: The righteous mind: Why good people are divided by politics and religion. Pantheon, New York (2012)

He, J., Guo, D., Zhai, S., Shen, M., Gao, Z.: Development of social working memory in preschoolers and its relation to theory of mind. Child Development **90**(4), 1319–1332 (2019). https://doi.org/10.1111/cdev.13025

Hofmann, W., Gschwendner, T., Friese, M., Wiers, R.W., Schmitt, M.: Working memory capacity and self-regulatory behavior: Toward an individual differences perspective on behavior determination by automatic versus controlled processes. J. Persona. Soc. Psychol. **95**(4), 962–977 (2008). https://doi.org/10.1037/a0012705

Jones, L.M., Finkelhor, D., Halter, S.: Child maltreatment trends in the 1990s: Why does neglect differ from sexual and physical abuse? Child maltreatment **11**(2), 107–120 (2006). https://doi.org/10.1177/1077559505284375

Kidd, D.C., Castano, E.: Reading literary fiction improves theory of mind. Science **342**(6156), 377–380 (2013). https://doi.org/10.1126/science.1239918

Kogut, T., Ritov, I.: The "identified victim" effect: An identified group, or just a single individual? J. Behavi. Deci. Mak. **18**, 157–167 (2005)

Leslie, A.M.: Pretense, autism, and the theory-of-mind module. Curr. Direct. Psychol. Sci. **1**, 18–21 (1992)

Mar, R.A., Oatley, K.: The function of fiction is the abstraction and simulation of social experience. Perspec. Psycholo. Sci. **3**(3), 173–192 (2008). https://doi.org/10.1111/j.1745-6924.2008.00073.x

Meyer, M.L., Spunt, R.P., Berkman, E.T., Taylor, S.E., Lieberman, M.D.: Evidence for social working memory from a parametric functional MRI study. Proce. Nation. Acad. Sci. **109**(6), 1883–1888 (2012). https://doi.org/10.1073/pnas.112107710

Pinker, S.: The better angels of our nature: Why violence has declined. Viking, New York (2011)

Pennycook, G., Bago, B., McPhetres, J.: Science beliefs, political ideology, and cognitive sophistication. J. Experim. Psychol. Gene. **152**(1), 80–97 (2023). https://doi.org/10.1037/xge0001267

Pennycook, G., Rand, D.G.: The psychology of fake news. The Trends in Cognitive Science **25**(5), 388–402 (2021). https://doi.org/10.1016/j.tics.2021.02.007

Stanovich, K.E.: Who is rational? Studies of individual differences in reasoning. Erlbaum (1999)

Stanovich, K.E.: Distinguishing the reflective, algorithmic, and autonomous minds: Is it time for a tri-process theory? In: Evans, J.St.B.T., Frankish, K. (eds.) In two minds: Dual processes and beyond, pp. 55–88. Oxford: Oxford University Press (2009)

Stanovich, K.E., West, R.F.: Individual differences in rational thought. J. Experim. Psychol. Gene. 127(2), 161–188 (1998). https://doi.org/10.1037/0096-3445.127.2.161

Strack, F., Deutsch, R.: Reflective and impulsive determinants of social behavior. Person. Soc. Psychol. Rev. 8(3), 220–247 (2004). https://doi.org/10.1207/s15327957pspr0803_1

Swami, V., Voracek, M., Stieger, S., Tran, U.S., Furnham, A.: Analytic thinking reduces belief in conspiracy theories. Cognition 133(3), 572–585 (2014). https://doi.org/10.1016/j.cognition.2014.08.006

Verkuyten, M., Schlette, A., Adelman, L., Yogeeswaran, K.: Deliberative thinking increases tolerance of minority group practices: Testing a dual-process model of tolerance. J. Experim. Psychol. Appl. Adva. online Pub. (2023). https://doi.org/10.1037/xap0000429M.

Verkuyten, M., Yogeeswaran, K., Adelman, L.: Does deliberative thinking increase tolerance? Political tolerance toward individuals with dual citizenship. Social Cognition 40(4), 396–409 (2022). https://doi.org/10.1521/soco.2022.40.4.396

Wrangham, R.W.: Hypotheses for the evolution of reduced reactive aggression in the context of human self-domestication. Frontiers in Psychology 10(1914), 1–11 (2019). https://doi.org/10.3389/fpsyg.2019.01914

Yama, H.: Morality and contemporary civilization: A dual process approach. In: Yama, H., Salvano-Pardieu, V. (eds.) Adapting human thinking and moral reasoning in contemporary society, pp. 92–114. IGI Global (2020)

# Neuropsychology and Interaction

# Mindfulness Is in the Eye of the Machine

Léa Lachaud[1,2](✉) [iD], Geoffrey Tissier[1], and Ugo Ballenghein[2] [iD]

[1] UPL, Univ Paris 8, CHArt, 93526 Saint-Denis, France
`geoffrey.tissier04@univ-paris8.fr`
[2] Univ Paris Est Créteil, CHArt, 94380 Bonneuil, France
`{lea.lachaud,ugo.ballenghein}@u-pec.fr`

**Abstract.** Mindfulness can be defined relative to mind-wandering, while the former is associated with focused attention and enhanced monitoring of conscious experience, the latter corresponds to a process of distraction. The aim of this human-machine interaction study was to investigate the ocular correlates of the state of mindfulness by comparing it with mind-wandering and resting-state. To this end, experienced meditators and nonmeditators performed a point-fixation task while carrying out three different actions: meditating, reflecting on a philosophical question, and resting. Meditating induced mindfulness, reflecting induced mind-wandering, and resting-induced a resting-state. Eye movement recordings revealed a decrease in microsaccade amplitude and velocity during the meditation task, compared with the other two tasks. Participants also blinked more during the reflecting task than during the other two tasks, especially those in the experienced meditator group. These results suggest that microsaccades are indicators of sustained attention, and blinking of distraction, meaning that it may be possible to detect mind-wandering episodes versus states of mindfulness. Detection of this episodes will be used to develop a biofeedback device to learn mindfulness meditation.

**Keywords:** mindfulness · mind-wandering · eye movement

## 1 Introduction

The aim of this preliminary study is to investigate the ocular correlates of mindfulness and mindwandering, with a view to designing a gaze-controlled human-computer interaction interface. The algorithmic identification of the subjective states of mindfulness and mindwandering could, in the future, enable the design of a biofeedback device to learn meditation. Mindfulness differs from mind-wandering in attentional processes. Whereas mindfulness is associated with focused attention and increased monitoring of conscious experience [40], mindwandering is linked to a process of distraction that occurs when attention strays from the main task in hand [52]. Eye movements are thought to be physiological markers of the different attentional processes associated with cognitive states

[27,36,48], but few studies have focused on the ocular correlates of mindfulness. Given that the practice of mindfulness involves focused attention exercises [40], which in turn increase attentional flexibility [30], we can assume that the physiological markers of the mindfulness state are similar to those of focused attention and attentional flexibility. The attentional control involved in mindfulness has already been found to be negatively correlated with catch-up and anticipatory saccades during a smooth pursuit task [34]. This control is characterized by focused attention [40], where attentional focus is directed toward a specific activity over a prolonged period [16]. During a visual attention task, the blink rate decreases when attention is engaged, and increases when attention is disengaged [50]. It appears to be higher under stress, and lower during tasks requiring sustained attention [21,50,53].

Mindfulness exercises based on focused attention foster concentration skills, reducing the impact of distractors [3,13,56]. One study found that the magnitude of microsaccades was negatively correlated with the level of concentration [7], meaning that it might be an indicator of focused attention. As specified earlier, the practice of mindfulness leads to attentional flexibility, by mobilizing the capacity for cognitive decoupling [18]. This is achieved through set-shifting exercises [44] and by increasing the focus of attention on constantly changing information [45]. Two studies showed that a high blink rate is associated with greater attentional flexibility in a set-shifting task [14,46,54]. However, they also showed that attentional flexibility is associated with an increase in distractibility. Distractibility may correspond to the mind-wandering state [41][1]. The practice of mindfulness is thought to raise dopamine levels [32], and high dopamine levels have been found to increase blinking during a resting task [14,26,46,54]. Mindfulness may therefore be correlated with a high blink rate. In addition, individuals who have a high blink rate when they are in a state of relaxation appear to be better able than individuals with a low blink rate to engage in attentional switching [14,26,46,54]. However, another study unexpectedly found that experienced meditators had a lower blink rate than naive meditators [33]. Experienced meditators also produced fewer eye movements during a breath-focused attention meditation task than during an induced mind-wandering task [44]. Furthermore, the more experienced the meditators, the fewer eye movements they produced in both mindfulness and mind-wandering tasks [44].

## 1.1 Induction and Comparison of Different Mental States

The tasks in which individuals engage generate specific mental states [24]. These correspond to their internal subjective experiences of thoughts, emotions, perceptions, intentionality, information processing, attention, memory, reasoning, and decision-making [24,42,55]. Yarbus (1967) [57] was one of the first to use

---

[1] A state of distraction can occur during mindfulness practice. It is characterized by self-generated mental activity that arises spontaneously, and is associated with the individual's concerns and hopes, rather than with immediate perceptions of the environment [8,9].

saccades and fixations to discriminate between different mental states during a visual task. This author found that gaze path was influenced by the instructions given to participants. Subsequent studies have supported this finding, showing that eye movements are systematically influenced by the task in which the individual is engaged, the emotional valence of the stimuli and, by extension, the individual's mental state [2,24,55].

**Induction of the State of Mindfulness.** Breath counting involves concentrating on one's breathing and the sensations that accompany it [29,39]. Studies have shown that this type of exercise can constitute a behavioral measure of mindfulness [17,25,39], in addition to the self-report measures with which it is correlated. It is relevant for both experienced and novice meditators, and can be used to distinguish between the two [39]. The ability to count one's breaths has been associated with greater metaconsciousness, reduced mind-wandering, and less distraction [39]. This measure operationalizes mindfulness, insofar as it depends on direct perception of the present moment experience and makes it possible to become aware of the occurrence of mind-wandering, thus facilitating a return of attention to the initial activity.

**Induction of the Mind-Wandering State.** Mind-wandering occurs when attention initially focused on a task in progress is diverted from that task. This process of distraction leads to errors in the execution of the initial task, as well as to more superficial representations of the external environment [52]. These characteristics make mind-wandering difficult to study, as individuals need to be aware of their own internal experiences in order to report them. The self-caught mind-wandering method involves asking participants to spontaneously indicate each time they realize that their attention is straying [10,52], while the probe-caught mind-wandering method involves stopping them during a task to ask them where their attention is currently directed [10,52]. By contrast, the instructed mind-wandering method [44] relies on the artificial induction of a mind-wandering episode. Although the resulting mind-wandering state may be slightly different, given that it does not occur spontaneously, this method improves experimental control by making it easier to objectify. Moreover, mind-wandering studied in the laboratory is often intentionally triggered by the participants themselves [51].

**Induction of the Resting-State.** Resting-state refers to the basic state of the brain when it is not actively oriented toward a goal. It is characterized by spontaneous, self-organized neuronal activity that occurs in several brain regions in the absence of sensory stimulation or a specific cognitive task. It has long been regarded as the brain's default mode, serving as a baseline (or reference) for studying different brain states [6]. The resting-state is characterized by intrinsic activity representing spontaneous cognition free from any voluntarily induced constraints [47]. The method used to induce this resting-state consists in staring at a fixation point or into space while doing nothing in particular [20]. The literature has shown that the brain activates the same areas during a resting-state with eyes closed or fixated on a point, or during passive viewing of stimuli [20].

## 1.2   Aim of the Study

The aim of the present study was to explore the ocular correlates of the mindfulness state. More specifically, we sought to discriminate this state from the states of induced mind-wandering and resting-state in experienced meditators versus nonmeditators by analyzing eye movements. The detection of mind-wandering episodes will be used to develop a biofeedback device to learn mindfulness meditation. We expected to uncover specific features of gaze behavior during a point-fixation (i.e., meditation) task intended to induce the mindfulness state, compared with two other tasks designed to induce mind-wandering and the resting-state. Given that all three mental states can be induced during a point-fixation task, we chose to exclusively study microsaccades, which indicate shifts in peripheral attention [15], and blinking.

The magnitude of microsaccades is negatively correlated with the level of concentration [7], and experienced meditators produce fewer eye movements during a focused attention meditation task [44]. Consequently, we expected to observe a lower magnitude of microsaccades during the meditation task versus the other two tasks, and among experienced meditators versus nonmeditators. As far as blinking is concerned, the results in the literature diverge. Several studies have shown that attentional flexibility is associated with an increase in blink rate during a relaxation task [14,26,46,54], but another found that experienced meditators had a lower blink rate than naive meditators [33]. Consequently, we expected the number of blinks to differ between the meditation task and the other two tasks, without knowing whether it would be higher or lower. We also expected to observe a difference between experienced meditators and nonmeditators.

The main function of the resting-state is to facilitate flexible, self-relevant mental explorations that serve to anticipate and assess future events [6]. In this sense, it can resemble the mind-wandering state [8,12,43]. We did not therefore necessarily expect to observe a difference between the mind-wandering state and the resting-state. By contrast, as reduced activation of brain areas associated with the resting-state has been observed during mind-wandering episodes among individuals who regularly practice mindfulness training [4,12], we expected to observe differences between experienced meditators and nonmeditators during the mind-wandering state.

Finally, to control for mindfulness expertise in both groups (meditators and nonmeditators), we administered the Five-Facet Mindfulness Questionnaire (FFMQ) [1]. This 39-item self-report questionnaire measures the trait mindfulness [1,23] that develops with mindfulness practice [3]. As it can be used to control exposure to mindfulness [39], we used it as a group validation tool. We predicted that meditators would achieve a higher FFMQ score, but did not expect this score to be correlated with eye movements, given the results of a similar study [34].

## 2    Method

### 2.1    Participants

A power analysis using G*Power 3 software [19] indicated that a total sample of 28 people would be needed to detect medium effects ($d = .5$) with power $(1 - \beta)$ set at 0.80 and $\alpha = .05$ using an F test (Repeated measures ANOVA, within-between interaction) for the effect of mindfulness, mind-wandering and resting-state.

Sixty-five adults took part in the study on a voluntary basis. The data of six were excluded, owing to insufficient eye movement data. The final sample therefore comprised 59 participants. All participants ($M_{age} = 41.88 years, SD = 12.58, range = 22 - 66$), were native French speakers, and identified themselves as belonging to one of two binary genders (32 women, 27 men). All had obtained a high-school diploma or higher, and had normal or corrected-to-normal eyesight. None had any cognitive pathology. All participants signed an informed consent form stating that they could stop the experiment at any time, and were debriefed at the end of the experiment.

Participants were divided into two groups according to their self-reported experience of mindfulness meditation: experienced meditators versus nonmeditators. The self-reported practice of participants in the experienced meditators group ($n = 26$) had to include meditation exercises focused on attending to the present moment, breathing, body sensations, thoughts, and the environment. The different types of accepted meditation practice had to include instructor-supervised mindfulness exercises (mindfulness-based stress reduction, mindfulness, Buddhism, vipassanā, Zen). Experienced meditators reported practicing mindfulness at least once a week for at least 2.5 years within a club or organization. On average, they had been practicing mindfulness for 8.46 years ($SD = 8.12$), and the mean duration of a session was 29.35 min ($SD = 15.44$). Participants in the control group of nonmeditators ($n = 33$) were not expected to report any experience of mindfulness-related practices, including meditation, yoga, tai chi, qigong or martial arts, but were expected to be potentially interested in this type of activity. The group formation methodology was similar to that used in several other studies [5, 11, 31, 34, 35].

### 2.2    Material

**Physiological Sensors.** We used an SMI RED500, a 500-Hz remote binocular eye tracker. The infrared camera was embedded in a Dell PP2210 22" LCD monitor (width: 47.6 cm; height: 29.5 cm; diagonal: 56 cm). The eye tracker was accurate to 0.4°, and its spatial resolution was 0.03°. The amplitude of head movement tolerated by the eye tracker was 40 x 20 cm, and the maximum tolerated velocity of head movement was 50 cm/s. There were five calibration points.

**Screen and Visual Support.** The monitor was covered by a rectangle of light-gray cardboard (width: 54.4 cm; height: 33.3 cm; diagonal: 63.5 cm) marked with a dark-gray fixation dot. This dot was positioned at the zenith of their gaze (i.e., x = 27.2 cm; y = 9.7 cm). A digital marker was also located at the same point on the screen, in order to define the correct zone of interest for the eye tracker. The purpose of the opaque cardboard cover was to allow participants to fixate a point on a nonluminescent support, in order to reduce variations and eye fatigue linked to screen radiation. A second screen was positioned to the left of the participant (width: 49.6 cm; height: 31 cm; diagonal: 58.7 cm). This displayed a series of questions that participants had to answer during the experiment. Participants responded using a mouse and keyboard.

**Software.** SMI iViewX$^{TM}$ (version 2.8) software and SMI Experiment Center$^{TM}$ (version 3.6) software were coupled for data acquisition and stimulus presentation. SMI BeGaze$^{TM}$ (version 3.6) software was used to save, pretest, and export the eye-tracker data. All statistical analyses were performed using Excel and JASP (version 0.17). Online questionnaires were completed using Google Form.

**Questionnaires.** Participants completed two online questionnaires on their own computers before coming to the laboratory. The first questionnaire collected demographic information. The second questionnaire was the French version [1] of the FFMQ scale for measuring trait mindfulness [23]. This questionnaire measured 5 facets of mindfulness on a 5-point Likert scale (1 = Never or very rarely true , 5 = Very often or always true): Observing (e.g. "When I'm walking, I deliberately notice the sensations of my body moving"), Describing (e.g. "I'm good at finding words to describe my feelings."), Acting with Awareness (e.g. "When I do things, my mind wanders off and I'm easily distracted"), Nonjudging of inner experience (e.g. "I believe some of my thoughts are abnormal or bad and I shouldn't think that way.") and Nonreactivity (e.g. "In difficult situations, I can pause without immediately reacting."). The sum of all items (39) corresponds to the mindfulness score.

**Instructions for Inducing the Three States.** Participants were given instructions on how to induce three different mental states while fixating the dot: mindfulness, mind-wandering, and resting-state. The mindfulness state was induced by a breath counting task [17,25,29,39]. The instructions were: "Concentrate on your breathing. You are going to count each breathing cycle. One inhalation + one exhalation = 1 cycle." The mind-wandering state was induced by three philosophical questions adapted to an audience with at least a high-school diploma[2]. These were philosophy essay questions including concepts covered in the philosophy syllabus for the last year of high school [46]. The questions were as follows: "Does culture make us more human?", "What's the point

---

[2] This method was preferred to that of recalling autobiographical episodes, as it reduces the risk of strong emotional activation in participants. It also homogenizes the mind-wandering that is elicited, as all undergo the same induction process.

of explaining a work of art?", and "Is freedom the absence of constraint?" The resting-state was induced by the following instruction: "Please fixate the dot while doing nothing in particular."

## 2.3   Procedure

Before starting the experiment, participants were asked to respond to a demographic questionnaire (gender, age, profession, level of education, presence of strabismus, level of meditation practice) and the FFMQ. On their arrival in the laboratory, they signed a free and informed consent form. Participants were then seated approximately 60 cm from the eye tracker [28,37,38] embedded in the monitor. After calibrating the eye tracker, the cardboard rectangle with the fixation dot was placed over the screen. Participants were instructed to fixate the dot. These instructions were displayed on the second screen, and provided orally by the experimenter: "In all the trials, you must stare attentively at the dot in front of you while performing the requested mental activity." Participants performed three 3-minute trials for each task: breath counting (mindfulness state), philosophical reflection (mind-wandering state), and resting task (resting-state). The order of the trials was randomized. Both groups (experienced meditators vs. nonmeditators) performed the same protocol.

## 2.4   Data Selection and Analysis

**Success for Each Trial.** For each participant, we retained one of the three trials for each task (i.e., one meditation trial, one mind-wandering trial, and one resting-state trial). This selection was based on two criteria: the quality of the eye-tracker recording, and participants' responses to questions asked after each trial. More specifically, we selected the trials in the following order of priority: 1) best recording quality, 2) greatest ease of performance, and 3) greatest interest. To this end, participants were asked about the ease with which they had performed the task, and their interest in the task. Regarding the mind-wandering and resting-state trials, we asked participants to describe in a few sentences what they had been thinking about during the task, to check that they had actually performed the required tasks.

**Eye Behavior Indicators.** Two eye behavior indicators were selected for this study: microsaccades (number, duration, amplitude, velocity, and peak velocity) and blinks (number and frequency). As recommended in the literature, we defined microsaccades as saccades with a maximum duration of 239 ms [22]. We averaged the velocity values of all the eye movements identified as microsaccades. Peak velocity, also known as maximum velocity, corresponds to the highest velocity reached by a microsaccade. We averaged the peak velocity values of all the microsaccades [49].

**Statistical Analyses.** Linear regressions were performed to investigate whether demographic measures and trait mindfulness (FFMQ score) had an influence on eye movement behavior during the three tasks. Repeated-measures analyses of variance (ANOVAs) were used to test the effects of different mental states on eye behavior in the two groups (experienced meditators vs. nonmeditators). Contrast analyses were then performed to refine our understanding of these results. These tests compared eye behavior during the three tasks (meditation, mind-wandering, resting-state), and according to group (experienced meditators vs. nonmeditators) (mixed design)[3].

## 3　Results

### 3.1　Sociodemographic Analysis

Descriptive analyses were performed on the participants' sociodemographic data. When we ran multiple linear regressions, we found no direct influence of trait mindfulness (FFMQ score) on gaze behavior in the three mental states (mindfulness, mind-wandering, and resting-state). However, experienced meditators achieved a higher FFMQ score ($M = 152, SD = 12.5$) than nonmeditators ($M = 135, SD = 18.8$), $t(57) = 3.98, p < .001, d = 1.04$. This result suggests that the grouping was valid, as experienced meditators had higher trait mindfulness than nonmeditators.

### 3.2　Effects of Different Mental States on Eye Movements

We ran seven ANOVAs on different eye movement indicators, including blinks and microsaccades. More specifically, we analyzed whether the following variables could be indicators of mental state (meditation vs. mind-wandering vs. resting-state) and group (experienced meditators vs. nonmeditators): (i) numbers of microsaccades, (ii) duration of microsaccades, (iii) amplitude of microsaccades, (iv) numbers of blinks, (v) frequency of blinks, (vi) velocity of microsaccades and (vii) peak velocity of microsaccades. The three independent variables were the three mental states (mindfulness, mind-wandering, and resting-state). The gaze behavior of the two experimental groups (experienced meditators and nonmeditators) was compared according to these three states.

**Analysis of Microsaccades: Duration, Amplitude, and Velocity.** Neither mental state nor group had an effect on the duration and number of microsaccades ($F = .09$ and $F = .56$). By contrast, participants' microsaccades were of a lower mean amplitude during the meditation task ($M = 2.9, SD = 2.4$) than during the mind-wandering ($M = 3.9, SD = 2.9$) and resting-state ($M = 3.7, SD = 3.3$), $F(2, 114) = 7.3, p < .001, d = .3$, tasks. Contrast analyses

---

[3] The data are available on the online repository at https://osf.io/w3qhf/?view_only=f31bdb4e351040d19c5047c8481dc727.

revealed significant differences between mindfulness and both mind-wandering and resting-state, $t(114) = 3.7, p < .001, 95\%$ $CI[0.4, 1.3])$, between mindfulness and resting-state, $t(114) = 2.6, p = .009, 95\%$ $CI[0.18, 1.26])$, and between mindfulness and mind-wandering, $t(114) = 3.7, p < .001, 95\%$ $CI[0.47, 1.56])$ (Fig. 1).

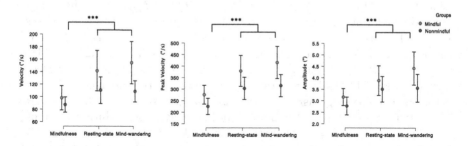

**Fig. 1. Velocity, peak velocity and amplitude of microsaccades according to group and task.** Vertical bars denote 95% CI., *** $p < .001$.

Mean velocity of microsaccades was also lower during the meditation task ($M = 92, SD = 59.3$) than during the mind-wandering ($M = 129, SD = 85$) and resting-state ($M = 124, SD = 114$) tasks, $F(2, 114) = 7.1, p < .001, d = .39$. Contrast analyses revealed significant differences between mindfulness and both mind-wandering and resting-state, $t(114) = 3.7, p < .001, 95\%$ $CI[16.9, 55.2]$, between mindfulness and resting-state, $t(114) = 3.5, p = .004, 95\%$ $CI[11.1, 55.3]$, and between mindfulness and mind-wandering, $t(114) = 3.5, p < .001, 95\%$ $CI[17, 61]$. Mean peak velocity of microsaccades was also lower during the meditation task ($M = 247, SD = 176$) than during the mind-wandering (M = 360, SD = 230) and resting-state ($M = 336, SD = 274$) tasks, $F(2, 114) = 11.5, p < .001, d = .43$. Contrast analyses revealed significant differences between mindfulness and both mind-wandering and resting-state, $t(114) = 4.7, p < .001, 95\%$ $CI[59.7, 146.6]$, between mindfulness and resting-state, $t(114) = 3.5, p < .001, 95\%$ $CI[40, 141]$, and between mindfulness and mind-wandering, $t(114) = 4.5, p < .001, 95\%$ $CI[65, 165]$.

**Analysis of Blinks: Number and Frequency.** Two ANOVAs showed that the number and frequency of blinks were influenced by both mental state and group. The total number of blinks was higher in experienced meditators than in nonmeditators, $F(2, 57) = 4.24, p = .044, d = .48$. During the mindfulness task, experienced meditators averaged 49.9 ($SD = 45.4$) blinks, while nonmeditators averaged 32.6 ($SD = 28.1$). During the resting-state task, experienced meditators averaged 53.9 ($SD = 45.0$) blinks, while nonmeditators averaged 37.5 ($SD = 31.0$). Finally, during the mind-wandering task, experimented meditators averaged 70.9 ($SD = 57.1$) blinks, while nonmeditators averaged 44.9 ($SD = 37.9$). A significant difference was also noted between the three mental states, $F(2, 114) = 9.22, p < .001, d = .34$, with means of 56.4 ($SD = 48.6$)

for the mind-wandering task, 44.7 $(SD = 38.4)$ for the resting-state task, and 40.2 $(SD = 37.4)$ for the meditation task. More specifically, four contrast analyses revealed that the number of blinks was higher during the mind-wandering task than during the resting-state task, $t(114) = 4.1, p < .001, 95\%$ $CI[24.651, 8.713]$, higher during the mind-wandering task than during the meditation task, $t(114) = 3.03, p = .003, 95\%$ $CI[20.177, 4.239]$, higher during the mind-wandering task than during the meditation and resting-state tasks, $t(114) = 4.15, p < .001, 95\%$ $CI[21.346, 7.544]$, and higher during the mind-wandering and resting-state tasks than during the meditation task, $t(114) = 3.03, p = .003, 95\%$ $CI[3.677, 17.479]$. Another contrast analysis revealed a higher number of blinks for experienced meditators versus nonmeditators during the mind-wandering task, $t(84) = 12.41, p = .018, 95\%$ $CI[4.616, 42.275]$. For experienced meditators, the number of blinks was higher during the mind-wandering task than during the resting-state task, $t(114) = 2.81, p = .006, 95\%$ $CI[28.9, 5.04]$, the meditation task, $t(114) = 3.49, p < .001, 95\%$ $CI[32.9, 9.1]$, and both the meditation and resting-state tasks, $t(114) = 3.64, p < .001, 95\%$ $CI[29.3, 8.65]$. For nonmeditators, the number of blinks was also higher during the mind-wandering task than during the resting-state task, $t(114) = 2.31, p = .022, 95\%$ $CI[1.78, 22.94]$, and both the resting-state and meditation tasks, $t(114) = 2.14, p = .034, 95\%$ $CI[19.07, 0.74]$ (Fig. 2).

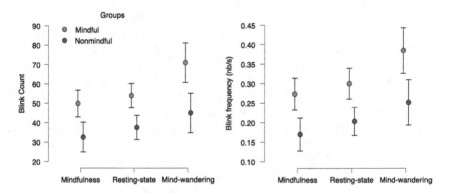

**Fig. 2. Blink count and blink frequency according to group and task.** Vertical bars denote 95% CI.

Total blink frequency was higher for experienced meditators than for nonmeditators, $F(1, 57) = 4.17, p = .046, d = .48$. During the meditation task, mean blink frequency was 0.27 $(SD = 0.25)$ for experienced meditators, and 0.17 $(SD = 0.16)$ for nonmeditators. During the resting-state task, mean blink frequency was 0.30 $(SD = 0.26)$ for experienced meditators, and 0.20 $(SD = 1.17)$ for nonmeditators. Finally, during the mind-wandering task, mean blink frequency was 0.38 for experienced meditators $(SD = 0.32)$, and 0.25 $(SD = 0.22)$ for nonmeditators. A significant difference was also noted between the three mental states, $F(2, 114) = 9.12, p < .001, d = .35)$, with means of 0.31 $(SD = 0.27)$ for the mind-wandering task, 0.25 $(SD = 0.22)$ for the resting-state task, and 0.21 $(SD = 0.21)$ for the meditation task. More specifically, four contrast analyses showed that blink frequency was higher during the mind-wandering task than during the resting-state task, $t(114) = 2.87, p = .005, 95\%$ $CI[0.112, 0.021]$, higher during the mind-wandering task than during the meditation task, $t(114) = 4.17, p < .001, 95\%$ $CI[0.143, .051]$, higher during the mind-wandering task than during the meditation and resting-state tasks, $t(114) = 4.07, p < .001, 95\%$ $CI[0.121, 0.04]$, and higher during the mind-wandering and resting-state tasks than during the meditation task, $t(114) = 3.161, p = .002, 95\%$ $CI[0.024, 0103]$. Contrast analyses also revealed a higher blink frequency in experienced meditators versus nonmeditators during the mind-wandering task, $t(85) = 2.19, p = .031, 95\%$ $CI[0.013, 0.25]$. In the experienced meditators group, blink frequency was higher during the mind-wandering task than during both the resting-state and meditation tasks, $t(114) = 3.26, p = .001, 95\%$ $CI[0.039, 0.030]$. It was also higher during the mind-wandering task than during the meditation task, $t(114) = 3.22, p = .002, 95\%$ $CI[0.043, 0.180]$. Finally, in the nonmeditator group, blink frequency was higher during the mind-wandering task than during both the resting-state and meditation tasks, $t(114) = 2.45, p = .016, 95\%$ $CI[0.118, 0.012]$.

## 4    Discussion

The aim of the present study was to compare the ocular correlates of mindfulness, mind-wandering, and the resting-state in experienced meditators versus nonmeditators. In other words, we sought to discriminate between these three states in each group on the basis of eye movements. All participants performed lower amplitude microsaccades during the meditation task than during the other two tasks (mind-wandering and resting-state). Microsaccade velocity and peak velocity were also lower during the meditation task than during the other two tasks. By contrast, there were no significant difference between either the mind-wandering and resting-state tasks, or the meditating and nonmeditating groups. This suggests that regardless of mindfulness expertise, meditation activity and, by extension, the mindfulness state, decreases the magnitude of microsaccades. This result is in line with the literature, which has shown that the higher the level of concentration, the lower the amplitude of microsaccades [7]. Focused attention exercises carried out during mindfulness activities [40] increase the level of attention paid to a task and, by extension, the level of concentration [3,13,56]. Our results showed that participants engaged in mindfulness did indeed concentrate more on the task in hand, notably by increasing their visual attentional focus onto the fixation point. Similarly, the higher amplitude and velocity of microsaccades during the mind-wandering and resting-state tasks may have reflected greater distraction [41,45]. A previous study reported lower saccadic amplitude during breath-focused attention meditation than during a mind-wandering induction task [44]. Furthermore, the more experienced the meditators, the fewer the eye movements they made during both the meditation and mind-wandering tasks. Our study yielded similar findings on microsaccade amplitude and velocity across all participants. However, we did not observe any difference between experienced meditators and nonmeditators. This may be because the meditation performed in our study lasted only half as long as that performed in Matiz et al. [44]. Our observations were therefore partly in line with this earlier study.

In addition, we observed lower microsaccadic amplitude and velocity during the meditation task than during the resting-state task. This can be explained by the fact that the resting-state can generate an unconscious and involuntary mind-wandering state, insofar as it constitutes the brain's default mode [20]. One study has suggested that mind-wandering is associated with the default network of brain regions activated during the resting-state [43]. Thus, when they were instructed to do nothing in particular, participants probably sought to distract their thoughts, which explains why there was the same difference between the meditation state and the mind-wandering state as between the meditation state and the resting-state.

The other two indicators we considered were the number and frequency of blinks. Results showed that these indicators were influenced by both task and group. Specifically, (i) The number and frequency of blinks in the mind-wandering state were higher for experienced meditators than for nonmeditators, (ii) for both meditators and nonmeditators, the number and frequency of blinks in the mind-wandering state were higher than those in the two other tasks (mindfulness and resting state). These results are partly in line with those of a previous study [33] that found that, compared with mind-wandering, mindfulness was associated with a reduction in blinking. The lower frequency of blinking during the meditation task than during the mind-wandering task was particularly marked in experienced meditators. This observation can be interpreted in the light of another study [50] that reported a decrease in blinking when attention is engaged. This result therefore suggests that the mindfulness state engages greater attention than during the mind-wandering state, especially in experienced meditators [3,13,56].

In addition, a positive correlation has previously been observed between blinking and attentional flexibility, itself associated with increased distractibility [54]. The practice of mindfulness increases attentional flexibility [30], and mind-wandering refers to a state of distraction [52]. Thus, the increase in blink frequency we observed during the mind-wandering task may have been due partly to distractibility in nonmeditators, and partly to the combination of distractibility and attentional flexibility in experienced meditators. That said, a decrease in blinking has previously only been reported in long-term meditators versus nonmeditators in a resting-state condition [33]. This difference was not observed in our study. One possible explanation is that the experienced meditators in our study had slightly less experience (8.46 years vs. 10.1 years) than those in Kruis et al. (2016) [33]. It is thus possible that the stable changes in basic striatal dopaminergic functioning responsible for the reduction in blinking [14,33] can only be achieved through intensive, long-term mindfulness practice [33].

# 5   Conclusion

The aim of the present study was to compare gaze indicators of mindfulness, mind-wandering and resting-state in experienced mindfulness mediators versus nonmeditators. Microsaccades and eye blinks were examined to differentiate between these states. Results showed that all participants performed microsaccades of lesser amplitude during meditation than during the mind-wandering and resting-state tasks. Moreover, the velocity and peak velocity of microsaccades were also lower during meditation. By contrast, there was no difference between the mind-wandering and resting-state tasks. These results suggest that meditation, independent of mindfulness expertise, reduces the amplitude and velocity of microsaccades, which can therefore be regarded as an indicator of sustained attention. Results also showed that for both groups (experienced meditators and nonmeditators), the number and frequency of blinks were higher during the mind-wandering task than during the meditation and resting-state tasks. However, experienced meditators had a higher number and frequency of blinks than nonmeditators during mind-wandering. These results suggest that the mind-wandering state leads to an increase in blinking, linked to distractibility and attentional flexibility, and that mindfulness practice can modulate these effects. They also suggest that it is possible to detect mind-wandering episodes, relative to a mindfulness state. The use of these indicators to differentiate between states associated with this practice could contribute to the development of biofeedback devices aimed at facilitating mindfulness learning. A further step would be to combine these indicators with other measures, such as variations in pupil size and heart rate.

**Acknowledgements.** We are grateful to Pr. J. Baratgin for his advice, F. Gonzalez for her contribution to this study, and E. Portier for proofreading.

# References

1. Baer, R.A., et al.: Construct validity of the five facet mindfulness questionnaire in meditating and nonmeditating samples. Assessment **15**(3), 329–342 (2008). https://doi.org/10.1177/1073191107313003
2. Ballenghein, U., Megalakaki, O., Baccino, T.: Cognitive engagement in emotional text reading: concurrent recordings of eye movements and head motion. Cogn. Emot. **33**(7), 1448–1460 (2019). https://doi.org/10.1080/02699931.2019.1574718
3. Bishop, S.R., et al.: Mindfulness: a proposed operational definition. Clin. Psychol. Sci. Pract. **11**(3), 230 (2004). https://doi.org/10.1093/clipsy.bph077
4. Brewer, J.A., Worhunsky, P.D., Gray, J.R., Tang, Y.Y., Weber, J., Kober, H.: Meditation experience is associated with differences in default mode network activity and connectivity. Proc. Natl. Acad. Sci. **108**(50), 20254–20259 (2011). https://doi.org/10.1073/pnas.1112029108
5. Brown, D., Forte, M., Dysart, M.: Differences in visual sensitivity among mindfulness meditators and non-meditators. Percept. Mot. Skills **58**(3), 727–733 (1984). https://doi.org/10.2466/pms.1984.58.3.727

6. Buckner, R.L., Andrews-Hanna, J.R., Schacter, D.L.: The brain's default network: anatomy, function, and relevance to disease. Ann. N. Y. Acad. Sci. **1124**, 1–38 (2008). https://doi.org/10.1196/annals.1440.011

7. Buettner, R., Baumgartl, H., Sauter, D.: Microsaccades as a predictor of a user's level of concentration. In: Davis, F., Riedl, R., vom Brocke, J., Léger, P.M., Randolph, A. (eds.) Information Systems and Neuroscience. Lecture Notes in Information Systems and Organisation, vol. 29, pp. 173–177. Springer, Cham (2019). https://doi.org/10.1007/978-3-030-01087-4_21

8. Callard, F., Smallwood, J., Golchert, J., Margulies, D.: The era of the wandering mind? twenty-first century research on self-generated mental activity. Front. Psychol. **4**, 891 (2013). https://doi.org/10.3389/fpsyg.2013.00891

9. Christoff, K., Irving, Z.C., Fox, K.C., Spreng, R.N., Andrews-Hanna, J.R.: Mind-wandering as spontaneous thought: a dynamic framework. Nat. Rev. Neurosci. **17**(11), 718–731 (2016). https://doi.org/10.1038/nrn.2016.113

10. Cunningham, S., Scerbo, M.W., Freeman, F.G.: The electrocortical correlates of daydreaming during vigilance tasks. J. Ment. Imag. **24**, 61–72 (2012)

11. Daubenmier, J., Sze, J., Kerr, C.E., Kemeny, M.E., Mehling, W.: Follow your breath: respiratory interoceptive accuracy in experienced meditators. Psychophysiology **50**(8), 777–789 (2013). https://doi.org/10.1111/psyp.12057

12. Davidson, R., Schuyler, B.: Neuroscience du bonheur. Rev. Québécoise de Psychol. **38**(1), 39–64 (2017). https://doi.org/10.7202/1040069ar

13. Deikman, A.J.: Experimental meditation. J. Nerv. Ment. Dis. (1963). https://doi.org/10.1097/00005053-196304000-00002

14. Dreisbach, G., et al.: Dopamine and cognitive control: the influence of spontaneous eyeblink rate and dopamine gene polymorphisms on perseveration and distractibility. Behav. Neurosci. **119**(2), 483 (2005). https://doi.org/10.1037/0735-7044.119.2.483

15. Engbert, R., Kliegl, R.: Microsaccades uncover the orientation of covert attention. Vision. Res. **43**(9), 1035–1045 (2003). https://doi.org/10.1016/s0042-6989(03)00084-1

16. Esterman, M., Rothlein, D.: Models of sustained attention. Curr. Opin. Psychol. **29**, 174–180 (2019). https://doi.org/10.1016/j.copsyc.2019.03.005

17. F Wong, K., AA Massar, S., Chee, M.W., Lim, J.: Towards an objective measure of mindfulness: replicating and extending the features of the breath-counting task. Mindfulness **9**(5), 1402–1410 (2018). https://doi.org/10.1007/s12671-017-0880-1

18. Farrar, S.T., Yarrow, K., Tapper, K.: The effect of mindfulness on cognitive reflection and reasoning. Mindfulness **11**(9), 2150–2160 (2020). https://doi.org/10.1007/s12671-020-01429-z

19. Faul, F., Erdfelder, E., Lang, A.G., et al.: G*power 3: a flexible statistical power analysis program for the social, behavioral, and biomedical sciences. Behav. Res. Methods **39**, 175–191 (2007). https://doi.org/10.3758/BF03193146

20. Gusnard, D.A., Raichle, M.E.: Searching for a baseline: functional imaging and the resting human brain. Nat. Rev. Neurosci. **2**(10), 685–694 (2001). https://doi.org/10.1038/35094500

21. Haak, M., Bos, S., Panic, S., Rothkrantz, L.J.: Detecting stress using eye blinks and brain activity from EEG signals. In: Proceeding of The 1st Driver Car Interaction and Interface (DCII 2008), pp. 35–60 (2009)

22. Hauperich, A.K., Young, L.K., Smithson, H.E.: What makes a microsaccade? a review of 70 years of research prompts a new detection method. J. Eye Move. Res. **12**(6), 13 (2020). https://doi.org/10.16910/jemr.12.6.13

23. Heeren, A., Douilliez, C., Peschard, V., Debrauwere, L., Philippot, P.: Cross-cultural validity of the five facets mindfulness questionnaire: adaptation and validation in a French-speaking sample. Eur. Rev. Appl. Psychol. **61**(3), 147–151 (2011). https://doi.org/10.1016/j.erap.2011.02.001

24. Henderson, J.M., Shinkareva, S.V., Wang, J., Luke, S.G., Olejarczyk, J.: Predicting cognitive state from eye movements. PLoS ONE **8**(5), e64937 (2013). https://doi.org/10.1371/journal.pone.0064937

25. Isbel, B., Stefanidis, K., Summers, M.J.: Assessing mindfulness: experimental support for the discriminant validity of breath counting as a measure of mindfulness but not self-report questionnaires. Psychol. Assess. **32**(12), 1184 (2020). https://doi.org/10.1037/pas0000957

26. Jongkees, B.J., Colzato, L.S.: Spontaneous eye blink rate as predictor of dopamine-related cognitive function-a review. Neurosci. Biobehav. Rev. **71**, 58–82 (2016). https://doi.org/10.1016/j.neubiorev.2016.08.020

27. Just, M.A., Carpenter, P.A.: A theory of reading: from eye fixations to comprehension. Psychol. Rev. **87**(4), 329 (1980)

28. Kaakinen, J.K., Ballenghein, U., Tissier, G., Baccino, T.: Fluctuation in cognitive engagement during reading: evidence from concurrent recordings of postural and eye movements. J. Exp. Psychol. Learn. Mem. Cogn. **44**(10), 1671 (2018)

29. Kabat-Zinn, J.: Full Catastrophe Living: Using the Wisdom of Your Body and Mind to Face Stress, Pain, and Illness. Dell Publishing, New York (1990)

30. Kang, Y., Gruber, J., Gray, J.R.: Mindfulness and de-automatization. Emot. Rev. **5**(2), 192–201 (2013). https://doi.org/10.1177/1754073912451629

31. Khalsa, S.S., Rudrauf, D., Damasio, A.R., Davidson, R.J., Lutz, A., Tranel, D.: Interoceptive awareness in experienced meditators. Psychophysiology **45**(4), 671–677 (2008). https://doi.org/10.1111/j.1469-8986.2008.00666.x

32. Kjaer, T.W., Bertelsen, C., Piccini, P., Brooks, D., Alving, J., Lou, H.C.: Increased dopamine tone during meditation-induced change of consciousness. Brain Res. Cogn. Brain Res. **13**(2), 255–259 (2002). https://doi.org/10.1016/s0926-6410(01)00106-9

33. Kruis, A., Slagter, H.A., Bachhuber, D.R., Davidson, R.J., Lutz, A.: Effects of meditation practice on spontaneous eyeblink rate. Psychophysiology **53**(5), 749–758 (2016). https://doi.org/10.1111/psyp.12619

34. Kumari, V., et al.: The mindful eye: smooth pursuit and saccadic eye movements in meditators and non-meditators. Conscious. Cogn. **48**, 66–75 (2017). https://doi.org/10.1016/j.concog.2016.10.008

35. Lachaud, L., Jacquet, B., Baratgin, J.: Reducing choice-blindness? An experimental study comparing experienced meditators to non-meditators. Eur. J. Inv. Health Psychol. Educ. **12**(11), 1607–1620 (2022). https://doi.org/10.3390/ejihpe12110113

36. Lai, M.L., et al.: A review of using eye-tracking technology in exploring learning from 2000 to 2012. Educ. Res. Rev. **10**, 90–115 (2013). https://doi.org/10.1016/j.edurev.2013.10.001

37. Lemercier, A.: Développement de la pupillométrie pour la mesure objective des émotions dans le contexte de la consommation alimentaire. Ph.D. thesis, Paris 8 (2014)

38. Lemercier, A., Guillot, G., Courcoux, P., Garrel, C., Baccino, T., Schlich, P.: Pupillometry of taste: methodological guide-from acquisition to data processing-and toolbox for matlab. Quant. Methods Psychol. **10**(2), 179–195 (2014). https://doi.org/10.20982/tqmp.10.2.p179

39. Levinson, D.B., Stoll, E.L., Kindy, S.D., Merry, H.L., Davidson, R.J.: A mind you can count on: validating breath counting as a behavioral measure of mindfulness. Front. Psychol. **5**, 1202 (2014). https://doi.org/10.3389/fpsyg.2014.01202

40. Lutz, A., Slagter, H.A., Dunne, J.D., Davidson, R.J.: Attention regulation and monitoring in meditation. Trends Cogn. Sci. **12**(4), 163–169 (2008). https://doi.org/10.1016/j.tics.2008.01.005

41. Malinowski, P.: Neural mechanisms of attentional control in mindfulness meditation. Front. Neurosci. **7**, 8 (2013). https://doi.org/10.3389/fnins.2013.00008

42. Marshall, S.P.: Identifying cognitive state from eye metrics. Aviat. Space Environ. Med. **78**(5 Suppl), B165–B175 (2007)

43. Mason, M.F., Norton, M.I., Van Horn, J.D., Wegner, D.M., Grafton, S.T., Macrae, C.N.: Wandering minds: the default network and stimulus-independent thought. Science **315**(5810), 393–395 (2007). https://doi.org/10.1126/science.1131295

44. Matiz, A., Crescentini, C., Fabbro, A., Budai, R., Bergamasco, M., Fabbro, F.: Spontaneous eye movements during focused-attention mindfulness meditation. PLoS ONE **14**(1), e0210862 (2019). https://doi.org/10.1371/journal.pone.0210862

45. Miyake, A., Friedman, N.P., Emerson, M.J., Witzki, A.H., Howerter, A., Wager, T.D.: The unity and diversity of executive functions and their contributions to complex "frontal lobe" tasks: a latent variable analysis. Cogn. Psychol. **41**(1), 49–100 (2000). https://doi.org/10.1006/cogp.1999.0734

46. Müller, J., Dreisbach, G., Brocke, B., Lesch, K.P., Strobel, A., Goschke, T.: Dopamine and cognitive control: the influence of spontaneous eyeblink rate, drd4 exon iii polymorphism and gender on flexibility in set-shifting. Brain Res. **1131**, 155–162 (2007). https://doi.org/10.1016/j.brainres.2006.11.002

47. Raichle, M.E., Snyder, A.Z.: A default mode of brain function: a brief history of an evolving idea. Neuroimage **37**(4), 1083–1090 (2007). https://doi.org/10.1016/j.neuroimage.2007.02.041

48. Raptis, G.E., Fidas, C.A., Avouris, N.M.: Using eye tracking to identify cognitive differences: a brief literature review. In: Proceedings of the 20th Pan-Hellenic Conference on Informatics. vol. 21, p. 6. Association for Computing Machinery, New York, NY, USA (2016). https://doi.org/10.1145/3003733.3003762

49. Rayner, K.: Eye movements in reading and information processing: 20 years of research. Psychol. Bull. **124**(3), 372–422 (1998). https://doi.org/10.1037/0033-2909.124.3.372

50. Sakai, T., et al.: Eda-based estimation of visual attention by observation of eye blink frequency. Int. J. Smart Sens. Intell. Syst. **10**(2), 1–12 (2017). https://doi.org/10.21307/ijssis-2017-212

51. Seli, P., Risko, E.F., Smilek, D., Schacter, D.L.: Mind-wandering with and without intention. Trends Cogn. Sci. **20**(8), 605–617 (2016). https://doi.org/10.1016/j.tics.2016.05.010

52. Smallwood, J., Schooler, J.: The restless mind. Psychol. Bull. **132**, 946–958 (12 2006). https://doi.org/10.1037/0033-2909.132.6.946

53. Stern, J.A., Boyer, D., Schroeder, D.: Blink rate: a possible measure of fatigue. Hum. Factors **36**(2), 285–297 (1994). https://doi.org/10.1177/001872089403600209

54. Tharp, I.J., Pickering, A.D.: Individual differences in cognitive-flexibility: the influence of spontaneous eyeblink rate, trait psychoticism and working memory on attentional set-shifting. Brain Cogn. **75**(2), 119–125 (2011). https://doi.org/10.1016/j.bandc.2010.10.010

55. Trumbo, M.C., Armenta, M.L., Haass, M.J., Butler, K.M., Jones, A.P., Robinson, C.S.H.: Real time assessment of cognitive state: research and implementation challenges. In: Schmorrow, D., Fidopiastis, C. (eds.) Foundations of Augmented Cognition: Neuroergonomics and Operational Neuroscience. LNCS, vol. 9744, pp. 107–119. Springer, Cham (2016). https://doi.org/10.1007/978-3-319-39952-2_12
56. Wenk-Sormaz, H.: Meditation can reduce habitual responding. Altern. Ther. Health Med. **11**(2), 42–58 (2005)
57. Yarbus, A.L.: Eye movements during perception of complex objects. In: Eye Movements and Vision, pp. 171–211. Springer, Boston (1967). https://doi.org/10.1007/978-1-4899-5379-7_8

# Action Boundary in 2D-Cyberspace: A Critical Review of the Action Boundary Perception Tasks

Kévin Bague[1] and Éric Laurent[2]([envelope])

[1] University Paris 8, Saint-Denis, France
[2] University of Franche-Comté, Besançon, France
eric.laurent@univ-fcomte.fr

**Abstract.** In everyday life, adaptive behavior depends on the ability to perceive action possibilities. The field of research on affordances has addressed how individuals perceive action possibilities through the coupling between the perceiver's capacity and environmental features. However, with the increasing digitalization of our environment, people usually need to perceive action boundaries in a 2D-cyberspace. Consequently, researchers need new tasks to operationalize action boundary perception in 2D-cyberspace. The current article aims to propose a critical review of two action boundary perception tasks: the Perception-Action Coupling Task and the Action Boundary Perception Online Task, as assessment tools of the virtual action-boundary perception in digital environment. First, we present the historical background and context of the emergence of these tasks. Then, we review the studies that have employed these tasks. Finally, we conclude by discussing the potential of these tasks, as well as providing a critical description of the limitations that future research will need to address.

**Keywords:** perception · affordance · human-computer interaction · virtual environment · perception and action

## 1 Introduction

Performing an action adequately requires perceiving whether this action is possible or not (Fajen and Turvey, 2003). For example, when we want to cross a road while a car is approaching, we must perceive if we are fast enough to cross before the car passes. In everyday life, similar situations are often present, in which adaptive behaviour depends on the ability to perceive action boundary (Bhargava et al., 2020; Chemero and Turvey, 2007; Harrison et al., 2016; Tosoni et al., 2021). For several decades, the field of affordance study has addressed this question. It proposes that the boundary between what is perceived as possible and impossible depends on the coupling between the perceiver's action capabilities (*e.g.,* walking speed) and the properties of the environment (*e.g.,* car speed) (Fajen and Turvey, 2003).

Nevertheless, the last decade have been characterized by a transformation of our environment. It has become increasingly digital and virtual. A significant portion of our actions now requires human-computer interactions, through the use of various technologies (*e.g.*, traditional computer, smartphone, digital television). In other words, we often have to perceive action boundary in a 2D-cyberspace. As a result, research should embrace this digital transformation and study the perception of action boundaries in a digital environment. The study of action boundary perception in 2D-cyberspace seems to be very important. It has been shown that multiscale factors (Laurent, 2014) can influence the perception of action possibilities (*e.g.*, for affective factors see Bague and Laurent, 2023; Graydon et al., 2012; Vegas and Laurent, 2022). These factors could also disturb the perception of action possibilities in a digital environment. Thus for optimal development and use of digital devices, the influence of such factors has to be taken into account. However, this cannot be achieved without developing new and adapted tasks to study the perception of action boundary in 2D-cyberspace.

Two promising tasks are the digital versions of the Action Boundary Perception Task proposed by Smith and Pepping (2010). The current article aims to propose a critical review of these tasks. We first describe the historical background of these tasks and then expose a review of research that have used them. Finally, we provide a critical discussion of the potential strengths of these tasks, as well as its current limitations, with the aim of suggesting directions for future research.

## 2  Historical Background

Traditionally, in the affordance field, perceived action boundary is operationalized by the determination of the perceived critical point. It corresponds to the ratio between the maximal dimensions (*e.g.*, height, distance) of an environmental layout and the biomechanical feature of the perceiver relevant for the action. For example, the perceived reachability boundary could be determined by the ratio between the maximal distance at which an individual can reach an object and their arm length (*i.e.*, a dimension of an environmental layout that depends on the perceiver's biomechanical feature relevant to reach a target) (Bague and Laurent, 2023; Carello et al., 1989).

Smith and Pepping (2010) provided another operationalization. They proposed to study the perception of action boundaries through the Initiation Time (IT) (*i.e.*, the time between stimulus presentation and the start of the response movement). They hypothesized that IT is a function of how easily an action can be perceived as possible or impossible. In experiment 1[1], participants had to judge whether they could fit or post a real ball through virtual apertures of different sizes. The ball's size was 30 mm, and the size of the apertures ranged from 3 to 60 mm in increments of 3 mm. The authors defined 20 ratios by dividing the size of the aperture by the ball's size. For the first 10 ratios (from 0.1 to 1), the ball could not pass through the aperture (*i.e.*, the aperture did not afford ball posting). For the next 10 ratios (from 1.1 to 2), the ball could pass through the aperture (*i.e.*, aperture afforded ball posting). Smith and Pepping (2010) showed that the distribution of IT as a function of postability ratios followed a curvilinear pattern. The peak of IT (i.e., the longest IT) corresponded to the ratio of the perceived action

---

[1] In the experiment 1, authors did not exactly describe the equipment that allows to measure IT. In the experiment 2, Smith & Pepping (2010) have used a light gate sensor to detect the initiation of the ball movement.

boundary. IT decreased for ratios close to the tails of the distribution. In other words, IT was longer for ratios close to the perceived action boundary. Furthermore, when participants performed the task multiple times, the perceived action boundary was updated. It became more accurate. The authors observed that the peak of IT synchronized with this updated perceived boundary. Therefore, IT would index the perception of action boundaries. Assessing the perception of action boundary through IT is very interesting in digital context. IT can be easily measured in computerized tasks. In other words, an IT from a computerized version of the task by Smith and Pepping (2010) (*i.e.,* with virtual ball and apertures displayed in a digital screen) would allow assessing perception of action boundary in 2D-cyberspace.

Connaboy's team has proposed a tablet adaptation of Smith and Pepping's (2010) task. This adaptation has been called the Perception-Action Coupling Task (PACT) (Connaboy, Johnson, et al., 2019; Connaboy, LaGoy, et al., 2019; Eagle et al., 2021; Johnson et al., 2019). The organizing principle of the PACT is similar to that of Smith and Pepping's (2010) task. Participants are required to indicate whether a virtual ball can pass through a virtual aperture. The PACT contains 8 ratios ranging from 0.2 to 1.8 (in increments of 0.2). One cycle of the PACT corresponds to 16 trials for each ratio. For each trial, participants began with their finger on a start button. To respond, they had to move their fingers towards a virtual joystick. If they estimated the ball could pass through the aperture, they oriented the joystick upwards to guide the ball into the aperture. If they estimated the ball could not pass through the aperture, they oriented the joystick downwards to move the ball away from the aperture. At the end of the trial, they would rest their finger on the start button. Response time was assessed. It included the reaction time (*i.e.,* the time interval between the stimulus presentation and the removal of the finger from the start button), movement time (*i.e.,* the time interval between the finger removal from the start button and the finger contact with virtual joystick), and "initiation time"[2] (*i.e.,* the time to execute the answer with the joystick, corresponding to the interval between the finger contact with virtual joystick and completing of the response). Accuracy is computed by the ratio between number of correct responses and number of trials, multiplied by 100. The main differences with the princeps task are that both the ball and the opening are virtual in the PACT. Additionally, in the PACT, both the ball and the aperture vary in size (diameters ranging from 10–60 mm and 18–44 mm respectively), whereas in the Smith and Pepping's (2010) task, the size of the ball remains fixed. Thus there are ratios ranging from 0.2 to 1.8. The ratio 1.0 is excluded in the PACT. Finally, in the PACT, the definition of the IT, as defined by Smith and Pepping (2010), is different. In the former, it corresponds to the interval between the finger contact with virtual joystick and completing of the response. In the latter, it corresponds to the time between the stimulus presentation and the start of the movement. This terminological ambiguity can lead to confusion, especially as IT is a prominent measure in the operationalization of Smith and Pepping (2010).

We have developed the Action Boundary Perception Online Task (ABP-OT) (see Laurent et al., 2023). This task is inspired by the PACT in the sense that it is entirely

---

[2] We put the initiation time as measured by the Connaboy's team within quotation marks to avoid any confusion with the aforementioned IT as assessed by Smith & Pepping (2010). The two are associated with different definitions (see below for more details).

digital and computerized. However, it is a fully adaptable web-based version that closely resembles the Smith and Pepping's (2010) task. The ABP-OT is coded with PsyToolkit (Stoet, 2010, Stoet, 2017). It can be administered online on any computer with an internet connection and a mouse. In the ABP-OT, participants have to launch a mouse movement from the center of a gray virtual ball having a fixed diameter in order to click on "yes" or "no" buttons, to indicate whether this ball could pass through a white virtual aperture of variable diameter. We proposed twenty ratios from 0.1 to 2.0 in increments of 0.1. The ABP-OT allows the assessment of the IT, following Smith and Pepping's (2010) definition. That is the time interval between the stimulus presentation and the initiation of mouse movement is measured as the IT. Using the same definition as Smith and Pepping (2010) avoids any terminological confusion. It is very important because IT is the main outcome that are supposed to index the perception of affordance. Furthermore, the ABP-OT allows the computation of accuracy. It corresponds to the ratio between number of correct responses and number of trials, multiplied by 100.

## 3 Empirical Review

### 3.1 Validation Studies

Connaboy, Johnson, et al. (2019) have conducted an experiment to assess the systematic bias, the reliability and within-subject variability associated with the PACT. They found that to avoid differences between sessions and cycles or interaction between the two, it is not necessary to remove cycles for reaction and movement times. However, it is necessary to remove one cycle for accuracy and three cycles for "initiation time" to avoid systematic bias. In other words, three cycles would be required to obtain a stable outcomes (Connaboy, Johnson, et al., 2019). This recommendation to avoid systematic bias could be adapted according to the purpose of the study. If a study includes analyses on the "initiation time" and/or the total response time (including the "initiation time"), then it is necessary to remove three cycles. If a study includes analyses on the accuracy but not on the "initiation time" and/or the total response time, then it is necessary to only remove one cycle. If a study does not include analyses on the accuracy, the "initiation time" and/or the total response time, then it is not necessary to remove three cycles. When the test-rest reliability is computed without these three first cycles, the PACT is characterized by a satisfactory reliability for all outcomes from the first cycle (*i.e.,* the first cycle immediately after the three suppressed cycles) (Connaboy, Johnson, et al., 2019). Indeed, intraclass correlation coefficients range from .71 to .98 for all outcomes considered (Connaboy, Johnson, et al., 2019). Finally, when the three first cycles are suppressed, the PACT needs two cycles to reach a stable within-participants variability, as assessed with the coefficient of variation using typical error (Connaboy, Johnson, et al., 2019). Connaboy, Johnson, et al., (2019) concluded that "with a three-cycle familiarization period, the PACT demonstrates no systematic bias, good reliability, and within-subject variability that is consistent with expected values, requiring only one 5-min cycle of testing." (page 82).

Furthermore, Johnson et al. (2019) proposed to remove (a) 0.2 and 1.8 ratios because they can be considered redundant with 0.4 and 1.6 ratios respectively; (b) 12 trials of each aperture in each cycle (*i.e.,* keep only 4 trials for each aperture in a cycle).

They showed that such modifications do not alter the reliability and within-participant variability. These modifications allow for a reduction in the experiment's duration while maintaining a high-quality task (Johnson et al., 2019).

### 3.2 Influence of Contextual Factors

Perceived action boundary is dynamical and recalibrated based on the individual's overall state at multiple scales (biomechanical, physiological, affective, neurological, etc.) (Eagle et al., 2021; Fajen and Turvey, 2003; Johnson et al., 2019). The PACT has been used to assess the recalibration or the calibration failure of the perceived action boundary according to several factors. It is a central issue, because erroneous perception of action boundaries can lead to risky behaviors or inaction (Connaboy, Johnson, et al., 2019; Connaboy, LaGoy, et al., 2019; Eagle et al., 2021; Johnson et al., 2019; LaGoy et al., 2020, 2022).

We found two studies that assessed the relationship between sport-related concussion (SRC) and perceived action-boundary as assessed by the PACT. Eagle et al. (2021) showed that athletes with a SRC (dating on average 264 days ago) had longer movement, reaction and total response times, in comparison with athletes without SRC, for all ratio considered and each ratio separately. Athletes with a SRC were also less accurate than athlete without SRC for three ratios (0.2, 0.4 and 0.6). These findings suggest that SRC is associated with a disturbance of action boundary perception. In a second study, Eagle et al. (2020) partially confirmed these finding. Among athletes aged 12 to 18 years, those with a SRC (dating a maximum of 21 days ago) were less accurate in the PACT for all ratios combined, and for the ratios 0.4, 0.8, 1.4, 1.6, compared to those without a SRC. However, the authors reported no difference between these two groups for the different time outcomes.

Two studies have assessed the influence of stressful environment on PACT performances. At the end of a 30-day NASA mission in an isolated and confined environment, participants had shorter total response time (LaGoy et al., 2020). Furthermore, during this mission, participants with high anxiety and those with low cortisol-DHEA ratio were faster than participants with lower anxiety and those with high cortisol-DHEA ratio (LaGoy et al., 2020). Finally, participants with the lower total sleep time were faster but also less accurate than participants with the higher total sleep time (LaGoy et al., 2020). The last findings seem to be consistent for the accuracy and inconsistent for the response times with those found in another study assessing the effect of a simulated military operational stress scenario, characterized by sleep deprivation and restricted caloric intake, on PACT performances (LaGoy et al., 2022). In this study, total response time, reaction time, and movement time increased with the mission duration, especially during typical sleeping period (LaGoy et al., 2022). Overall, response times during typical sleeping period were longer than during typical wakeful period (LaGoy et al., 2022). The accuracy followed the same data pattern (LaGoy et al., 2022). Likewise, Connaboy, LaGoy, et al. (2019) have shown that after a sleep deprivation of one night, participants exhibited a decrease in accuracy and longer movement and reaction times in the PACT.

With the ABP-OT, Laurent et al. (2023) showed that the perception of virtual action boundaries in a 2D-cyberspace is different between younger adults ($M_{age} = 26.63$, $SD_{age} = 11.69$) and older adults ($M_{age} = 68.36$, $SD_{age} = 5.48$). They showed that older participants have a higher IT than younger participants for all ratios considered, and are less accurate for ratios close to the action boundary.

## 4   Critical Discussion

In this article, we have identified two computerized action boundary perception tasks: the PACT and the ABP-OT. We have also reviewed studies using these tasks. These two tasks, based on the non-computerized task by Smith and Pepping (2010), have the merit of offering ambitious perspectives for the study of action boundaries perception in the digital environment. They have the advantage of being relatively simple tasks, quick and easy to administer. Studies relying on the PACT have demonstrated the suitability of the PACT to various settings. The ABP-OT has been fully adapted for web-based use, expanding its accessibility. We could also envision the use of these tasks in conjunction with physiological measurement tools (e.g., eye-tracking, fMRI, EEG), to study the physiological substrates of action boundary perception. Finally, the PACT and the ABP-OT allow us to study the multi-level factors involving adaptive recalibration of the perception of virtual action boundaries in digital environment, necessary to avoid risky behaviors or inaction. The PACT and the ABP-OT have already allowed studying of the role of sleep deprivation, aging, and concussions. However, it is necessary to further broaden the framework in a multi-scale perspective to better grasp the complexity of action boundary perception in a digital environment and for a deeper understanding of interactions among the involved factors.

The PACT has benefited from validation studies. However, the quality assessment is limited by frequentist statistics. For example, Connaboy, Johnson, et al. (2019) stated an absence of systematic bias from non-significant $p$ values. However, non-significant $p$ values does not allow quantifying evidence in favor of the null statistical hypothesis (*e.g.*, Amrhein et al., 2019), unlike Bayesian inference (*e.g.*, Rouder et al., 2016). It is a commendable initiative to have proposed validation studies. However, future studies should strengthen review process by using Bayesian analyses. The ABP-OT has not yet benefited from validation studies. It could be a future endeavor to employ similar approaches to that used by Connaboy, Johnson, et al. (2019) to ensure the quality of the task. The advantage of the ABP-OT is that its quality assessment could be based on a large number of participants, thanks to its web-based features. Finally, the two tasks could also reinforce each other. Indeed, it could be interesting to assess the convergent validity between the two tasks.

A major limitation concerns the measurement of response times. We identified two issues. Firstly, there is terminological ambiguity that can lead to confusion. According to Smith and Pepping (2010) response time consists in both the IT (*i.e.*, the time between stimulus presentation and the start of the response movement) and the movement time (MT) (*i.e.*, the time between the start of the response movement and the end of the response). The ABP-OT follows this IT definition in its design, while the PACT does not. The same terms refer to different outcomes across these tasks. Connaboy, Johnson, et al. (2019) stated that the IT in the PACT corresponds to the total response time (*i.e.*, the sum of reaction time, movement time, and original "initiation time"). This definition is not entirely consistent with the princeps definition. This can be problematic since the IT is a prominent measure in the operationalization of Smith and Pepping (2010). In Connaboy, LaGoy, et al. (2019) the IT corresponds to reaction and movement times. In Eagle et al. (2020) reaction time corresponds to the time for affordance perception. Thus, definitions of IT which are supposed to index the perception of affordance, are heterogenous in the field. Likewise, within PACT studies there are various methodologies. For example, in studies dealing with concussions, Eagle et al. (2020) used the shortened version of the PACT following Johnson et al. (2019), while Eagle et al. (2021) used the classical version following Connaboy, Johnson, et al. (2019). These subtle variations can be discrete but may pose a threat to the comparability of results across studies. Both studies included the analysis of SRC on outcomes for all ratios combined, without including the same ratios. This may account for the difference in findings between the two studies. There is a clear need for harmonization of definitions and method across research. Secondly, construct validity needs to be evaluated for both tasks. Smith and Pepping (2010) warns that a major limitation of their operationalization concerns construct validity. In other words, does the IT genuinely assess the ease with which an action boundary is perceived? According to Smith and Pepping (2010), the response to the question "can the ball pass through the opening?" could be based on the perception of action boundaries, but it could also be based on a size comparison between the two stimuli. The former mobilize a process of action capability perception ("am I able to pass the ball through the opening?"). The latter involves a process of comparative judgment of extrinsic features without emphasizing action capabilities ("is the ball circle bigger or smaller than the opening circle?"). Future research should address this critical question.

The last reflection is conceptual and concerns the concept of affordance. The PACT and the ABP-OT are attempts to adapt experimental paradigms from the field of affordances to study the perception of action possibilities in a digital environment. For this purpose, they assess the perception of the action boundary, which is also used in classical paradigms of affordance studies. However, this classical paradigm also assesses the intrinsic scaling of the perceived action boundary, meaning the coupling of an perceiver's characteristic with an environmental feature. In contrast, in the PACT and ABP-OT, the ratios do not include a perceiver's feature but rather two environmental feature (a ball on which the participant is supposed to have control on and an aperture). The question is whether we can speak of affordances in these tasks in the sense of the ecological approach to perception. This issue may also require reflection on the evolution of the concept of affordance in human-computer interaction. Specifically, for future studies, we draw attention to how the observer could appropriate, embody, or become one with the computer equipment. For example, when you need to cross the street before a car arrives, your walking speed depends on your bodily capabilities. In contrast, when you need to perceive if you can click on a target (e.g., in a video game) before a given time, it depends not only on your arm speed but also on your mouse's speed. The question is, to what extent do we embody and become one with these familiar objects, and how are they integrated into our perception.

# 5   Concluding Remarks

Following the increasing digitalization of our interactions with the world, understanding the perception of action possibilities in the digital environment becomes crucial. The PACT and ABP-OT are promising tasks that have already (a) operationalized the perception of action boundaries in a digital environment; (b) investigated the influence of various factors on the perception of action boundaries (*e.g.,* aging, sleep deprivation). However, further research is still needed. First, it is essential to further improve the quality of these tasks both methodologically and conceptually. This involves, for example, clarifying and harmonizing the terminology used for response times assessment, as well as ensuring a comprehensive understanding of the perceptual processes involved. Specifically, researchers need to discern whether participants are making size-based stimulus comparisons or assessing their own action capabilities when answering questions: "Can the ball pass through the opening?" By addressing these considerations, the tasks can be refined to provide more precise and accurate insights into the perception of action boundaries in digital environments. Second, it would be interesting to (a) continue exploring the initial empirical data showing the influence of multilevel factors on the perception of action boundaries to better grasp the complexity of action boundary perception in a digital environment; (b) determine the physiological substrates *(e.g.,* with EEG, eye-tracking, fMRI) underlying the perception of action boundary in digital environement. Finally, this article, while raising questions about both embodiment and embededness with computers, also draws our attention to the opportunity of testing hypotheses related to how we perceive affordances in our modern and technology-mediated environment.

# References

Amrhein, V., Greenland, S., Mcshane, B.: Scientists rise up against statistical significance. Nature **567**(7748), 305–307 (2019). https://doi.org/10.1038/d41586-019-00857-9

Bague, K., Laurent, É.: Depressive symptoms and affordance perception: the case of perceived reachability boundary. Psychonomic Bulletin & Review. Advance Online Publication (2023). https://doi.org/10.3758/s13423-022-02242-6

Bhargava, A., et al.: Revisiting affordance perception in contemporary virtual reality. Virtual Reality **24**, 713–724 (2020). https://doi.org/10.1007/s10055-020-00432-y

Carello, C., Grosofsky, A., Reichel, F.D., Solomon, H.Y., Turvey, M.: Visually perceiving what is reachable. Ecol. Psychol. **1**(1), 27–54 (1989). https://doi.org/10.1207/s15326969eco0101

Chemero, A., Turvey, M.T.: Gibsonian affordances for roboticists. Adapt. Behav. **15**(4), 473–480 (2007). https://doi.org/10.1177/1059712307085098

Connaboy, C., Johnson, C.D., et al.: Intersession reliability and within-session stability of a novel perception-action coupling task. Aeros. Medi. Human Perform. **90**(2), 77–83 (2019). https://doi.org/10.3357/AMHP.5190.2019

Connaboy, C., LaGoy, A.D., et al.: Sleep deprivation impairs affordance perception behavior during an action boundary accuracy assessment. Acta Astronaut. **166**, 270–276 (2019). https://doi.org/10.1016/j.actaastro.2019.10.029

Eagle, S.R., et al.: Utility of a novel perceptual-motor control test for identification of sport-related concussion beyond current clinical assessments. J. Sports Sci. **38**(15), 1799–1805 (2020). https://doi.org/10.1080/02640414.2020.1756675

Eagle, S.R., Nindl, B.C., Johnson, C.D., Kontos, A.P., Connaboy, C.: Does concussion affect perception-action coupling behavior? action boundary perception as a biomarker for concussion. Clini. J. Sport Medi. Offi. J. Canadian Acad. Sport Medi. **31**(3), 273–280 (2021). https://doi.org/10.1097/JSM.0000000000000731

Fajen, B.R., Turvey, M.T.: Perception, categories, and possibilities for action. Adapt. Behav. **11**(4), 276–278 (2003). https://doi.org/10.1177/1059712303114004

Graydon, M.M., Linkenauger, S.A., Teachman, B.A., Proffitt, D.R.: Scared stiff: The influence of anxiety on the perception of action capabilities. Cogn. Emot. **26**(7), 1301–1315 (2012). https://doi.org/10.1080/02699931.2012.667391

Harrison, H.S., Turvey, M.T., Frank, T.D.: Affordance-based perception-action dynamics: A model of visually guided braking. Psychol. Rev. **123**(3), 305–323 (2016). https://doi.org/10.1037/rev0000029

Johnson, C.D., et al.: Action boundary proximity effects on perceptual-motor judgments. Aeros. Medi. human Perform. **90**(12), 1000–1008 (2019). https://doi.org/10.3357/AMHP.5376.2019

LaGoy, A.D., et al.: Differences in affordance-based behaviors within an isolated and confined environment are related to sleep, emotional health and physiological parameters. Acta Astronaut. **176**, 238–246 (2020). https://doi.org/10.1016/j.actaastro.2020.06.034

LaGoy, A.D., et al.: Combined effects of time-of-day and simulated military operational stress on perception-action coupling performance. Chronobiol. Int. **39**(11), 1485–1497 (2022). https://doi.org/10.1080/07420528.2022.2125405

Laurent, E.: Multiscale Enaction Model (MEM): The case of complexity and "context-sensitivity" in vision. Front. Psychol. **5**, 1–16 (2014). https://doi.org/10.3389/fpsyg.2014.01425

Laurent, E., Cichon, L., Vegas, C.: Aging alters virtual action-boundary perception in 2D-cyberspace. Book of abstracts of the 63rd Annual Meeting of the Psychonomic Society **27**(November), 263 (2022)

Rouder, J.N., Morey, R.D., Verhagen, J., Swagman, A.R., Wagenmakers, E.J.: Bayesian Analysis of Factorial Designs. Psychological Methods **22**(2), 304–321 (2016). https://doi.org/10.1037/met0000057

Smith, J., Pepping, G.-J.: Effects of affordance perception on the initiation and actualization of action. Ecol. Psychol. **22**, 119–149 (2010). https://doi.org/10.1080/10407411003720080

Stoet, G.: PsyToolkit - A software package for programming psychological experiments using Linux. Behav. Res. Methods **42**(4), 1096–1104 (2010)

Stoet, G.: PsyToolkit: a novel web-based method for running online questionnaires and reaction-time experiments. Teach. Psychol. **44**(1), 24–31 (2017)

Tosoni, A., Altomare, E.C., Brunetti, M., Croce, P., Zappasodi, F., Committeri, G.: Sensory-motor modulations of EEG event-related potentials reflect walking-related macro-affordances. Brain Sci. **11**(11), 1506 (2021). https://doi.org/10.3390/brainsci11111506

Vegas, C., Laurent, É.: Mood influences the perception of the sitting affordance. Atten. Percept. Psychophys. **84**(1), 270–288 (2022). https://doi.org/10.3758/s13414-021-02419-6

# Physiological Anxiety Recognition

Beatriz Guerra[1]($\boxtimes$) and Raquel Sebastião[2]

[1] Department of Physics (DFis), University of Aveiro, 3810-193 Aveiro, Portugal
beatrizguerra@ua.pt

[2] Institute of Electronics and Informatics Engineering of Aveiro (IEETA), Department of Electronics, Telecommunications and Informatics (DETI), Intelligent Systems, Associate Laboratory (LASI), University of Aveiro, 3810-193 Aveiro, Portugal
raquel.sebastiao@ua.pt

**Abstract.** Anxiety is currently increasing in human daily life. Studies aimed to deepen the understanding of it, to minimize its negative impact on people's lives, have gained significant importance. In this context, the main focus of this work is to study the use of several physiological signals (electrocardiogram, electrodermal activity, and blood volume pulse) to predict the level of anxiety felt by a subject using four different approaches. These involve the use of a different number of features that are selected as more informative, and by training two classification models, with different properties. For the chosen approaches, the obtained results are compared and analysed to understand which performs better, i.e., which has a greater ability in anxiety recognition, being the anxiety level of each participant assessed through the application of the STICSA questionnaire. When selecting 20 features to train a Linear Discriminant Analysis model, an accuracy and precision of over 70% were achieved. This strategy presented the best performance as this model surpasses, for all the used metrics, the results obtained when using the models based on a decision tree. The encouraging obtained results sustained the feasibility of the use of simultaneous different physiological signals to train models for predicting the level of anxiety.

**Keywords:** Anxiety · Classification · Decision Tree · Filtering · Linear Discriminant Analysis · Physiological Signals

## 1 Introduction

Today, more and more people are suffering from anxiety, stress and depression. It is a growing problem and has become an inevitable part of everyone's daily life, as it not only affects those who suffer from it but also all those around them such as family and friends. According to the Mental Health Foundation, in 2022/23 there was an increase in the percentage of women from 21.8% to 37.1% and men from 18.3% to 29.9% reporting high levels of anxiety compared to data from 2012 to 2015 [1].

Growing over time, with an increasing number of people suffering from it, anxiety has been receiving increasingly attention. Considering all the expenses involved, and the long road and burden that anxiety brings with it, there is a need for the creation of

J. Baratgin et al. (Eds.): HAR 2023, LNCS 14522, pp. 249–262, 2024.
https://doi.org/10.1007/978-3-031-55245-8_17

healthcare policies and plans to help everyone in need. In addition, it is important to understand and study the effects of anxiety on people to help them control and reduce it, and, consequently, reduce the financial burden and personal and social damage of this increasingly widespread symptom.

This work has as its main objective to predict the level of anxiety through the use of different types of physiological signals collected under a neutral state without the elicitation of any condition. This will help deepen the understanding of alterations in physiological signals that are due to anxiety. Moreover, the prediction performance obtained sustains the development of methods and wearable devices for continuously and accurately assessing this symptom and, in some way, helping the subject to manage and control it.

For that purpose, different physiological signals were filtered and several features were extracted for each type of signal.

After an introduction, Sect. 2 presents related works on the topic, while Sect. 3 describes the data and the methods used in this study. Finally, and before the conclusions and suggestions for future directions, the results are presented and discussed in Sects. 4 and 5, respectively.

## 2 Literature Review

Recent studies have devoted efforts to the study of physiological signals for the identification of anxiety and stress, allowing a deepened understanding of related alterations.

*Ancillon et al.* [2] performed a review of several works, published between the years of 2012 and 2022, in which various machine learning algorithms are used for the analysis of anxiety through biosignals, primarily electrocardiogram (ECG) and electrodermal activity (EDA), collected non-invasively. Based on these, the authors intend to understand which strategies present the best results in the recognition of anxiety through characteristics or patterns observed in the collected signals. In most of the articles reviewed, anxiety has been categorised into 2 or more levels, such as low and high or low, moderate and high, respectively, and different induction protocols were been implemented. The authors conclude that machine learning methods based on random forest and neural networks are the most widely used, and the ones that provide the most accurate results. In addition, they also recommend the simultaneous use of the abovementioned signals, ECG and EDA, as a source of data for the recognition of anxiety.

Comparing clinical data from 60 adolescent girls diagnosed with some or several anxiety-associated illnesses and/or major depressive disorders with 53 healthy girls, serving as controls, the authors of the work [3] found that patients had significantly lower Heart Rate Variability (HRV) values than the control group. With similar purposes, the authors of the study [4] concluded that HRV is lower for individuals suffering from depressive disorders compared to healthy individuals. Considering that it is widely established that low values of HRV are a risk factor for the onset of problems associated with cardiovascular disease, the authors place as important future work, to understand the benefits that antidepressants bring to the subjects concerning heart diseases.

For continuous, accurate, and quantitative stress assessment, the authors of the work [5] collected ECG and electromyogram (EMG) signals from 34 healthy participants under several simulated stress states while reporting, through a questionnaire, the level of stress felt. The stages that were intended to provoke more anxiety/stress in the students involved mathematical tests and tests using the Stroop effect, with a time limit and with noise around them. Regarding signals analysis, from the ECGs 20 features related to HRV were extracted, while from the EMG, besides muscle activity, 9 time and frequency features were computed. Using a fuzzy model to assess stress continuously based on the extracted features, the authors demonstrated a relationship between the increasing difficulty of the mathematical tests and the stress level of the subjects and a significant difference between the relaxation phases and the phases implemented to provoke anxiety. The obtained results also show a high correlation between the estimation of stress and the stress reported by the subjects, as well as high percentages in the calculation of the accuracy of stress detection. Moreover, the authors sustain that the proposed method can be beneficial in practical situations, such as when students are driving or taking exams, to control their stress and anxiety levels by implementing it in various everyday devices.

Addressed driving stress, *Liu* and *Du* [6] used EDA signals, from eleven drivers, of the MIT drivedb database [7]. The three rating levels were divided into low, medium, and high, and each participant was subjected to three different driving conditions: at rest, on the open road (highway), and city driving. Using eighteen features extracted from five-minute segments for each driving period, the authors relied on linear discriminant analysis to distinguish between the three distinct levels of stress, obtaining an average discrimination rate of 81.82%. This approach supports the feasibility of using only one type of signal, in contrast to the different and multiple signals used in several existing methodologies in the literature [8].

*Zamkah et al.* [9] review the existing literature regarding sweat biomarkers that can be used, in the future, in wearable sensors to improve stress and emotion detection, concluding that cortisol is presented as the main stress biomarker. They concluded, however, that the antibody-based technique is the most widely used when it comes to detecting this hormone in sweat. Notwithstanding, other components possibly present in sweat, such as metabolites and antistress hormones, and volatile organic components (VOCs) are also pointed out as possible stress biomarkers.

## 3  Data and Methods

This section describes the physiological signals used and the anxiety levels assessment, as well as the methodology for anxiety prediction based on the extracted features.

After ethical approval CED-UA - 24-CED/2021, 37 volunteers, 23 females and 14 males, belonging to the university community, aged between 19 and 25 y.o. (average age of 21.36 y.o. and standard deviation of 1.27 y.o), participated were enrolled in this study.

Before starting the protocol, the experimental procedure was explained to the participants and after removing the remaining doubts, the declaration of informed consent was signed. Afterward, the individuals had to answer questionnaires regarding their personality attributes, health state, and anxiety. Regarding anxiety, the self-report Portuguese version of the State-Trait Inventory for Cognitive and Somatic Anxiety questionnaire (STICSA) [10, 11] was implemented. Table 1 shows the obtained STICA scores for all the participants as well as the respective categorization into anxiety levels.

**Table 1.** Categorization of the participants into anxiety levels according to the STICSA anxiety scores.

| Anxiety levels | STICSA scores | | | | | | | | | | | | | | | | | |
|---|---|---|---|---|---|---|---|---|---|---|---|---|---|---|---|---|---|---|
| LOW | 26 | 26 | 27 | 32 | 25 | 23 | 29 | 25 | 24 | 30 | 26 | 29 | 31 | 27 | 24 | 25 | 24 | 32 | |
| HIGH | 35 | 40 | 33 | 40 | 34 | 34 | 33 | 59 | 37 | 36 | 50 | 51 | 35 | 46 | 33 | 47 | 48 | 38 | 48 |

As suggested in [11], based on the median of the anxiety scores of all participants (Mdn = 33), two groups of anxiety levels were considered: a group (N = 19) with higher anxiety level (Mdn = 38) and a group (N = 18) with low anxiety level (Mdn = 26). This allowed to balance the data, as shown in Fig. 1.

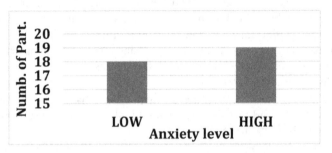

**Fig. 1.** Anxiety levels for all participants.

The number of participants engaged in the study was mainly chosen considering constraints such as the time, equipment availability and human resources available for participant recruitment and data collection. Those were the main reasons for relying on a homogeneous population with similar characteristics, such as the professional activity (students) and ages (we also tried to balance the gender in our sample, but we did not succeed as we have a greater number of female students than male in the Biomedical Engineering course). After categorizing the participants into the defined anxiety levels, the effect size was estimated, using 1000 bootstrap replications, through Cohen's d, obtaining a high value (−2.313) and a narrow confidence interval]−2.9598, −1.684[ which does not cross zero, suggesting a substantial and statistically significant effect.

The data collection comprises a 5-min baseline recording while the participant was watching an emotionally neutral documentary regarding natural world themes, which were classified by a research team of psychologists based on the emotion they elicit, allowing them to establish a neutral psychological state. Along with the video visualization, ECG, EDA (in the hand palm), and blood volume pulse (BVP), among others, were recorded at 1000 Hz and using minimally invasive equipment [12]. Thereafter, frequency domain analysis was performed to understand which frequencies were most appropriate for each signal, and signals were filtered accordingly. To extract features, the filtered signals were then processed using the *Neurokit2*, a Python open source toolbox for physiological signal processing[1]. While processing the ECG, the location and amplitudes of the peaks of P, Q, R, S, and T waves, the P onsets and T offsets, were computed. Concerning the EDA, the Skin Conductance Response (SCR) SCR peaks, and associated components as onsets, height, amplitude, rise time, and recovery time, were extracted. Finally, the location and amplitude of BVP peaks were computed. Figure 2 presents a scheme of the time series of features, characterizing each signal, that were computed using the processing functions of *Neurokit2*.

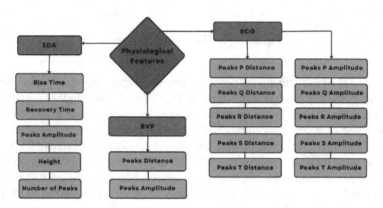

**Fig. 2.** Representative scheme of the time series features computed from each signal.

### 3.1 Distribution Analysis

A distribution analysis was performed to assess the possible effect of anxiety levels on physiological signals. Violin plots allowed inferring differences in the distribution of the corresponding data for both considered levels.

For the ECG, the location of the P, Q, R, S, and T peaks were used to compute time series regarding the amplitude and the distance between consecutive peaks. The differences respecting the distribution of the distance between consecutive peaks in both anxiety levels are shown in Fig. 3 (left). Nevertheless, a similarity in the violins for the different peaks is also observed, which can be explained by the fact that when the electrical impulse follows a normal path, the electrocardiogram waves, composed of the different peaks, follow a practically constant pattern called sinusoidal rhythm. In this rhythm, the waves occur in sequence and without any interruption, and their amplitude, order, and duration remain constant. The figure on the right shows the violin plots for the distribution of time series with peaks' amplitude. In the case of P and Q peaks, and since the values of the amplitudes reached by these are small, compared to those achieved by the others, the distinction between levels is not noticeable. However, for R peaks, the low anxiety level shows lower amplitude values than the high anxiety level. The amplitude of S peaks also presented a clear distinction between the two anxiety levels, achieving greater values for the high anxiety level. Finally, T peaks show a less evident differentiation in the amplitude distributions.

---

[1] https://neuropsychology.github.io/NeuroKit/.

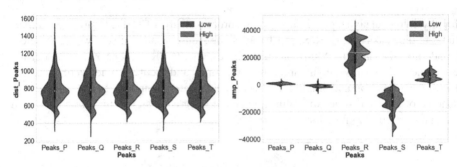

**Fig. 3.** Violin plots of ECG features.

For the EDA, the distribution of peaks' amplitude, height, number of peaks, rise time, and recovery time was evaluated and shown in Fig. 4. For the amplitude of the peaks, it is visible the different distributions along the two levels of anxiety, being noticeable, for the high anxiety level, a large concentration of data in a region of restricted amplitude when compared with the relative distribution corresponding to the low anxiety level. These results also stand for height and rise time. Regarding recovery time, for lower anxiety, the highest data density is found for higher values, as opposed to higher level of anxiety. This indicates that for participants with higher anxiety, the time corresponding to the period in which there is an abrupt decrease in amplitude after reaching the maximum amplitude corresponding to the peak is shorter. In the case of the violin plot that highlights the distinction for the case of the number of peaks in the EDA signal, a similar distribution is observed between anxiety levels.

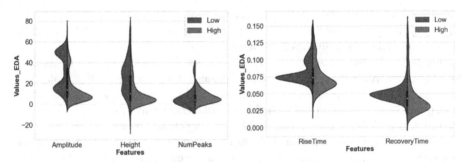

**Fig. 4.** Violin plots of EDA features.

Regarding the BVP, violin plots for the distances between consecutive peaks and their amplitude are shown in Fig. 5. From the visual analysis of the distance between consecutive peaks, a distinct distribution of data between the anxiety levels is perceptible. There is a more elongated distribution and a slightly higher value for the median, represented by the white dot in the centre, in the graph that corresponds to high anxiety. For peaks' amplitude, no significant distinction is apparent between anxiety levels.

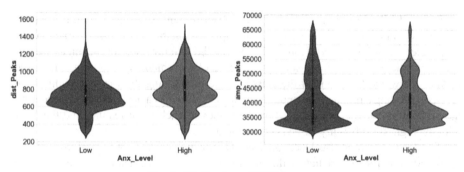

**Fig. 5.** Violin plots of BVP features.

From the results obtained, in general, considerable differences in the distribution of the time series were observed regarding the two anxiety levels used, supporting that it can, indeed, have an impact on the characteristics of the physiological signals.

## 3.2  Anxiety Classification

Considering that the length of the time series of the same feature was significantly different among participants, several time series were not considered for anxiety classification, namely the time series regarding P, Q, and T peaks (as those are considerably more susceptible to detection errors when compared to R and S peaks), and the number of EDA peaks.

Thereafter, four different metrics were computed from the respective time series of the remaining features obtained: mean, median, standard deviation, and range. After initial analysis, five participants were removed since three of them had no values for the EDA signal while the other two, had widely different values when compared with those obtained for the rest of the participants for the same type of signal. In addition, for a small number of participants, outlier values were replaced through the imputation of the median values of the remaining participants with the same anxiety level.

Training and testing of this final dataset were performed using the leave-one-out cross-validation (LOOCV) strategy, where the total number of participants minus one is used for the training set, and the participant that is left out forms the test set. This strategy is run until all samples are evaluated as a test set. The choice of LOOCV method was due to the low number of samples present.

As a second step, data was standardized by subtracting the mean and dividing by the standard deviation, reducing all features to a common scale.

The next step entails the selection of the most important features. The goal is to reduce the size of the data by eliminating the less important features, that is, to remove non-informative or redundant predictors to improve the performance of the classification model. This was achieved through the *SelectKBest* method that chooses the best K features with the highest scores, being K the number of features to select. This search is done only on the training set. Once the best features are found, only these will be used in the test set, that is, the test set will contain only the features that were selected based on the training set.

The first ML algorithm used was Linear Discriminant Analysis (LDA). It has been widely used in many applications being an important method for feature extraction and for problems involving classification [13]. Using the training data, for the binary case, it works by calculating summary statistics for each class, which are used to maximize the distance between the means of the classes while minimising the variance within each class, developing a probabilistic model per class. It classifies a new example considering the probability for each class based on the values of each input feature, and the class with the highest probability value is chosen and assigned [14].

The second method was Decision Tree (DT), which consists of several instructions and conditions for the predictors that partition the data. A decision tree starts from a root node from which new subsets are generated, based on a feature value. This splitting process continues from the generated subsets until leaf nodes are created, which have the important function of assigning the class to the sample in question [14].

Thereafter, four prediction models were trained by fitting the model in the training set, for both situations where the number of features selected was distinct. Then, using the *predict* function, each model was used to make the desired prediction of the anxiety level for every participant using a new set of data, the test set, which in this case is the participant left out of the training set.

The prediction ability of the learnt models was assessed through the following performance metrics: accuracy, precision, recall, and F1 score. Accuracy establishes the correspondence between the classes that are correctly predicted and the observed ones. Precision represents the rate with which the classifier is capable of correctly classifying a sample as being positive when in reality it is positive, while recall, is the model's ability to classify samples that are positive as positive. Finally, the F1 score is calculated taking into account these last two metrics, precision and recall, thus combining both the positive and the negative results of the classifier.

# 4 Results

This study aimed to, through the LOOCV strategy, learn ML models, using two distinct algorithms, to evaluate the predictive efficacy of the participants' anxiety level based on features from three types of physiological signals. These were collected while the participants were seated in a neutral condition. As previously mentioned, the study accounts for a total of 37 participants, 5 of whom were removed due to significant discrepancies or even the absence of values regarding the signals used.

## 4.1 Feature Selection

Feature selection considered the selection of two dissimilar numbers of features, 10 features in the first approach and 20 in the second. From each, using the LOOCV strategy, distinct models using two different ML algorithms were learned, which is schematically represented in Fig. 6.

Feature selection was implemented to eliminate irrelevant features that are non-informative or redundant, thus improving the quality of the data extraction process. After building a loop that went through all the participants for this study, it was possible

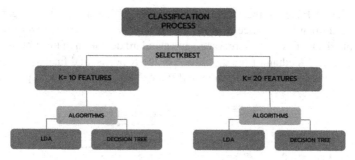

**Fig. 6.** Diagram representing the different prediction strategies.

to build a final matrix that matched each feature to the total number of times it was selected. In this study, and as can be seen in the scheme of Fig. 6, two initial paths were followed before the implementation of the algorithms.

Considering the case that 10 features were selected for each participant, from the total of 40, 16 features were chosen at least one time, that is, for at least 1 participant they were selected as one of the principal features. To illustrate this, Fig. 7 shows the features that were picked for at least half of the participants, i.e., that were selected at least 16 times. It reveals the great importance that 6 features have as they were selected for all participants demonstrating their relevance for predicting anxiety, such as the mean calculated from the distance between consecutive peaks for the BVP signal and the range obtained from the distance between the S and R peaks. The features relating to the standard deviation calculated from the amplitude of the S peaks and from the distance between R and S peaks were also always selected.

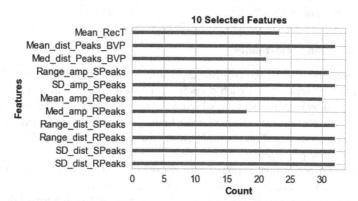

**Fig. 7.** Features selected most of the time for the entire set of participants when k = 10.

Regarding the selection of 20 features, of the total number of features, only 3 were never selected. Figure 8 shows the features that were selected for at least half of the participants, allowing to conclude that 11 features were consistently selected each time the loop was realised, being selected for all 32 participants. From the EDA signal, only the feature that corresponds to the mean of the recovery time was always chosen.

Regarding the BVP signal, the mean and median of the distance between consecutive peaks revealed great importance as well. Ultimately, the features that were constantly selected from the ECG signal were: range and standard deviation of the amplitude of the S peaks, mean and median of the amplitude of the R peaks, and finally, the range and the standard deviation calculated from the distance between the R and S peaks.

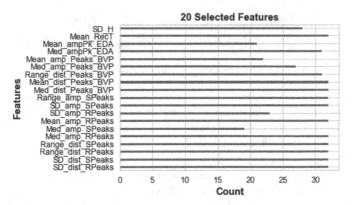

**Fig. 8.** Features selected most of the time for the entire set of participants when k = 20.

By analysing the results for the number of selected features, it is possible to conclude that all the features chosen when only 10 features were selected by the *SelectKBest* function were also selected when the search went to 20 features for each participant. Furthermore, all these features were selected 32 times, even if in the initial situation this had not happened for some of them, that is, they were selected as informative and non-redundant features for all participants in the second strategy.

## 4.2  Classification Performance

For both cases and each of the two ML algorithms, it was possible to make four predictions of the anxiety level of each participant. From these and with the true level of anxiety reported by each participant, which was established through the STICSA-Trait questionnaire, four confusion matrices were built. These allow to compute the performance for each case, based on a simple cross-tabulation between the classes that are observed and those predicted.

The obtained performance, taking into consideration the number of features that were selected and the ML algorithms used, is shown in Fig. 9.

Through the use of only 10 features, it is possible to visualize that for all the metrics used, the models learned by the LDA algorithm led to slightly better results, compared to those obtained for the decision tree-based models, obtaining values higher than 60% for all metrics, reaching a value close to 80% for recall. For the case of the graph on the right of Fig. 9, similar conclusions can be drawn. In this case, 20 features were used, and higher values were also reached when classifying from the LDA-based models.

Comparing the two graphs of Fig. 9, it is interesting to verify that for a larger number of extracted features, the LDA models led to higher values for all metrics used: accuracy,

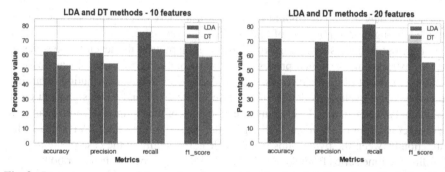

**Fig. 9.** Percentage values of the performance metrics for the situation where 10 (left) and 20 (right) features were selected.

precision, recall, and F1 score, compared to those obtained when only 10 features are used. For the DT-based models, the percentage values of the measures are lower for the case where 20 features are used, excluding the recall, which remains identical in both cases.

Considering these results, it is advisable to use LDA-based models, trained with a larger number of features, to predict anxiety levels since present significantly better results when compared with the other three strategies that were tested. Moreover, the statistical significance of the outperformance of LDA over DT was assessed through the McNemar test. The obtained result from the McNemar test (p-value = 0.0228) indicates evidence to reject the null hypothesis that the LDA is less accurate than DT.

## 5   Discussion

*Pourmohammadi* and *Maleki* [5] also used different types of physiological signals to obtain a reliable and accurate index for stress. The authors considered two stress scores and the signals used were subjected to similar steps to those performed in the present study. Through the implementation of an appealing fuzzy model, a fairly high correlation between the predictions and what was actually felt by the subjects was obtained. This reinforces the importance and feasibility of using different signals in the study of stress and anxiety. Also, in the review work done by *Ancillon et al.* [2] articles are presented where different types of signals were used and in which DT models are frequently adopted. Commonly, with DT, and with the simultaneous utilization of both ECG and EDA signals, good results were achieved.

The conclusions of these authors support the use of presented physiological features but did not sustain when it comes to the use of the DT-based model. As in the present work, DT was outperformed in all metrics by the LDA model and using 20 features, the LDA model performed significantly better than the DT model.

## 6   Conclusions and Further Work

In this work, four different classification strategies were applied, including two different classification algorithms, with the aim of predicting the level of anxiety of each participant using data from three physiological signals: ECG, EDA, and BVP. For this purpose, anxiety was categorised into high anxiety and low anxiety levels using the assessment provided through the application of the STICSA questionnaire.

A distribution analysis was performed using violin plots of the several time series of features extracted in the signal processing step. Through these results, it was possible to observe, in most of the features used, clear differences in data distribution regarding the two anxiety levels considered. In the case of the ECG and EDA signals, a clear distinction between anxiety levels can be seen in most of the computed features through the distinct location of the areas with higher data density. Regarding BVP signals, the distinction between the two groups is not so clear, especially in the case of the feature regarding the time series from the amplitude reached by the peaks since, for both levels of anxiety, a quite symmetric violin was obtained.

Furthermore, regarding the prediction of anxiety, 10 and 20 features were selected from the total number of features. For each case, the classification models LDA and DT were trained with LOOCV strategy, and the evaluation of the four approaches implemented was performed by computing accuracy, precision, recall, and f1-score metrics.

When selecting the least redundant and most informative features, it was notable, both when 10 and 20 features were selected, and taking into account that in the corresponding graphics only those features that were chosen at least 16 times were represented, that a large proportion of these was selected for all participants. It is possible to conclude that the features related to the ECG signal and the BVP signal had a higher selection rate when compared to those collected from the EDA signal. Compared to the conclusions drawn from the violin plots, in the case of the EDA signal, this situation was not expected since an obvious distinction was visible between the groups of participants with different anxiety levels, in the majority of features related to this signal.

From the analysis of the performance metrics, it is possible to conclude that, in general, they were quite positive. Nevertheless, better results were obtained when the LDA-based models were applied, performance metrics higher than 60% when 10 features were selected, and equal or higher than 70% when selecting 20 features. Overall, the performance obtained when implementing the DT method was poorer, due to the consistently lower results for the four performance metrics compared to the LDA method.

In view of the above, using the LDA-based models trained with 20 selected features, it was possible to predict anxiety levels with reasonable accuracy and f1-score, of around 75%. These encouraging results demonstrate that it is feasible to predict anxiety levels through the use of features extracted from three distinct physiological signals collected under a neutral condition. This work is also a step further to contribute to objective assessment of anxiety levels based on physiological signals.

To support the conclusions obtained and to reinforce the viability of the ML algorithms employed, a larger data collection would be beneficial. Additionally, it could be considered the use of only one of the physiological signals to compare the prediction performance obtained from it, as performed in the study conducted by *Liu* and *Du* [6]. These authors used only features from the EDA signal, collected on the foot, achieving an average rate of 81.82% to discriminate into 3 levels. Furthermore, bearing in mind that the results obtained from the present study were obtained from signals collected in a situation considered normal, a neutral condition without the elicitation of any stimuli, the importance of using physiological signals for the effective recognition of anxiety levels and the management of its impact on human daily life becomes visible. This relationship within physiological signals may bring benefits and, after understanding the most appropriate wearable physiological biomarkers and the most effective techniques and models for detecting and classifying anxiety, it may support the development of devices accessible to everyone with the ability to notify, in real-time, the anxiety state of the subject, and provide appropriate recommendations to avoid the undesirable health outcomes, helping with coping strategies to manage anxiety.

**Acknowledgments.** This work was funded by national funds through FCT –Fundação para a Ciência e a Tecnologia, I.P., under the Scientific Employment Stimulus CEECIND/03986/2018 (R.S.) and CEECINST/00013/2021 (R.S.). This work is also supported by the FCT through national funds, within the R&D unit IEETA/UA (UIDB/00127/2020) and under the project EMPA (2022.05005.PTDC).

# References

1. Anxiety: statistics. https://www.mentalhealth.org.uk/explore-mental-health/mental-health-statistics/anxiety-statistics, accessed 21 May 2023
2. Ancillon, L., Elgendi, M., Menon, C.: Machine learning for anxiety detection using biosignals: a review. Diagnostics **12**(8), MDPI (2022). https://doi.org/10.3390/diagnostics12081794
3. Blom, E.H., Olsson, E.M., Serlachius, E., Ericson, M., Ingvar, M.: Heart rate variability (HRV) in adolescent females with anxiety disorders and major depressive disorder, Acta Paediatrica. Int. J. Paediatrics **99**(4), 604–611. Wiley (2010). doi: https://doi.org/10.1111/j.1651-2227.2009.01657.x
4. Licht, C.M.M., et al.: Association between major depressive disorder and heart rate variability in the netherlands study of depression and anxiety (NESDA). Arch Gen Psychiatry **65**(12), 1358–1367. American Medical Association (2008). https://doi.org/10.1001/archpsyc.65.12.1358
5. Pourmohammadi, S., Maleki, A.: Continuous mental stress level assessment using electrocardiogram and electromyogram signals. Biomedical Signal Processing and Control **68**. Elsevier (2021). https://doi.org/10.1016/j.bspc.2021.102694
6. Liu, Y., Du, S.: Psychological stress level detection based on electrodermal activity. Behavioural Brain Research **341**, 50–53. Elsevier (2018). https://doi.org/10.1016/j.bbr.2017.12.021
7. Healey, J.A., Picard, R.W.: PHYSIONET, Stress Recognition in Automobile Drivers (drivedB). (2008). https://physionet.org/content/drivedb/1.0.0/, accessed 08 Jun. 2023
8. Healey, J.A., Picard, R.W.: Detecting stress during real-world driving tasks using physiological sensors. IEEE Transactions on Intelligent Transportation Systems **6**(2), 156–166. Institute of Electrical and Electronics Engineers Inc. (2005). https://doi.org/10.1109/TITS.2005.848368
9. Zamkah, A., et al.: Identification of suitable biomarkers for stress and emotion detection for future personal affective wearable sensors. Biosensors **10**(4). MDPI (2020). https://doi.org/10.3390/bios10040040
10. Mendes, A.: STICSA: Análise Psicométrica Numa Amostra De Estudantes Universitários Portugueses. M.S. thesis. DEP, Universidade de Aveiro, Aveiro, PT (2018)

11. Barros, F., Figueiredo, C., Brás, S., Carvalho, J.M., Soares, S.C.: Multidimensional assessment of anxiety through the State-Trait Inventory for Cognitive and Somatic Anxiety (STICSA): From dimensionality to response prediction across emotional contexts. PLOS ONE **17**(1). Public Library of Science (2022). https://doi.org/10.1371/journal.pone.0262960

12. Silva, P.: Using the electrocardiogram for pain classification under emotion elicitation. M.S. thesis. DFIS, Universidade de Aveiro, Aveiro, PT (2022)

13. Ye, J., Janardan, R., Li, Q.: Two-dimensional linear discriminant analysis. In: Advances in Neural Information Processing Systems (NIPS 2004) **17**, 1569–1576. MIT Press (2004)

14. Flach, P.: Machine Learning: the Art and Science of Algorithms That Make Sense of Data. Cambridge University Press, MA, USA (2012)

# Physiological Characterization of Stress

Diogo Esteves[1] and Raquel Sebastião[2(✉)]

[1] Department of Physics (DFis), University of Aveiro, 3810-193 Aveiro, Portugal
diogomsesteves@ua.pt
[2] Institute of Electronics and Informatics Engineering of Aveiro (IEETA), Department of Electronics, Telecommunications and Informatics (DETI), Intelligent Systems Associate Laboratory (LASI), University of Aveiro, 3810-193 Aveiro, Portugal
raquel.sebastiao@ua.pt

**Abstract.** This study analyses and characterizes changes in features extracted from physiological signals in the presence of affective states, namely, neutral, stress, and amusement. With a focus on the stress condition, for this purpose a statistical analysis was performed on various features extracted from ECG (electrocardiogram), EMG (electromyography), EDA (electrodermal activity), and RESP (respiration) signals from the WESAD dataset. This dataset provides data from 15 healthy participants regarding three affective conditions. Concerning the ECG features, from the statistical analysis, it was possible to observe a significant decrease in the interval between consecutive R peaks, meaning that there was an increase in heart rate. The EMG signal showed a significant increase in muscle activation. Regarding the EDA, it was noticed that during the stress condition, the production of sweat increased, leading to greater skin conductivity. Lastly with respect to the RESP signal, although no significant difference was observed in regards to changes in the inspiration and expiration durations, the increase in the standard deviation leads inferring an increase in the irregularity of the breathing pattern of the participants during the stress condition.

**Keywords:** ECG · EDA · EMG · RESP · Stress

## 1 Introduction

Stress is an ever-growing problem in our society. According to the World Health Organization, due to the recent pandemic, the global prevalence of anxiety and depression increased by 25.6% and 27.6% respectively after the first year of isolation [1].

Stress has been shown to negatively impact an individual's mental health, as prolonged exposure to stress leads to depression and anxiety [2]. Additionally, stress has also been shown to increase the risks of other health complications including, cardiovascular disease, hypertension, diabetes, obesity, and heart attack or stroke [2–4]. Due to the relation between stress and a worse quality of life and shorter life expectancy, recent studies have proposed to create systems capable of detecting stress.

J. Baratgin et al. (Eds.): HAR 2023, LNCS 14522, pp. 263–277, 2024.
https://doi.org/10.1007/978-3-031-55245-8_18

In an attempt to improve future stress detection systems, the aim of this study is to analyze and characterize changes in features extracted from physiological signals that are related to the stress condition, including electrocardiogram (ECG), electromyography (EMG), electrodermal activity (EDA), and respiration (RESP).

Stress refers to the response of the mind and body to any demand or threat that exceeds the individual ability to cope effectively, triggering several physiological, emotional, and cognitive changes [5]. Previous studies have established a relationship between stress and emotion models, like the circumplex model. The axes of this two-dimensional space are related to valence and arousal. The valence axis refers to if the emotion is perceived as positive or negative, while the arousal axis indicates the level of activation related to the emotion [6, 7]. Stress has been mapped in the circumplex map as a high arousal and negative valence [6].

The fast response to sudden stress, regulated by the Sympathetic-Adrenal Medullary axis [8], is mainly assessed through indirect measurements of the sympathetic nervous system (SNS) response, by analysing physiological changes associated with the SNS activation.

Variations in the heart activity, obtained through ECG, increases in the secretion of sweat, observed using EDA, increases in muscle excitability, observable through EMG, and changes in the breathing pattern, among others had been described as associated with stress [6, 8, 9].

After an introduction, Sect. 2 presents related works, while Sects. 3 and 4 describe the data and the methodology used in this study, respectively. Section 5 is devoted to the analysis of the obtained results. Before final remarks on the impact of stress on rationality and suggestions for further work, conclusions are presented and discussed respecting the literature.

## 2 Literature Review

The publicly available WESAD dataset [10] had been used to assess physiological stress indicators, by performing statistical analysis and classification modelling on bio-physiological data [11] The proposed analysis was done on statistical features (mean, standard deviation, and median) extracted from the heart rate (HR), heart rate variability (HRV) and the interval between consecutive R peaks (RRI), both extracted from the ECG signal, EMG signal, phasic component of the EDA signal, and the respiration rate (RspR), extracted from the RESP signal. From the statistical tests, the study concluded that the RspR was the best indicator of stress, among the ones considered in this study, and found that the EMG and EDA features showed no statistically significant difference between the stress and baseline states. Logistical regression was utilized for the classification modelling evaluated using leave-on-out cross-validation. The results from the classification model confirmed that the RspR features had the highest specificity and sensitivity for stress detection, also finding that the junction of the Resp features with the HR and HRV features achieved almost the same performance as all six parameters, achieving a sensitivity and specificity of 0.89 and 0.72 vs 0.86 and 0.86, respectively.

Also, ECG and EMG signals were exploited to derive a continuous personalized stress detection method by employing a fuzzy-based model [12]. The signals were collected from 34 healthy participants subjected to the Stroop colour and mathematical tests with three difficulty levels. The proposed model calculated a stress index from twenty features extracted from the ECG signal and nine from the EMG signal. Paired t-tests were used for the statistical analysis of the stress indexes obtained throughout the study, finding a statistical difference between the rest and all of the stress phases. Additionally, the correlation between the indexes obtained and the self-reports of each participant was calculated, obtaining a correlation of over 90%. The fuzzy-based model proposed in this study was implemented through two steps. The first step consisted of fuzzy clustering, which established membership values between the data and five clusters, one cluster per stage of the protocol. The second step implemented a fuzzy inference system that calculated numerical weights associated with each membership value, and finally, using both the membership values and their corresponding numerical weights the mental stress indexes were calculated. Lastly, the indexes were converted into labels and used for two-label (stress vs no stress) and three-label (low stress vs medium stress vs high stress) classification tasks, obtaining an accuracy of 96.7% and 75.6%, respectively.

Several measures of RESP and ECG signals from 39 participants, obtained using a wearable device, were used in a three-level (no stress, moderate stress and high stress) stress detection method [13]. The stress conditions were elicited by using the Montreal Imaging Stress Task protocol, adjusted to better fit the measures required for the study and to include factors to better simulate workplace stress. Relevant features were extracted from both signals, and feature selection was performed using Random Forest. From four classifiers (SVM – Support Vector Machines, LDA – Linear Discriminant Analysis, Adaboost and kNN – k-Nearest Neighbours), the best performance was reached with the SVM, with feature selection, obtaining an accuracy of 84% for the three-class (no-stress vs moderate-stress vs high-stress) classification task and 94% for binary stress detection (stress vs no-stress).

Multi-stress levels detection was also evaluated through the use of EMG and ECG to train SVMs [14]. The ECG and EMG signals were collected from 34 healthy participants, and the stress condition was induced by employing the Stroop colour test with the addition of mental arithmetic and a stressful environment. The model with the highest accuracy used both signals and achieved an accuracy of 100%, 97.6%, and 96.2% for four (no tress vs low stress vs medium stress vs high stress), three (no-stress vs moderate-stress vs high stress) and two (no-stress vs stress) level stress detection, respectively. Additionally, while comparing the results from the models utilizing only the EMG or ECG signals it was possible to observe that their performance was similar.

## 3   Dataset

The present work used the publicly available dataset WESAD [10], obtained through two wearable devices: the Empatica E4 and the RespiBAN Professional. The Empatica E4 is a wrist-worn device capable of recording BPV (64 Hz), EDA (4 Hz), ACC (32 Hz), and TEMP (4 Hz). The RespiBAN Professional is a chest-worn device capable of recording ACC and RESP and was connected to four additional sensors to record ECG, EMG,

EDA, and TEMP. All signals obtained with the RespiBAN Professional were sampled at 700 Hz.

In this work, only the ECG, EDA, EMG and RESP signals, obtained with the RespiBAN Professional, were analyzed. Regarding the sensors that were additionally connected to the chest-worn device and that will be considered in this work, the ECG was collected using three electrodes. The EDA was measured on the rectus abdominis due to the elevated number of sweat glands present in that area. Lastly, the EMG was collected on the upper trapezius muscle on both sides of the spine. Figure 1 shows the placement of the electrodes as described.

This dataset was obtained from 15 participants, with a mean age and standard deviation of $27.5 \pm 2.4$ years old, twelve male and three female subjects. All participants were graduate students from the University of Siegen in Germany.

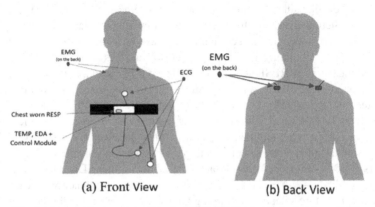

**Fig. 1.** Sensor placement used in the WESAD Dataset (adapted from [11]).

The study where the WESAD dataset was collected had three main objectives. The first was to provide a publicly available multimodal dataset for wearable stress and affect detection. The second was to close the gap between studies on stress and emotions by providing a dataset with three affective states: neutral, stress and amusement. Lastly, to create a benchmark for stress and affect detection by a large number of well-known features and common machine learning methods to identify these affective states.

The protocol for collecting the WESAD dataset had the goal of inducing mentioned three affective states. To elicit these states, the participants were subjected to 3 different conditions. The baseline condition consists of a 20-min record while the participants were reading neutral magazines. In the amusement condition, data were recorded while the participants were watching a selection of funny videos, with a total duration of around 7 min. The stress condition, with a total duration of approximately 10 min, was recorded while the subjects were submitted to a modified version of the Trier Social Stress Test (TSST), formed by two distinct tasks to elicit stress [15, 16]. The first is the preparation (3 min) and delivery of 5 min speech, without using the preparation's notes. The speech was made in front of a three-person panel acting as human resources specialists and was intended to boost participants' career options with a focus on their traits, either strengths or weaknesses. The second task, with a duration of 5 min, was also performed in front of the panel and involved the participant performing a mental arithmetic task by counting from 2023 to zero, with steps of 17 (and starting over from 2023 in case of a mistake).

After the TSST the participants had a rest period of 10 min. In addition, two meditation sessions of 7 min each were added. The meditations were performed after amusement and stress conditions to bring the participant closer to a neutral affective state by having the participant sit and follow a breathing exercise. In total, the study had an expected duration of two hours, comprising the total duration of the conditions as well as instructing the participant regarding the protocol, informed consent signing, responses to questionnaires, and preparation, placement and removal of the equipment and sensors.

To accurately gauge the psychological responses, the WESAD protocol account for the following control procedures:

- Participants were randomly assigned to two sequences of the protocol, by swapping the placement of the defined conditions, to mitigate the order effects that might influence participants' responses (as represented in Fig. 2).
- Each condition involved a standardized activity, ensuring that participants experienced similar stimuli, which allows for meaningful comparisons.
- Before engaging in the different affective conditions, a baseline assessment was recorded, allowing to establish initial psychological states.
- To ensure the elicitation of different conditions, after each condition, ground truth was obtained through participants' self-reports by the application of several questionnaires (PANAS – to assess positive and negative affect, shortened STAI – for anxiety evaluation, and SAM – to evaluate valence and arousal). Moreover, after the stress condition, a shortened SSSQ questionnaire was also applied.

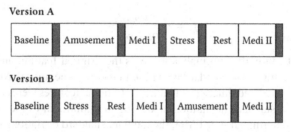

**Fig. 2.** The two versions of the protocol applied to obtain the WESAD database [10].

# 4   Methodology for Data Visualization and Analysis

Before performing feature extraction, the raw signals were filtered to remove noise, baseline drift and artefacts. For all the signals, except RESP, a fourth-order Butterworth filter was used with different frequency bands, which were chosen based on the characteristics of the signals and possible types of noise, observed through the analysis of the signals in time and frequency domains. The RESP signals were pre-processed as described in [17], by performing linear detrending, followed by the application of a fifth-order low-pass Butterworth filter. Thereafter, the signals for each subject were segmented according to the procedure, removing the time between conditions. After segmentation, every participant has a baseline, stress, amusement and two meditation segments for all signals used.

## 4.1   Feature Extraction

To extract features, the filtered signals were processed using the *Neurokit2*, a Python open source toolbox for physiological signal processing[1].

From ECG, the position of P, Q, R, S and T peaks were identified and therefore the distance between consecutive peaks (P, Q, R, S and T) was calculated. Regarding EMG, the amplitude, the location of onsets and offsets, and activity were computed. The EDA was separated into the tonic component, corresponding to slow and long-term fluctuations, and the phasic component, associated with sudden and rapid changes. From this component, it was extracted the position, height, amplitude, onsets, recovery time, and rise time for each peak. For RESP, the inspiration and expiration phases were estimated.

To compare the responses of participants across conditions, inter-participant variability was minimized by reducing the effect of the initial state. Thus, these extracted time series features were normalized to the respective baseline, accordingly to:

$$y' = \frac{y - \mu_{baseline}}{\sigma_{baseline}},$$

where $y'$ is the feature after normalization, $y$ is the original feature, and $\mu_{baseline}$ and $\sigma_{baseline}$ are the average and standard deviation in the baseline, respectively.

From these normalized time series features, additional features were computed. From ECG, the median, standard deviation, maximum and minimum of the height of the R peaks were also determined. For EMG, the median, standard deviation, and root-mean-square error were computed for both the filtered EMG and amplitude, and the total number of activations was also obtained. From EDA, the median and range of the phasic and tonic components were calculated, as well as the number of peaks and the area under the phasic component. For RESP, the duration of each inspiration and expiration was computed, from which the total duration of each breath and the ratio between the duration of the phases of each respiration cycle were calculated. Thereafter, the median and standard deviation were obtained.

Table 1 shows all of the extracted features used.

---

[1] https://neuropsychology.github.io/NeuroKit/.

**Table 1.**  List of extracted features and respective abbreviations.

| Signal | Feature name | Abbreviation |
|---|---|---|
| ECG | R-R interval | RRI |
| | P-P interval | PPI |
| | Q-Q interval | QQI |
| | S-S interval | SSI |
| | T-T interval | TTI |
| | Median R peaks height | MRH |
| | Standard deviation R peaks height | SDRH |
| | Max R peaks height | RHmax |
| | Min R peaks height | RHmin |
| EMG | Median of the EMG signal | MEMG |
| | Standard deviation of the EMG signal | SDEMG |
| | Root mean square error of the EMG signal | RMSE |
| | Median of the EMG amplitude | MEMGamp |
| | Standard deviation of the EMG amplitude | SDEMGamp |
| | Root mean square error of the EMG amplitude | RMSA |
| | Normalized number of activations during conditions | NNAct |
| EDA | SCR peak amplitude | SCRamp |
| | SCR peak height | SCRheight |
| | Median of the tonic component | MTonic |
| | Median of the phasic component | MPhasic |
| | Range of the phasic component | PhasicRange |
| | Range of the tonic component | TonicRange |
| | Normalized number of SCR peaks | NNSCR |
| | Area under the phasic component | AUSCR |
| | Rise time of SCR peaks | RiseTime |
| | Recovery time of SCR peaks | RecTime |
| RESP | Median expiration duration | MedianEx |
| | Median inspiration duration | Medianin |
| | Median total breath duration | MedianTotal |
| | Median ratio between inspiration and expiration | MedianRIE |
| | Standard deviation of expiration duration | SDEx |
| | Standard deviation of inspiration duration | SDin |
| | Standard deviation of total breath duration | SDTotal |
| | Standard deviation of ratio between inspiration and expiration | SDRIE |

## 4.2  Univariate Statistical Analysis

Visual and statistical analyses of the distribution of each feature were performed.

The distribution of each feature was studied using boxplots, a graphical plot that allows the visualization of descriptive statistics in a concise way allowing for a quick interpretation and understanding of the data. The boxplot displays a 5-summary consisting of the minimum and maximum values, the upper and lower quartiles, and the median.

Typically, the boxplot is constructed by distributing the data into four subsets of equal size, delimited by the values of the quartiles and the maximum and minimum values. If extreme outliers are present in the data, the upper and lower limits of the boxplot may be defined as a multiple of the upper and lower quartile instead of the maximum and minimum values, respectively. The outliers are then represented outside the boxplot. Lastly, the median is marked as a line dividing the data into two subsets of equal size [18]. Additionally, the mean can also be represented in the boxplot.

As the goal of this work is to associate the extracted features with stress or affective states, a statistical analysis was performed, using the Python scikit-posthoc package[2] and considering a significance level of 5% for hypothesis testing.

Two statistical tests were used to identify significant statistical differences between the several conditions of the protocol, for each feature. Firstly, the non-parametric Friedmann test was performed to detect the presence of any significant statistical difference between the five conditions (the use of a non-parametric statistical test was decided considering the small number of participants in the dataset). In the case of the existence of a significant difference, the post hoc Nemenyi test was employed to identify which pairs of conditions were significantly different. Regarding the time series features, the statistical analysis was performed considering the median value of each feature, for each participant, in each condition. Since a large number of outliers were observed for various time series features, the median was chosen instead of the mean, as it is less influenced by the presence of outliers.

## 5 Results

The results of the statistical analysis are shown for the features from which a significant statistical difference was observed. The distribution of the features is analyzed through boxplots, with the additional representation of the mean of the feature for each condition (yellow triangle), and a statistical analysis matrix with the computed p-values, obtained using the post hoc Nemenyi tests, between the different conditions is also presented.

### 5.1 ECG

The distance between two consecutive R peaks and the heart rate provide similar information. As presented in Fig. 3, the boxplots and statistical results matrix both show that the interval between consecutive R peaks was smallest during the TSST, meaning that during this condition the heart rate of the participants increased, as expected. When comparing non-stress conditions, it is clear that there are no significant changes in the heart rate of the participants (as can be observed in the statistical results Fig. 3 – right).

The observations made for the distance between consecutive R peaks also applies to the other distances between the consecutive peaks (P, Q, S and T). From the remaining ECG features, de median and standard deviation of the R peaks height showed no significant changes in between conditions, while the maximum R peaks height had a significant increase during the TSST compared to the meditation condition. Lastly, the minimum R peak height showed a significant decrease between stress and both meditations and amusement conditions.

[2] https://scikit-posthocs.readthedocs.io/en/latest/intro.html.

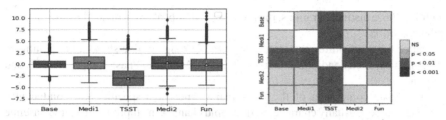

**Fig. 3.** Boxplot (left) and statistical results matrix (right) of the RRI feature.

## 5.2 EMG

From the seven features extracted from the EMG signal, two did not show statistically significant differences in the Friedmann test, namely the RMSA and the standard deviation of the EMG amplitude.

Regarding the number of EMG activations for each condition, although there is no significant difference between the baseline and the TSST, it is possible to observe that the number of activations diminished during meditation and amusement conditions, as shown in Fig. 4 (right). The post-hoc tests identified significant differences between the TSST and meditation1 and between TSST and amusement conditions, as shown in Fig. 4 (left). These differences could indicate that during meditation and amusement conditions the participants relaxed, leading to a decrease in muscle excitability.

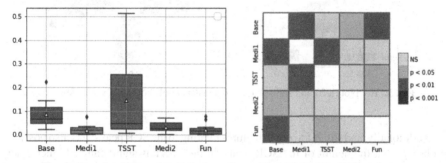

**Fig. 4.** Boxplot (left) and statistical results matrix (right) of the NNAct feature.

The remaining features (MEMG, SDEMG, MEMGamp and RMSE) showed significant differences between stress and meditation conditions. Additionally, both the RMSE and the SDEMG indicated statistically significant changes between amusement and both baseline and stress conditions. Lastly, it should be pointed out that no extracted EMG feature presented significant variations between baseline and stress conditions.

## 5.3 EDA

Due to a lack of SCR activity during the baseline of a significant number of participants, the EDA features weren't normalized concerning the baseline. Also, as long as the conditions have distinct durations, the number of SCR peaks was divided by the duration of the condition.

Out of the ten features extracted from the EDA signal, five showed no significant differences between conditions, namely, SCRamp, MPhasic, PhasicRange, AUSCR and RecTime.

The phasic component of the EDA mainly measures specific skin conductance responses (SCRs), usually characterized as rapid changes in amplitude from the baseline level (tonic component) value to the peak of the response. Thus, commonly SCR peaks occur in response to externally presented stimuli, such as stressful or surprise events, activating the SNS and causing an increase in perspiration and skin conductivity. By analyzing the boxplot and the results of the statistical tests, shown in Fig. 5, it is possible to notice that the number of SCR was highest during the TSST condition. When comparing the other conditions, the number of peaks was similar and significantly lower than when in stress, showing that the TSST protocol was able to activate the SNS. Lastly, as there are no statistically significant differences between the other four conditions, it is possible to infer that the participants were able to return to their baseline state after the realization of the stress condition.

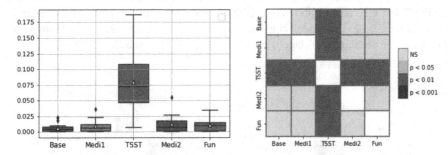

**Fig. 5.** Boxplot (left) and statistical results matrix (right) of the NNSCR feature.

Additionally, and still referring to features relating to the SCR peaks, it is also possible to observe that the height of the SCR peaks in the TSST is significantly higher when compared to the other conditions, except for Meditation2, as displayed in the boxplot of Fig. 6. The statistical results from the post-hoc tests also confirm that the TSST is significantly different from the remaining conditions, except for Meditation2, as shown in Fig. 6 - right.

The remaining EDA features presented a statistical difference between stress and both meditations and amusement conditions. The range of the tonic component and both the rise time and recovery time of the SCR were the only EDA features unable to establish significant differences between baseline and TSST conditions. Finally, the rise time of the SCR and the median of the tonic component didn't show a significant variation between baseline and amusement conditions.

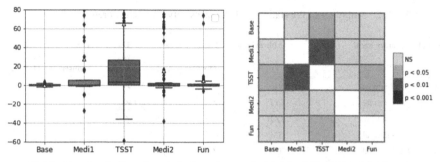

**Fig. 6.** Boxplot (left) and statistical results matrix (right) of the SCRheight feature.

## 5.4 RESP

When analyzing the statistical results for the RESP signal, none of the median features showed a significant difference between conditions. However, for all of the standard deviation features (SDIn, SDEx, SDTotal, SDRIE), the post-hoc tests showed a statistically significant difference between baseline and stress conditions, indicating that the breathing patterns of the participants became irregular during stress since the deviation from the mean durations increased, as can be observed, in Fig. 7, for the standard deviation of the ratio between inspirations and expirations.

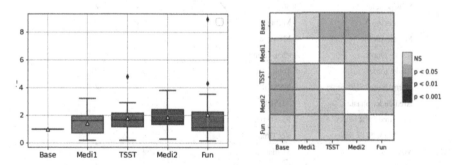

**Fig. 7.** Boxplot (left) and statistical results matrix (right) of the SDRIE feature.

## 5.5 Summary Results

Table 2 presents a summary of the statistical results for all features, including the results of the comparison between stress and remaining conditions. Additionally, it also contains a summary of the differences between amusement and baseline conditions. Lastly, the difference in statistical results between meditation-1 and meditation-2 was also included. The statistical results between the conditions mentioned above are presented as ✓; if a statistically significant difference was observed between the conditions, and was left out blank if no statistical difference could be established between the conditions.

**Table 2.** Summary of the statistical differences between the different conditions.

| Signal | Feature name | Abbreviation | Basel. vs TSST | Basel. vs Amusem. | TSST vs Medit. | TSST vs Amusem. | Medit.1 vs Medit. 2 |
|---|---|---|---|---|---|---|---|
| ECG | R-R interval | RRI | ✓ | | ✓ | ✓ | |
| | P-P interval | PPI | ✓ | | ✓ | ✓ | |
| | Q-Q interval | QQI | ✓ | | ✓ | ✓ | |
| | S-S interval | SSI | ✓ | | ✓ | ✓ | |
| | T-T interval | TTI | ✓ | | ✓ | ✓ | |
| | Median R height | MRH | | | | | |
| | StD R height | SDRH | | | | | |
| | Max R height | RHmax | | | ✓ | | |
| | Min R height | RHmin | | | ✓ | ✓ | |
| EMG | Median EMG | MEMG | | | ✓ | | |
| | StD EMG | SDEMG | | ✓ | ✓ | ✓ | |
| | Root mean square error EMG | RMSE | | ✓ | ✓ | ✓ | |
| | Median amp | MEMGamp | | | ✓ | | |
| | StD amp | SDEMGamp | | | | | |
| | Root mean square error amp | RMSA | | | | | |
| | Norm. Num. Act | NNAct | | ✓ | ✓ | ✓ | |
| EDA | SCR peak amplitude | SCRamp | | | | | |
| | SCR peak height | SCRheight | ✓ | | ✓ | ✓ | |
| | Median tonic | MTonic | ✓ | | ✓ | ✓ | |
| | Median phasic | MPhasic | | | | | |
| | Range phasic | PhasicRange | | | | | |
| | Range tonic | TonicRange | | ✓ | ✓ | ✓ | |
| | Norm. Num. SCR peaks | NNSCR | ✓ | | ✓ | ✓ | |
| | Area under phasic | AUSCR | | | | | |
| | Rise time SCR peaks | RiseTime | | ✓ | ✓ | ✓ | |
| | Rec. Time SCR peaks | RecTime | | | | | |
| RESP | Median exp | MedianEx | | | | | |
| | Median insp. | Medianin | | | | | |
| | Median total breath | MedianTotal | | | | | |
| | Median insp.-exp. Ratio | MedianRIE | | | | | |
| | StD insp. | SDE | ✓ | | | | |
| | StD exp | SDI | ✓ | | | | |
| | StD Total | SDTotal | ✓ | | | | |
| | StD insp.-exp. Ratio | SDRIE | ✓ | | | | |

# 6  Conclusions and Discussion

Using the WESAD dataset, this work analyzed the relation between stress and physiological features extracted from ECG, EMG, EDA and RESP.

From the ECG, HR was shown to be a good predictor of stress as it was possible to observe a statistically significant decrease in the interval between consecutive R peaks during the stress condition, providing a clear indication of an increase in the SNS activity, in comparison to the other conditions, including baseline and amusement conditions. On the other hand, the extracted features concerning the height of the ECG peaks showed almost no correlation with stress, with the exceptions of the maximum and minimum heights of the ECG, both showing a statistical difference between stress and meditation conditions, and a significant difference between stress and amusement for the minimum height.

Regarding EMG features, no statistical difference was obtained between baseline and stress conditions for any of the features. However, EMG standard deviation, root mean square, and the number of activations showed a significant statistical difference between amusement and both stress and baseline conditions. Lastly, the median, standard deviation and root mean square of the EMG signal, median of the amplitude, and number of activations presented a significant change between the stress and, at least, one of the meditation conditions. The analysis of the EMG features showed a high decrease in muscle excitability during meditation and amusement conditions, possibly showing that the participants relaxed during these phases.

Concerning the EDA, the best stress indicators appear to be the median of the tonic component and the number of SCR peaks, as for both, a significant difference was obtained between stress and all of the other conditions. From the remaining EDA features, only the rise time of the SCR peaks and the tonic range seem to have significant changes when comparing stress with meditation conditions, and the amusement with both stress and baseline conditions.

Lastly, from the extracted RESP features only the standard deviations of inspiration, of expiration, and total durations, and the ratio between the inspiration and expiration durations showed statistical differences between baseline and stress conditions.

In addition, none of the features extracted from the four signals revealed differences between meditation conditions, indicating that both were performed in similar circumstances.

The obtained results are in accordance with the conclusions from previous works [6, 8–14], showing that physiological features are associated with stress. Except for the EMG signal, several features from ECG, EDA and RESP presented significant differences among stress and baseline conditions.

Also using the WESAD dataset, *Iqbal et al.* [11] found that respiration rate was the best indicator of stress and that statistical features extracted from EMG and EDA phasic component showed no statistically significant difference between the stress and baseline states. These conclusions are sustained by the results of the present study. Out of 4 from 8 RESP features and out of 5 from 9 ECG features were able to distinguish between stress and baseline. However, as peaks intervals from ECG are highly related in between, RESP presented a greater ability to provide distinctive features. For the EMG, no statistical difference was obtained between baseline and stress conditions for any of the considered features. Regarding EDA features, the median of the phasic component (MPhasic) showed no difference among the conditions, while the number of SCR peaks and SCR height, not used in the compared work, showed significant differences between baseline and stress conditions.

Moreover, we also identify features with the ability to characterize amusement and meditation conditions. In contrast to the results regarding baseline and stress conditions, ECG features were able to discriminate between stress and amusement and stress and mediation conditions. On the other hand, the RESP signal reveals to be the most promising to distinguish between baseline and stress.

## 7    Final Remarks and Further Research

Stress impacts our daily life in many ways, posing several challenges to our overall well-being, including physical and mental health, and constraining social interactions. Often, stress exacerbates our emotional reactivity and limits our reasoning, hindering our capability of objective judgments and leading to an impairment of cognitive functions, such as rational thinking, decision-making, and problem-solving.

Being a response to any demand or threat that exceeds the individual ability to cope effectively, stress triggers the "fight or flight" response. As a consequence, blood flow diverges to regions of the brain with critical roles in evaluating and reacting to potential threats. Therefore, the brain's prefrontal cortex, responsible for logical and rational thinking, may receive less blood flow, affecting its capability.

Thus, stress management is of utmost importance to ensure our rationality.

The results obtained in this study are coherent with previous literature and with the physiological understanding of SNS, sustaining that efforts should be devoted to attaining stress recognition, based on physiological data obtained through wearable devices, and providing real-time strategies to cope with stress.

The reduced number of participants in the WESAD dataset hinders the generalization of these results. Thus, Bayesian inference should be considered to provide a deeper understanding of the overall strength of evidence for or against the presence of differences across conditions.

Besides the number of participants, during both versions of the protocol half of the participants were randomly chosen to perform the tasks while standing, and the other half performed them seated. This parameter was not considered in this analysis because no indication was given regarding which participants were standing and which were sitting. Moreover, this protocol considered two meditation conditions and only one state for the remaining conditions (rest, amusement, or stress), which adds an experimental bias. Thus, considering these limitations of the WESAD dataset, for future work, a new data acquisition protocol should be attained and a greater number of participants should be engaged. Lastly, a stress classification model must be addressed to obtain either binary (stress vs no-stress) or multiclass (considering various levels of stress) stress detection.

**Acknowledgements.** This work was funded by national funds through FCT –Fundação para a Ciência e a Tecnologia, I.P., under the Scientific Employment Stimulus CEECIND/03986/2018 (R.S.) and CEECINST/00013/2021 (R.S.). This work is also supported by the FCT through national funds, within the R&D unit IEETA/UA (UIDB/00127/2020).

# References

1. Mental Health and COVID-19 : Early evidence of the pandemic's impact, Scientific brief, World Health Organization,. Available: https://www.who.int/publications/i/item/WHO-2019-nCoV-Sci_Brief-Mental_health-2022.1, accessed: 10 Dec. 2022
2. Chrousos, G.P.: Stress and disorders of the stress system. Nat. Rev. Endocrinol. **5**, 374–381 (2009). https://doi.org/10.1038/NRENDO.2009.106
3. Epel, E.S., et al.: More than a feeling: A unified view of stress measurement for population science. Front. Neuroendocrinol. **49**, 146–169 (2018). https://doi.org/10.1016/J.YFRNE.2018.03.001
4. Cohen, S., Janicki-Deverts, D., Miller, G.E.: Psychological stress and disease. J. Am. Med. Assoc. **298**(14), 1685–1687 (2007). https://doi.org/10.1001/jama.298.14.1685
5. Fink, G.: Stress: Definition and history. Encycl. Neurosci. 549–555 (2009). https://doi.org/10.1016/B978-008045046-9.00076-0
6. Schmidt, P., Reiss, A., Duerichen, R., Van Laerhoven, K.: Wearable affect and stress recognition: a review. Sensors **19**, 4079 (2019). https://doi.org/10.3390/s19194079
7. Russell, J.A.: A circumplex model of affect. J. Pers. Soc. Psychol. **39**(6), 1161–1178 (1980). https://doi.org/10.1037/h0077714
8. Smets, E., De Raedt, W., Van Hoof, C.: Into the wild: the challenges of physiological stress detection in laboratory and ambulatory settings. IEEE J. Biomed. Heal. Informatics **23**(2), 463–473 (2019). https://doi.org/10.1109/JBHI.2018.2883751
9. Bota, P.J., Wang, C., Fred, A.L.N., Plácido Da Silva, H.: A review, current challenges, and future possibilities on emotion recognition using machine learning and physiological signals. IEEE Access **7**, 140990–141020 (2019). https://doi.org/10.1109/ACCESS.2019.2944001
10. Schmidt, P., Reiss, A., Duerichen, R., Van Laerhoven, K.: Introducing WeSAD, a multimodal dataset for wearable stress and affect detection. In: Proc. 2018 Int. Conf. Multimodal Interact. (ICMI 2018), pp. 400–408 (2018). https://doi.org/10.1145/3242969.3242985
11. Iqbal, T., et al.: A sensitivity analysis of biophysiological responses of stress for wearable sensors in connected health. IEEE Access **9**, 93567–93579 (2021). https://doi.org/10.1109/ACCESS.2021.3082423
12. Pourmohammadi, S.M.: A: Continuous mental stress level assessment using electrocardiogram and electromyogram signals. Biomed. Signal Process. Control **68**, 102694 (2021). https://doi.org/10.1016/j.bspc.2021.102694
13. Han, L., Zhang, Q., Chen, X., Zhan, Q., Yang, T., Zhao, Z.: Detecting work-related stress with a wearable device. Comput. Ind. **90**, 42–49 (2017). https://doi.org/10.1016/j.compind.2017.05.004
14. Pourmohammadi, S., Maleki, A.: Stress detection using ECG and EMG signals: A comprehensive study. Comput. Methods Programs Biomed **193** (2020). https://doi.org/10.1016/j.cmpb.2020.105482
15. Kudielka, B.M., Hellhammer, H., Kirschbaum, C.: Ten years of research with the trier social stress test. Soc. Neurosci. 56–83 (2007)
16. Allen, A.P., Kennedy, P.J., Dockray, S., Cryan, J.F., Dinan, T.G., Clarke, G.: The trier social stress test: principles and practice. Neurobiol. Stress **6**, 113–126 (2017). https://doi.org/10.1016/j.ynstr.2016.11.001
17. Khodadad, D., et al.: Optimized breath detection algorithm in electrical impedance tomography. Physiol. Meas. **39**(9), 094001 (2018). https://doi.org/10.1088/1361-6579/AAD7E6
18. Potter, K.: Methods for presenting statistical information: the box plot. Vis. Large Unstructured Data Sets **4**, 97–106 (2006)

# Artificial Agents and Interaction

# Prospective Memory Training Using the Nao Robot in People with Dementia

Kerem Tahan[(⊠)] and Bernard N'Kaoua

Université de Bordeaux, Bordeaux Population Health U-1219, 33000 Bordeaux, France
k.tahan@colisee.fr, bernard.nkaoua@u-bordeaux.fr

**Abstract.** The aim of the present study was to evaluate the ability of a robot to conduct cognitive training in elderly people with dementia. Fifty-two institutionalized elderly people were recruited and invited to participate in prospective memory training sessions, with or without a robot. They were divided into two equivalent groups in terms of age, level of education and MMSE score. Before and after the intervention phase, neuropsychological assessment was performed on each participant, including several cognitive and non cognitive (self-esteem) evaluations. Moreover, the sessions were recorded in order to compare the interaction behaviors of the 2 groups, using a validated observation grid. Results showed that: 1) the presence of the robot increases the interaction behaviors of the participants during the sessions (such as smiling, laughing, nodding, reaching out to others, talking, etc.); 2) That prospective memory training resulted in a significant increase of prospective memory performance, attentional abilities and executive functioning, but this improvement did not differ between the 2 groups These findings confirm the positive impact of a robot as a mediator of cognitive training, but suggest that further research is necessary to determine the effectiveness of these tools, in comparison with traditional training with humans.

**Keywords:** Aging · Dementia · Prospective Memory · Training · Robot · Randomised Control Trial

## 1 Problem

### 1.1 A Subsection Sample

Currently, more than 50 million people are living with dementia worldwide, and the number of people with dementia is projected to increase from around 57 million globally in 2015 to 152 million by 2050 (Nichols et al. 2022). Among the various neurodegenerative pathologies, Alzheimer's disease appears to be the most frequent cause of neurocognitive impairment, being the cause of approximately 70% of dementias.

Alzheimer's disease (AD) is characterized by memory loss, word-finding difficulties, executive dysfunction, and confusion in unfamiliar environments with secondary impairment of mood and quality of life (Wilson et al. 2011; Wilson et al. 1999).

Although studies of memory impairment in aging have primarily focused on retrospective memory, prospective memory difficulties have been shown to account for a

J. Baratgin et al. (Eds.): HAR 2023, LNCS 14522, pp. 281–295, 2024.
https://doi.org/10.1007/978-3-031-55245-8_19

large proportion of memory difficulties reported by older adults (Kliegel and Martin 2003). Prospective memory (PM) describes the ability to successfully plan and execute delayed intentions into the future (eg, Einstein and McDaniel 1990). Several studies have showed a general effect of age on PM performance (Schnitzspahn et al. 2009). In addition, people with mild cognitive impairment (MIC) perform PM tasks worse than healthy older people (McDaniel et al. 2011; Niedźwieńska and Kvavilashvili 2014; Shelton et al. 2016) and people with Alzheimer's disease (AD) perform worse on PM tasks than older people with MCI (Troyer & Murphy 2007).

Declines in PM functioning are of particular concern for older adults because of the strong links between PM and others cognitive abilities (Adda et al. 2008; Banville and Nolin 2000; Carlesimo et al. 2004; Knight et al. 2005; Schmitter-Edgecombe and Wright 2004; Titov and Knight 2000). Indeed, in different patient populations, a significant association was found between poor performance on PM tests and impaired planning (Shum et al. 2013), set-shifting abilities (Kumar et al. 2008) selective and divided attention (Pavawalla et al. 2012) and working memory.

Moreover, it has been shown that these PM difficulties have a much greater impact on the person's daily functioning and on the burden of their carers, compared to deficits in retrospective memory (Smith et al. 2000). This finding is probably due to the fact that PM is a key predictor of functional independence in Older Adults (Hering et al. 2018).

For example, a correlation has been shown between the Virtual Week (computerized assessment of PM) and the Instrumental Activities of Daily Life (Hering et al. 2018). Other PM assessments (the TBPM and EBPM) have been shown to be good predictors of age-related deficits in daily functioning (Sheppard et al. 2020). Poor performance in PM is associated with poor performance in the ecological "day-out" task, assessing daily functioning in people with MCI (Schmitter-Edgecombe et al. 2012). Finally, poor PM performance and self-reported PM failures are also associated with poorer quality of life in older adults (Woods et al. 2015). Given the consequences of PM decline on independent functioning, it is critical to explore and examine training conditions or interventions that may improve PM performance in older adults (Tse et al. 2020). Indeed, PM has been trained in people with healthy aging or people with dementia, sometimes with positive results (Kixmiller 2002) even if the repercussions on quality of life and activities of daily living remain to be clarified (Herring et al. 2014).

More recently, robots have been developed as a novel cognitive stimulation approach for the elderly, or those with general cognitive decline. This includes the use of Socially Assistive Robot (AR) for companionship (Marian and Bilberg 2008; Shibata 2012) exercise coaching (Feil-Seifer and Mataric 2005), assistance of daily living (Rudzicz et al. 2015) and to facilitate recreational activities (McColl et al. 2013; Tahan et al. 2023).

## 2   Assistive Robot Acceptance by Healthy Elderly and People with Dementia

Many animal-inspired robots (such as Paro, AIBO, etc.) have been used to maintain interest in the elderly and have yielded positive social and psychological outcomes compared to usual care (Tulsulkar et al. 2021; Yu et al. 2022). For example, in Liang et al.

(2017), the behavioral, affective and social responses of people with dementia interacting with the Paro robot were observed using a temporal sampling method. The presence of agitated behaviors (eg, repetitive behavior, wandering, fiddling), facial expressions (eg, smiling, sadness, fear), and social interactions (eg, talking to others, cooperation, reciprocity) were recorded during sessions where the researchers presented Paro, then passed him around so that each person could interact with him. Results showed that Paro significantly improved facial expressions and social interaction. Similar results were obtained by Moyle et al. (2017) and Hung et al. (2021).

Although there are fewer studies, humanoid-like robots (such as Nao, Pepper, etc.), are also used to interact with older people. In the Nakamura et al. (2021) study, for example, results showed that 60% of participants engaged in dialogue with mobile robots (twice as many as with stationary robots). Additionally, participants who did not engage in dialogue had low Mini-Mental State Examination (MMSE) scores.

In general, the conclusion of the studies is that social humanoid robots show promise in facilitating engagement in social interactions in people with dementia, but Martín Rico et al. (2020) insisted on the need for adapted and specific tools to assess the attitudes of people (and in particular people with dementia) towards robots.

## 3  Assistive Robot and Cognitive Training

In recent years, these robots have been used in a cognitive training context (Tanaka et al. 2012; Kim et al. 2015; Pino et al. 2019). For example, Pino et al. (2020) evaluated the effectiveness of human–robot interaction to reinforce therapeutic behavior and treatments in dementia people. The robot was programmed to perform some tasks from the usual memory training program. The subjects participated in sessions with the support of Nao or with the support of the psychologist. The data show that the training condition with Nao was associated with an increase in the patients' visual gaze and an enhancement of therapeutic behavior with, in some cases, a reduction in depressive symptoms.

## 4  Aim

Today, Social Assistive Robots are increasingly used to assist the elderly and the effectiveness of these devices must therefore be carefully assessed. In their review of the literature, Yu et al. (2022) conclude that the feasibility and acceptability of companion robotic animals is mixed and unclear. However, People with dementia enjoy the interaction with humanoid companion robots as, unlike animal robots, they are usually designed to speak although there can sometimes be problems with speech recognition. As a result, robots and particularly humanoid robots have been used as assistants in cognitive retraining protocols for people with dementia (Yuan et al. 2021). In this context, our study has 2 complementary aims:

– Evaluate and quantify interactions with a humanoid Assistive Robot of people with dementia (compared to usual care): very few studies have quantified the different interactions (between participants, between participants and robots, between participants and usual care) during cognitive retraining sessions, and compared these different

behaviors during sessions animated by a humanoid robot or by a usual care. In addition, our study is based on an observation grid (SOBRI) specifically dedicated to the observation of self-centered or other-centered behaviors of people with dementia;
- To compare, in a Controlled Random Trial design, a group of people with dementia trained by usual care with a group in which the training sessions are assisted by the Nao robot. The training task used is a prospective memory task with spaced retrieval. This technique consists of having the participant repeat the information to be retained, with gradually increasing intervals of information retention. In elderly people with cognitive decline or diagnosed with dementia, this method has been shown to be effective in improving prospective memory retrieval (Kinsella et al. 2007; Ozgis et al. 2009). The effect of these 2 types of intervention (usual care and usual care assisted by the Nao robot) will be evaluated on the cognitive and non-cognitive spheres.

## 5 Material and Method

### 5.1 Participants

Fifty-two institutionalized persons (58% women, age range 71–96 years) with Alzheimer disease elderly residing in two structures participated in the study (Table 1). Diagnoses were made by professionals.

Participants were randomly assigned to a robot-assisted cognitive training group (n = 26) or to usual-care group (n = 26). The 2 groups had a homogeneous male/female distribution, and did not differ for age, MMSE score (Folstein et al. 1975), and BDEA (Goodglass and Kaplan 1972; reading words and sentences) scores (respectively, $U = 333$; $p = .93$; $U = 325$; $p = .33$; $U = 268$; $p = .16$).

Participants with severe hearing or visual impairment, or behavioral impairment incompatible with the intervention were excluded from the study. Before the intervention, the study was presented to all participants so that they could decide if they wanted to participate. A consent form was completed by all persons wishing to participate in the study.

They were asked to participate in 8 training sessions on a prospective memory task with a frequency of two weekly sessions of 45 min. The intervention group was able to benefit from training conducted by the Nao robot and assisted by an usual care, while the control group performed the same intervention with the sole presence of an usual care.

### 5.2 Material

We used the Nao robot, which is one of the most popular humanoid robots in the world. It is widely used in research, in care services and in education. Its price is relatively affordable compared to most of the other robots on the market, which makes it a relevant robot for institutions looking for innovation in terms of activities. In addition, the software used to program the robot is simple to handle, which facilitates its use by professionals who are not qualified in robotics and programming. Regarding his physical characteristics, he is 58 cm tall and weighs 4.3 kg. It has 25 degrees of freedom allowing

**Table 1.** Descriptive characteristics of the two groups of participants

| Characteristics | Robot-assisted group | Usual care group |
|---|---|---|
| Age (years) | 71–96 (85,7) | 70–99 (87,5) |
| Gender | | |
| Male | 8 | 9 |
| Female | 18 | 17 |
| Dementia stage | | |
| Early | 16 | 16 |
| Middle | 10 | 10 |
| MMSE | 9–27 (20,5) | 11–28 (20,8) |
| BDAE (Boston Diagnostic Aphasia Examination: reading words & sentences) | Words: 29,23 Sentences: 8,96 | Words: 30 Sentences: 9,42 |
| Study level | | |
| Without diploma | 4 | 4 |
| Primary school certificate | 7 | 6 |
| College diploma | 15 | 16 |

it to move in all directions and adapt to the environment. It is also equipped with two 2D cameras, seven sensory sensors and four microphones and speakers.

All sessions conducted either with the usual caregiver or with Nao were videotaped, using 2 cameras placed at strategic locations to capture all the behavior of the participants (Fig. 1). Video data was viewed on a computer and manually analyzed by two independent reviewers. As soon as a behavior on the grid was observed, the evaluator stopped the video and marked it on a paper version of the ethogram. The onset time of each behavior was taken into account to see if both raters were observing the same behavior at the same time.

### 5.3 Acceptance of the Assistive Robot

The ethogram of observation (Mabire et al. 2016) consisted of 126 behaviors assigned to four categories: social interactions with other residents or with care staff (facial expressions, looking, verbal interactions, quasi- linguistic interactions, and interactive behaviors), self-centered behaviors (facial expressions, looking, non-directed verbalizations, sadness behaviors, stereotypical behaviors, comfort behaviors, inactivity behaviors, waiting behaviors, and movements) and unclassifiable behaviors. Moreover, the ethogram focused not only on behaviors directed at others but also on self-centered behaviors. The ethogram is presented in Fig. 2.

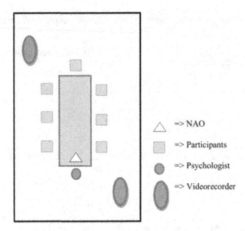

**Fig. 1.** A figure caption is always placed below the illustration. Short captions are centered, while long ones are justified. The macro button chooses the correct format automatically.

## 5.4 Cognitive Training

Before and after the intervention period, a neuropsychological assessment was carried out for each participant, including a memory, attentional and executive evaluation. The tests included: the envelope task (Huppert et al. 2000) for the assessment of prospective memory; the TMT-B for assessing mental flexibility, and the Modified Six Elements Test (Wilson et al. 1996) for assessing planning abilities. Selective attention and divided attention were assessed using computerized discrimination tasks (Rodda et al. 2011). Subjective measurements were also performed to provide additional information. First on memory with the PRMQ (Smith et al. 2000), which is a self-administered questionnaire with 8 items on prospective memory deficits in everyday life and 8 items on everyday memory difficulties retrospective. On a scale of 0 to 4 (from "Very often" to "Never"), the participants had to determine the frequency of appearance of certain memory difficulties. This tool has recently been validated in French with regard to its good psychometric characteristics (Guerdoux-Ninot et al. 2019) and would indeed constitute a useful questionnaire in the identification of memory complaints in the elderly. Additionally, we collected subjective data on participants' well-being with the Rosenberg Self-Esteem Scale (Rosenberg 1965).

## 5.5 Procedure

The spaced retrieval technique was used by Kinsella et al. (2007) to improve performance on a prospective memory task in which participants had to read a text aloud and remember to replace a target word (the name of the main character in the story) with another each time it appeared in the text. Immediately after the presentation of the instruction, the experimenter asked the participant: "What should you do while reading the text? ". If the participant was not sure, they were simply asked to answer "I don't know".

In our study, we adapted this procedure. Indeed, we explained to the participants that they were going to carry out a task of reading a text composed of 3 pages. First, the

**Self centered behaviors**

| | | |
|---|---|---|
| Facial expression | Positive | Smile / Laugh / Surprise |
| | Negative | Grimace / Frown / Raise eyes to the sky |
| Look | | Toward oneself / Blank stare / Outside/object (robot or people) |
| Non-directed verbalisation | Understandable – Positive | Talk / Whisper / Sing |
| | Understandable – Negative | speak loud / Yell / Insult |
| | Negative | Mumble / Rave / Scream |
| Sadness behaviours | | Outside/object (robot or people) / Fiddle with objects |
| Stereotypal behaviours | | Scrub the table / Swing / hitting oneself or hitting an object |
| Comfort behaviours | | Get up and sit down / Put on dentures/glasses / Readjust clothes or hair / Replace on chair / Yawn / Drink / Scratch |
| Inactivity behaviours | | To be withdrawn / Doze off / Sigh |
| Waiting behaviours | | Manipulate an objet / Wriggle on chair |
| Movements | Positive | Dance / To get closer |
| | Negative | Move away / Leave |

**Behaviours directed to others**

| | | |
|---|---|---|
| Facial expression | Positive | Smile / Laugh / Surprise |
| | Negative | Grimace / Frown / Raise eyes to the sky |
| Look | | At others / Mutual |
| Verbal interactions | Positive | Speak / Ask / Answer somebody / Humor |
| | Negative | Scream / Insult somebody |
| Quasi-linguistic interactions | Positive | "Hello" / "bye" / "Yes" with the head / Applause / Reach out to others |
| | Negative | Move away / Push someone away / "no" with hand / No with the head |
| Interactive behaviours | Positive and close | Take in arms / Touch somebody / Caress / Kiss / Bend / turn towards / Follow / Help / Leave with somebody |
| | Negative and agressive | Violence |

**Behaviours directed to speaker or robot**

| | | |
|---|---|---|
| Facial expression | Positive | Smile / Laugh / Surprise |
| | Negative | Grimace / Frown / Raise eyes to the sky |
| Look | | Mutual / At others |
| Verbal interactions | Positive | Ask / answer somebody / Humor |
| | Negative | Scream / Insult |
| Quasi-linguistic interactions | Positive | "Hello" / "Bye" / "Yes" with the head / Applause / reach out to others |
| | Negative | Distance / Push back / No with hand or the head |

**Fig. 2.** The ethogram of observation used in our study, adapted from Mabire et al. (2016)

experimenter (robot or human) gave the following instructions: "during your reading you will have to circle a target word and replace it with another but without writing it". This target word corresponded to the name of the main character of the story. To check that the instructions had been properly understood, each participant received a sheet with 5 suggestions on which was written the following question: "What word should you replace the target word of the story with?". The participants had to circle the word he had to use to replace the target word. This check was performed at intervals of 30 s, 1 min and 3 min. If the participant did not remember the answer, the experimenter made him reread the instruction and started again with the same time interval until the participant circled the correct answer. (e.g.: if a wrong answer is given at the 30-s interval, the experimenter has the instruction read again and the participant starts a 30-s interval again until the correct answer is obtained. Same process for the 1-min intervals and 3 min). The time interval was therefore increased only when the participant provided a correct answer. The entire procedure was repeated 3 times.

Following these three checks, the experimenter distributed the text to each participant and during their reading they had to circle the target word and remember to replace it with the word they had to circle during the instruction. At the end of the reading, the experimenter again distributed a sheet with 5 propositions on which the participants had to circle the word replacing the target word of the story.

Finally, a re-reading was carried out to all the participants by the robot or by the experimenter, and this re-reading included the word that the participants had to retain and replace. These tasks were proposed during each of the 8 retraining sessions with a text and a different target word at each session.

## 6 Results

### 6.1 Acceptance of the Assistive Robot

Table 2 presents the scores (means and standard deviations) of the behaviors observed (for the 3 categories identified (self-centered behaviors, participant/participant, interaction participants/animator) and the 2 conditions (animation by the robot or by the usual caregiver). For each indicator, and for each condition, non-parametric tests (Mann-Whitney U Test) were used to compare the averages of the group with robots with those of the groups with usual care.

**Interactions Resident/Resident:** Significant differences are present for the following indicators: smile ($U = 57.5$; $p = 0.002$), laugh ($U = 66$; $p = 0.01$); scream ($U = 94.5$; $p = 0.03$); answer somebody ($U = 63.5$; $p = 0.009$) and "yes" with the head ($U = 88.5$; $p = 0.03$). Smile, laugh, "yes" with the head and answer somebody were significantly more frequent with the robot group, while scream was more frequent with the caregiver group.

**Interactions Resident/Animator:** Significant differences are present for the following indicators: smile ($U = 63$; $p = 0.009$), laugh ($U = 59.5$; $p = 0.006$), "yes" with the head ($U = 55$; $p = 0.03$), humor ($U = 66.5$: $0.008$), reach out to others ($93.5$; $p = 0.01$). In any case, these behaviors were significantly more frequent with the robot compared to the caregiver group.

**Table 2.** Scores (means and standard deviations) of the behaviors observed on the different indicator (to facilitate reading, only the scores for which significant results are observed have been presented).

| Behaviours directed to others | | | | Behaviours directed to speaker or robot | | | | Self centered-behaviors | | | |
|---|---|---|---|---|---|---|---|---|---|---|---|
| Items | Group | Mean | p | Items | Group | Mean | p | Items | Group | Mean | p |
| Smile | Robot | 0.78 | 0.002 | Smile | Robot | 1.17 | 0.009 | Smile | Robot | 0.25 | 0.088 |
| | Usual care | 0.10 | | | Usual care | 0.46 | | | Usual care | 0.03 | |
| Laugh | Robot | 1.50 | 0.011 | Laugh | Robot | 1.61 | 0.006 | Laugh | Robot | 0.20 | 0.037 |
| | Usual care | 0.42 | | | Usual care | 0.64 | | | Usual care | 0.08 | |
| Answer somebody | Robot | 2.52 | 0.009 | Humor | Robot | 0.58 | 0.008 | Speak | Robot | 2.46 | 0.005 |
| | Usual care | 0.67 | | | Usual care | 0.16 | | | Usual care | 0.73 | |
| Scream | Robot | 0.01 | 0.039 | Scream | Robot | 0.01 | 0.090 | Sing | Robot | 0.15 | 0.061 |
| | Usual care | 0.11 | | | Usual care | 0.09 | | | Usual care | 0.00 | |
| «Yes» with head | Robot | 0.22 | 0.031 | «Yes» with head | Robot | 1.77 | 0.003 | Wander | Robot | 0.29 | 0.054 |
| | Usual care | 0.02 | | | Usual care | 0.44 | | | Usual care | 0.00 | |
| | | | | Reach out to others | Robot | 0.06 | 0.015 | Yawn | Robot | 0.03 | 0.011 |
| | | | | | Usual care | 0.00 | | | Usual care | 0.17 | |
| | | | | | | | | Sigh | Robot | 0.05 | 0.046 |
| | | | | | | | | | Usual care | 0.17 | |

**Self Centered-Behaviors:** Significant differences are present for the following indicators: laugh (U = 82; p = 0.03), speak (U = 57.5; p = 0.005), yawn (U = 76.5; p = 0.01) and sigh (U = 88; p = 0.04). Laugh and speak were significantly more frequent with the robot group, while yawn and sigh were more frequent with the caregiver group.

## 6.2 Prospective Memory Training

Table 3 presents the mean scores (and standard deviations) of the two groups for each of the neuropsychological measures collected before and after the intervention. The descriptive statistics show that the mean scores are all up slightly after the intervention period, with the exception of the PRMQ which decreases. This subjective measure evaluating the memory complaint, the more the value decreases, the less the memory complaint is pronounced.

**Table 3.** Mean scores (and standard deviations) in neuropsychological tests before intervention and after intervention, according to group.

| | Forward span | Backward span | Forward corsi | Backward corsi | Envelope task | MSET | TMT-B | Self esteem | PRMQ PM | PRMQ RM | PRMQ Total | Selective attention | Divided attention |
|---|---|---|---|---|---|---|---|---|---|---|---|---|---|
| Pre test | | | | | | | | | | | | | |
| Usual care | 5,12 (1,21) | 3,35 (0,892) | 4,54 (0,989) | 3,77 (1,03) | 0,5 (0,762) | 2,2 (1,98) | 8,88 (7,79) | 31,1 (4,57) | 15,4 (4,08) | 16 (3,77) | 31,3 (6,83) | 42.3 (4.9) | 35.3 (7.3) |
| Robot | 4,88 (0,816) | 3,19 (0,567) | 4,08 (0,862) | 3,58 (0,902) | 0,346 (0,485) | 2,36 (1,66) | 7,64 (5,7) | 29,7 (6,18) | 14,7 (5,23) | 16,8 (5,25) | 31,5 (9,88) | 45.2 (3.1) | 37.9 (5.7) |
| Post test | | | | | | | | | | | | | |
| Usual care | 5,81 (1,02) | 4,12 (0,816) | 5 (0,98) | 4,04 (0,958) | 1,12 (0,816) | 3,92 (1,66) | 14,1 (8,19) | 32 (4,22) | 14,2 (4,29) | 15,2 (4,03) | 29,9 (7,5) | 46.2 (1.9) | 40.6 (7.1) |
| Robot | 5,19 (0,749) | 3,65 (0,562) | 4,31 (0,884) | 3,62 (0,898) | 1,23 (0,71) | 4,36 (1,47) | 12,1 (8,05) | 31,8 (5,66) | 14,1 (4,2) | 15,9 (3,29) | 29,2 (6,58) | 46.9 (1.6) | 41.3 (6.0) |

The pre- and post-intervention neuropsychological data were analyzed using a two-way ANOVA, one with unpaired measures ("group" with two modalities control vs.

experimental) and the other with repeated measures ("intervention" factor with two modalities before vs. after). Statistical analyzes revealed a significant effect of the intervention for forward span task [F(1,50) = 13.24; p < 0.001], backward span task [F(1,50) = 23.97; p < 0.001], envelope task [F(1.50) = 39.73; p < 0.001], MSET [F(1.50) = 51.57, p < 0.001], selective attention [F(1,50) = 11.46, p < 0.003], divided attention [F(1,50) = 19.89, p < 0.001] and TMT- B [F(1,50) = 19.57; p < 0.001], indicating that the scores on these tests increased significantly after compared to before the intervention. The results of the self-esteem and memory complaint questionnaires revealed no significant effect of the intervention.

Finally, the analyzes revealed no group effect (difference robot vs usual care) and no interaction between the group and the intervention.

# 7  Discussion

The aim of our work was twofold: 1) to evaluate and quantify interactions with a humanoid Assistive Robot of people with dementia (compared to usual care); 2) and to compare, in a Controlled Random Trial design, a group of people with dementia trained by usual care with a group in which the training sessions are assisted by the Nao robot.

## 7.1  Interaction with Assistive Robot (Nao)

An important point of our study is that it evaluates the effectiveness of a humanoid robot, in a Controlled Random Trial design, during prospective memory training sessions, using an observation grid specifically designed to assess the behavior of people with dementia living in institutions. Indeed, the majority of studies in this field relate to the use of pet robots (PARO type) during more or less free interaction sessions of residents and staff with the robot. For example, Joranson et al. (2016) studied the behaviors of people with dementia during group activities with the Paro robot. The results showed that the use of Paro was associated with an increase in smiles or laughter directed at other participants. The authors concluded that the presence of Paro increased participants' social interactions and engagement. Wada and Shibata (2007) placed PARO in a public area of an institution for the elderly (for 2 months) and showed increased interaction between residents. The presence of the robot encouraged participants to communicate with each other and strengthened social ties over the two months.

Indeed, if the studies using an animal robot during free sessions are relatively numerous and show the benefits of the robot on many parameters, few studies have observed the behavior of people with dementia using a humanoid robot during rehabilitation sessions cognitive (prospective memory exercises in our study) with a Randomized Controlled Trial design (comparison robot/usual care). In this context, our results specify the behaviors that differ between the use of the robot and the usual care. It involves a wide range of positive and interactive social behaviors such as smiling, laughing, reaching out to others, talking, etc. The spheres concerned are both self-centered behaviors, resident/resident interactions, and resident/robot interactions. Our results confirm the impact of the robot on all facets of the social life of residents with dementia.

Finally, another important point of our study is to evaluate the interactions using a tool specifically dedicated to people with dementia. In their systematic review on social assistive robots for people with dementia, Yu et al. (2022) concluded that research in this area are sometimes difficult to compare or interpret due to the use of non-validated outcome instruments to collect interaction data. In our study, we used a tool developed by Mabire et al. (2016) which consists of a scientifically validated ethogram specifically dedicated to the study of the social interactions of institutionalized people with dementia.

## 7.2 Prospective Memory Training Intervention

In our study, a prospective memory training intervention carried out by usual care and an intervention carried out by usual care assisted by the Nao robot were compared on a set of cognitive measures in people with dementia. As the first result, prospective memory is improved as evidenced by the increase in scores on the envelope task observed in the 2 groups. This result is in agreement with other studies. For example, Farzin et al. (2018) evaluated the efficiency of a multicomponent prospective memory training in healthy older adults. The results showed that the training program was effective on the subjective and objective performance of prospective memory in healthy elderly people, also with positive effects on participants' activities of daily living. In their systematic review, Tse et al. (2022) identified forty-eight studies of prospective memory training in older adults, and 43% of these studies showed positive gains in performance.

In our study, in addition to prospective memory, post-intervention performance gains are transferred to other cognitive functions such as executive functions, selective attention, and divided attention. This result is consistent with work showing a significant association between poor performance on PM tests and impaired planning, set-shifting abilities, selective and divided attention, and working memory (Brooks et al. 2004). In the same way, Agogiatou et al. (2020) trained PM and investigated the transferability to other cognitive domains as well as the transfer of these benefits to ADLs in People with Mild Cognitive Impairment. At the end of the intervention, the experimental group (PM training) outperformed the control group in working memory, verbal fluency and ADL. All of these data confirm the link between PD and many other cognitive functions as well as the importance of PD for the daily functioning of the elderly.

A final objective of our study was to evaluate the interest of using a companion robot as a training mediator compared to usual care. Statistical analyzes revealed no significant effect of the experimental condition (control vs. experimental) on either cognitive or non-cognitive measures. Other studies have shown positive effects of robot-assisted training compared to usual care training. For example, the objective of the study by Park et al. (2021) was to assess the ability of multi-domain cognitive training programs, particularly robot-assisted training, to improve cognitive functions and depression in community-dwelling older adults with dementia mild cognitive impairment (MCI). One group consisted of 90 participants who would receive cognitive training and 45 who would receive no training (NI). The cognitive training group was randomly divided into two groups, 45 who received traditional cognitive training and 45 who received robot-assisted cognitive training.

For example, Park et al. (2021) assessed the ability of multi-domain cognitive training programs, particularly robot-assisted training, to improve cognitive function and depression in older adults with mild cognitive impairment (MCI). 90 participants would receive cognitive training and 45 participants would receive no training. The 90 participants were randomly divided into two groups, 45 who received traditional cognitive training and 45 who received robot-assisted cognitive training. Results show that robot group participants had significantly greater post-intervention improvement in memory, executive functions, and depression. The authors concluded that a 6-week robot-assisted cognitive training program can improve performance in global cognitive functions and depression in older adults with MCI. Pino et al., (2020) assessed the effectiveness of human-robot interaction in enhancing therapeutic behavior and treatment adherence in older adults. The subjects participated in sessions with Nao or with the psychologist. Training with Nao was associated with an increase in patients' visual gaze and improved therapeutic behavior with, in some cases, a reduction in depressive symptoms. Unexpectedly, significant changes in prose memory and measures of verbal fluency were detected. The authors conclude that further research on robotics in ecological contexts is needed to determine the extent to which they can effectively support clinical practice.

Investigations in the field are recent and dependent on technological advances but it will be necessary, in the future, to identify more clearly the parameters (characteristics of the training tasks, of the participants, etc.) which promote maximum efficiency when using a robot in the training of elderly people in good health or suffering from dementia.

**Acknowledgments.** This study is supported by the Colisée Group and more particularly by Doctor Vincent KLOTZ (medical director).

**Funding.** The authors have no funding to report.

**Data Availability.** The data supporting the findings of this study are available within the article and/or its supplementary material.

**Disclosure of Interests.** The authors have no competing interests to declare that are relevant to the content of this article.

# References

Adda, C.C., Castro, L.H., e Silva, L.C.A.M., de Manreza, M.L., Kashiara, R.: Prospective memory and mesial temporal epilepsy associated with hippocampal sclerosis. Neuropsychologia **46**(7), 1954-1964 (2008)

Agogiatou, C., Markou, N., Poptsi, E., Tsolaki, M.: Is it possible the training of prospective memory to enhance activities of daily living and executive function in people with mild cognitive impairment? A single-blind randomized controlled trial. Acta Sci. Med. Sci. **4**, 102–113 (2020). https://doi.org/10.31080/ASMS.2020.04.0747

Banville, F., Nolin, P.: Mémoire prospective: Analyse des liens entre les fonctions mnésiques et les fonctions frontales chez des adultes victimes d'un traumatisme craniocérébral. Rev. Neuropsychol. **10**, 255–279 (2000)

Brooks, B.M., Rose, F.D., Potter, J., Jayawardena, S., Morling, A.: Assessing stroke patients' prospective memory using virtual reality. Brain Inj. **18**, 391–401 (2004)

Carlesimo, G.A., Casadio, P., Caltagirone, C.: Prospective and retrospective components in the memory for actions to be performed in patients with severe closed-head injury. J. Int. Neuropsychol. Soc. **10**(5), 679–688 (2004)

Chang, W.L., Šabanović, S.: Interaction expands function: social shaping of the therapeutic robot PARO in a nursing home. In: 2015 10th ACM/IEEE International Conference on Human-Robot Interaction (HRI), pp. 343–350 (2015)

Einstein, G.O., McDaniel, M.A.: Normal aging and prospective memory. J. Exp. Psychol. Learn. Mem. Cogn. **16**(4), 717 (1990)

Farzin, A., Ibrahim, R., Madon, Z., Basri, H.: The efficiency of a multicomponent training for prospective memory among healthy older adults: a single-blind, randomized controlled within-participants cross-over trial. Am. J. Phys. Med. Rehabil. **97**(9), 628–635 (2018)

Feil-Seifer, D., Mataric, M.J.: Defining socially assistive robotics. In: 9th International Conference on Rehabilitation Robotics, 2005, ICORR 2005, pp. 465–468. IEEE, June 2005

Guerdoux-Ninot, E., Martin, S., Jailliard, A., Brouillet, D., Trouillet, R.: Validity of the French Prospective and Retrospective Memory Questionnaire (PRMQ) in healthy controls and in patients with no cognitive impairment, mild cognitive impairment and Alzheimer disease. J. Clin. Exp. Neuropsychol. **41**(9), 888–904 (2019)

Hering, A., Kliegel, M., Rendell, P.G., Craik, F.I., Rose, N.S.: Prospective memory is a key predictor of functional independence in older adults. J. Int. Neuropsychol. Soc. **24**(6), 640–645 (2018)

Herring, M.P., Lindheimer, J.B., O'Connor, P.J.: The effects of exercise training on anxiety. Am. J. Lifestyle Med. **8**(6), 388–403 (2014)

Hung, L., et al.: Exploring the perceptions of people with dementia about the social robot PARO in a hospital setting. Dementia **20**(2), 485–504 (2021)

Huppert, F.A., Johnson, T., Nickson, J.: High prevalence of prospective memory impairment in the elderly and in early-stage dementia: Findings from a population-based study. Appl. Cogn. Psychol.: Off. J. Soc. Appl. Res. Mem. Cogn. **14**(7), S63–S81 (2000)

Jøranson, N., Pedersen, I., Rokstad, A.M.M., Ihlebaek, C.: Change in quality of life in older people with dementia participating in Paro-activity: a cluster-randomized controlled trial. J. Adv. Nurs. **72**(12), 3020–3033 (2016)

Kixmiller, J.S.: Evaluation of prospective memory training for individuals with mild Alzheimer's disease. Brain Cogn. **49**(2), 237–241 (2002)

Kim, G.H., et al.: Structural brain changes after traditional and robot-assisted multi-domain cognitive training in community-dwelling healthy elderly. PLoS ONE **10**(4), e0123251 (2015)

Kinsella, G.J., Ong, B., Storey, E., Wallace, J., Hester, R.: Elaborated spacedretrieval and prospective memory in mild Alzheimer's disease. Neuropsychol. Rehabil. **17**(6), 688–706 (2007)

Kliegel, M., Martin, M.: Prospective memory research: why is it relevant? Int. J. Psychol. **38**(4), 193–194 (2003)

Knight, R.G., Harnett, M., Titov, N.: The effects of traumatic brain injury on the predicted and actual performance of a test of prospective remembering. Brain Inj. **19**(1), 19–27 (2005)

Liang, A., et al.: A pilot randomized trial of a companion robot for people with dementia living in the community. J. Am. Med. Dir. Assoc. **18**(10), 871–878 (2017)

Mabire, J.B., Gay, M.C., Vrignaud, P., Garitte, C., Vernooij-Dassen, M.: Social interactions between people with dementia: pilot evaluation of an observational instrument in a nursing home. Int. Psychogeriatr. **28**(6), 1005–1015 (2016)

Marian, N., Bilberg, A.: Experience with the behavior-based Robot development (2008)

Martín Rico, F., Rodríguez-Lera, F.J., Ginés Clavero, J., Guerrero-Higueras, Á.M., Matellán Olivera, V.: An acceptance test for assistive robots. Sensors **20**(14), 3912 (2020)

McColl, D., Louie, W.Y.G., Nejat, G.: Brian 2.1: a socially assistive robot for the elderly and cognitively impaired. IEEE Robot. Autom. Mag. **20**(1), 74–83 (2013)

Moyle, W., et al.: Use of a robotic seal as a therapeutic tool to improve dementia symptoms: a cluster-randomized controlled trial. J. Am. Med. Dir. Assoc. **18**(9), 766–773 (2017)

Nakamura, M., Ikeda, K., Kawamura, K., Nihei, M.: Mobile, socially assistive robots incorporating approach behaviour: requirements for successful dialogue with dementia patients in a nursing home. J. Intell. Robot. Syst. **103**, 1–11 (2021)

Nichols, E., et al.: Estimation of the global prevalence of dementia in 2019 and forecasted prevalence in 2050: an analysis for the Global Burden of Disease Study 2019. Lancet Public Health **7**(2), e105–e125 (2022)

Niedźwieńska, A., Rendell, P., Barzykowski, K., Leszczyńska, A.: Only social feedback reduces age-related prospective memory deficits in "Virtual Week." Int. Psychogeriatr. **26**(5), 759–767 (2014). https://doi.org/10.1017/S1041610214000027

Ozgis, S., Rendell, P.G., Henry, J.D.: Spaced retrieval significantly improves prospective memory performance of cognitively impaired older adults. Gerontology **55**(2), 229–232 (2009)

Palestra, G., Pino, O.: Detecting emotions during a memory training assisted by a social robot for individuals with Mild Cognitive Impairment (MCI). Multimedia Tools Appl. **79**(47–48), 35829–35844 (2020)

Park, E.A., Jung, A.R., Lee, K.A.: The humanoid robot Sil-Bot in a cognitive training program for community-dwelling elderly people with mild cognitive impairment during the COVID-19 pandemic: a randomized controlled trial. Int. J. Environ. Res. Public Health **18**(15), 8198 (2021)

Pavawalla, S.P., Schmitter-Edgecombe, M., Smith, R.E.: Prospective memory after moderate-to-severe traumatic brain injury: a multinomial modeling approach. Neuropsychology **26**(1), 91 (2012)

Pino, O., Palestra, G., Trevino, R., De Carolis, B.: The humanoid robot NAO as trainer in a memory program for elderly people with mild cognitive impairment. Int. J. Soc. Robot. **12**, 21–33 (2020)

Rodda, J., Dannhauser, T., Cutinha, D.J., Shergill, S.S., Walker, Z.: Subjective cognitive impairment: functional MRI during a divided attention task. Eur. Psychiatry **26**(7), 457–462 (2011)

Rosenberg, M.: Rosenberg self-esteem scale. J. Relig. Health (1965)

Rudzicz, F., Wang, R., Begum, M., Mihailidis, A.: Speech interaction with personal assistive robots supporting aging at home for individuals with Alzheimer's disease. ACM Trans. Accessible Comput. (TACCESS) **7**(2), 1–22 (2015)

Shelton, J.T., Lee, J.H., Scullin, M.K., Rose, N.S., Rendell, P.G., McDaniel, M.A.: Improving prospective memory in healthy older adults and individuals with very mild Alzheimer's disease. J. Am. Geriatr. Soc. **64**(6), 1307–1312 (2016)

Schmitter-Edgecombe, M., McAlister, C., Weakley, A.: Naturalistic assessment of everyday functioning in individuals with mild cognitive impairment: the day-out task. Neuropsychology **26**(5), 631 (2012)

Schnitzspahn, K.M., Kliegel, M.: Age effects in prospective memory performance within older adults: the paradoxical impact of implementation intentions. Eur. J. Ageing **6**, 147–155 (2009)

Sheppard, D.P., Matchanova, A., Sullivan, K.L., Kazimi, S.I., Woods, S.P.: Prospective memory partially mediates the association between aging and everyday functioning. Clin. Neuropsychol. **34**(4), 755–774 (2020)

Shibata, T.: Therapeutic seal robot as biofeedback medical device: Qualitative and quantitative evaluations of robot therapy in dementia care. Proc. IEEE **100**(8), 2527–2538 (2012)

Smith, G., Del Sala, S., Logie, R.H., Maylor, E.A.: Prospective and retrospective memory in normal ageing and dementia: a questionnaire study. Memory **8**(5), 311–321 (2000)

Tanaka, M., et al.: Effect of a human-type communication robot on cognitive function in elderly women living alone. Med. Sci. Monit. Int. Med. J. Exp. Clin. Res. **18**(9), CR550 (2012)

Titov, N., Knight, R.G.: A procedure for testing prospective remembering in persons with neurological impairments. Brain Inj. **14**(10), 877–886 (2000)

World Health Organization: World report on ageing and health. World Health Organization (2015).

# Dictator Game with a Robot in Children with Autism Spectrum Disorders: Sharing is Predicted by Positive Attributions Towards the Agent

Marion Dubois-Sage[1]([✉]) [ID], Yasmina Lembert[1] [ID], Frank Jamet[1,2,3] [ID],
and Jean Baratgin[1,2] [ID]

[1] Paris 8 University, 93526 Saint-Denis, France
`mariondbsg@hotmail.fr`
[2] Association P-A-R-I-S, 75005 Paris, France
[3] CY Cergy Paris Université, 95000 Cergy-Pontoise, France

**Abstract.** The increasing use of robots for individuals with Autism Spectrum Disorders (ASD), with the notable aim of supporting the development of social skills, prompts the question of the extent to which children with ASD interact with robots as with social agents. In the literature, resource-sharing tasks (e.g. dictator game) have already been used to study the perception of robots in typically developing children, but few studies have focused on sharing in children with ASD. 38 children aged 9 to 13 (28 typically developing children and 10 children with ASD) played the dictator game with a humanoid robot NAO and a human experimenter, in counterbalanced order. They also answered an adapted version of the Godspeed questionnaire designed to measure anthropomorphism. The results indicate that the quantity of shared resources is better predicted by the good personality score, which reflects the attribution of positive traits to the agent, than by the type of development: the higher the score, the greater the number of resources given by the participant. Contrary to our initial assumption, children with ASD tend to give fewer resources to the robot than to the human, unlike TD children who give the same amount to both agents. In the human experimenter condition, children with ASD and neurotypical children share similarly. Post-hoc analyses, however, indicate an effect of development type on good personality scores, with lower scores for ASD than neurotypical children. The attribution of positive traits to the agent could be the mediating variable between development type and sharing behavior.

**Keywords:** Child-robot interaction · Autism Spectrum Disorders · Anthropomorphism · Sharing behavior · Dictator game

## 1 Introduction

A great deal of research has been carried out on the prosocial behaviors of typically developing children, and in particular on their sharing behavior. However,

J. Baratgin et al. (Eds.): HAR 2023, LNCS 14522, pp. 296–322, 2024.
https://doi.org/10.1007/978-3-031-55245-8_20

despite their proven difficulties in social interactions, few studies have been carried out on the sharing behavior of children with autism spectrum disorders (ASD) [57]. These disorders are characterized in particular by deficits in communication and social interaction, as well as repetitive or restricted patterns of behavior, interests, and activities [1,55]. These children's prosocial behaviors, such as sharing with others, are likely to be affected by their social difficulties [76]. The first aim of the present study is to examine the sharing behavior of children with ASD towards others during a dictator game and to compare it with that of Typically Developing children (from now TD).

In addition, the increasing use of robots with children with ASD seems promising [12], particularly to promote social interaction [17,22,33,83]. Nevertheless, while numerous studies have determined how TD children perceive robots, few studies focus on the perception of robots by children with ASD. The literature indicates that TD children categorize the robot as an entity distinct from the human [37], but behave with it in the same way as they would with a human [18] and attribute human characteristics to robots, a phenomenon called anthropomorphism [9]. This suggests that TD children see robots as social agents. However, the child's type of development (TD children vs. children with ASD) could have an impact on how robots are perceived and interacted with. Indeed, the processing of robot features - its appearance and behavior - may vary according to the user's cognitive abilities [36]. As individuals with ASD may have cognitive impairments, they will not necessarily interpret cues in the same way as TD individuals [65]. Consequently, the second aim of this paper will be to specify the perception that children with ASD have of robots, based on both an implicit measure (sharing behavior towards the robot during the dictator game) and an explicit measure (an adapted version of the Godspeed questionnaire by [6]. These measures taken with a non-human agent will also enable us to study the social-cognitive abilities of children with ASD in a more indirect way.

In a review of the literature, we will first examine sharing behavior in an interaction with a human in children with ASD, and we will study sharing behavior in an interaction with a robot (only in TD children as we found no study focusing on children with ASD sharing with a robot), as well as the perceptions of robots by children with ASD. Then we will describe the experiment itself, presenting the methodology employed and the results obtained, which we will discuss before concluding.

### 1.1 Prosocial Behavior and Sharing in Autism

**Prosocial Behaviors.** Prosocial behavior can be defined as a voluntary act that aims to benefit someone else, without direct benefit to the individual. Three types of prosocial behaviors can be distinguished [23]: helping, comforting, and sharing. Helping behavior responds to an instrumental need (e.g., helping a person who has fallen to get up); comforting behavior responds to an emotional need (e.g., asking the person who has fallen if he or she is all right); and sharing behavior responds to a material need (e.g., offering to lend one's phone to the person who has fallen so that he or she can be picked up). Studies suggest a link between prosocial behavior and the Theory of Mind (ToM) which is the ability to attribute mental states to others, in autism and in typical development [76]. It is generally considered that people with ASD have a deficit in ToM [5]. Children with ASD, who show lower ToM abilities compared to TD individuals of the same age, would also perform fewer prosocial behaviors. In particular, they show an atypical pattern of social attention [11], which could be the consequence of lower social motivation [10]. Indeed, individuals' tendency to act prosocially (e.g. helping or sharing behavior towards others) would be linked to the motivation to create relationships with others [54]. According to *social motivation theory* [10], individuals with ASD show less social motivation than typically developing individuals. This deficit would prevent children with ASD from accessing rich social experiences, and would thus be social interaction disorders. Nevertheless, even if adolescents with ASD do indeed show less social motivation than TD peers (as reported by their parents) and are less prone to help the experimenter, no direct link between these variables was found [54]. Moreover, a review of the literature highlights contradictory results concerning the prosocial tendency of individuals with ASD.

Some studies indicate that ASD is associated with a decrease in prosocial behavior [43,54]. In the general population, people with more autistic traits are less prone to prosocial behavior than people with fewer autistic traits [34,82]. The decrease in prosocial behavior concerns collaborative tasks [43]. When an experimenter drops an object, adolescents with ASD help significantly less to pick up the object compared with TD adolescents [54]. However, this difference needs to be qualified, since a large proportion of autistic adolescents helped the experimenter (71.6% of autistic participants helped the experimenter, compared with 84.8% of non-autistic participants). Adolescents with ASD therefore spontaneously display prosocial behavior, albeit to a lesser extent than TD adolescents.

Others studies show similar [24] or even better performance than typically developing children [57]. Children with ASD would help as much as children with global developmental delay [43] and TD children [24]. However, children with ASD would be less likely to adapt their behavior to the interaction situation: they help as much when the experimenter has not dropped the object on purpose as when they voluntarily puts it on the ground, unlike TD children who help more when the experimenter has dropped the object involuntarily [24]. Children with autism also show a similar tendency to comfort others as TD children [24]. Finally, one study even notes a greater tendency to spontaneous help in children with ASD [57].

As the different categories of prosocial behavior have distinct developmental trajectories and characteristics, they are best analyzed independently of each other when looking at a population with ASD [25]. For this reason, in the present study, we focus solely on sharing behavior in children with ASD and TD children.

**Sharing.** Various paradigms have been used to evaluate sharing behavior in the literature, but they are generally based on resource distribution tasks from game theory (i.e. variants of the dictator or ultimatum game). In the ultimatum game, player 1 is allocated a certain resource (e.g. a sum of money), and must decide how much to keep for himself and how much to allocate to player 2. This second participant must then decide whether to accept or refuse the offer. If he refuses, neither individual receives the resource (see for example [7]). Some parameters of the task may vary (e.g. presence or absence of the recipient), but different versions generally give convergent results [30]. In the ultimatum paradigm, the participant can play the role of the proposer (player 1) or receiver (player 2). Studies can thus focus on the amount offered or accepted. The dictator game is based on the same principle as the ultimatum game. The only difference is that the recipient has no choice but to accept the offer. In this case, the participant only plays the role of proposer: He makes an offer to the receiver, which the latter is obliged to accept. The proposer therefore runs no risk of receiving nothing (unlike the ultimatum game, in which he can lose everything if the receiver refuses). The amount of shared resources indicates the individual's tendency to behave in a pro-social way (see [32]) and is related to empathy score (see [29] for an example).

Many of these tasks have been adapted for children. The resources provided are pleasant objects for the children, such as coins [66], stickers [52,73], or toys [73]. Studies examining sharing behavior in children with ASD have found conflicting results (see [61] for a review). Some studies suggest a lesser tendency for sharing behavior in ASD children compared to TD children [24,76]. Other studies argue conversely that children with ASD share in the same way as TD children whether they are faced with a puppet [30], a peer [66] or an adult [73]. They also show a similar preference for an equitable distribution of resources between themselves and the recipient [66,73], although one study shows that children with ASD give more resources to others [57].

The discrepancies in results could be explained at the methodological level. Firstly, the recipient is not necessarily present when resources are shared: many studies are based on recipients presented in the form of photos or videos [57,66] or even simply virtually [76]. Yet the presence of the recipient is likely to impact sharing towards him or her: children share more when the recipient is present in the room at the time of transmission rather than absent, and this is true regardless of their development type [73]. This underlines the need to develop more ecological experiences, with a physically present recipient.

Among the studies in which the resource recipient is physically present, another methodological difference can be highlighted: the type of recipient. Children with ASD show similar performance to TD children in a task involving a same-age recipient [30] but poorer performance in a sharing task with an adult recipient [24]. While TD children share as much with a peer as with an adult [74], it's possible that this is different in children with ASD, and that the presence of an adult hinders their performance.

Furthermore, we have seen that ASD and TD children can show prosocial behaviors towards a puppet [30]. We can ask whether prosocial behaviors apply to other non-human agents, such as robots. Robotic agents are relevant tools for studying cognitive and social abilities, as they provide interaction partners whose parameters can be modified according to the needs of the experimental conditions

[2–4, 35, 48–51, 64, 79]. In the research field of sharing, these tools would enable further investigation of the differences between individuals with ASD and TD individuals [20].

## 1.2 Prosocial Behavior and Anthropomorphism Toward a Robot

In typical development, individuals express prosocial behaviors towards robots, which can be considered anthropomorphism, i.e. the attribution of human characteristics to a non-human agent [72]. These attributions would be the result of a cognitive bias, which leads individuals to base themselves on human behaviors to explain the behaviors of non-human agents [15,16]. In adults, participants put in an ultimatum game situation with a robot and a human tend to behave in the same way with both agents [63,72]. TD children show prosocial behaviors towards robots: they share resources with a robot [52] and to the same extent as with a human [18]. Applying human norms of fairness to a non-human agent is then interpreted as a sign that the latter is considered a social agent, which amounts to anthropomorphism.

Nevertheless, it is important to distinguish between explicit measures of anthropomorphism, which generally correspond to questionnaires (e.g. the Godspeed Questionnaire Series; [6]), and implicit measures (e.g. dictator game), which refer to behavioral or indirect measures. These two types of measure can lead to different results [18,71]. At the explicit level (i.e., when children are explicitly asked what capabilities they attribute to the robot), children identify the robot as an entity distinct from a human. However, at the implicit level (i.e. when studying children's performance in the ultimatum game), children show similar behavioral responses towards the robot and the human. A comparable result is observed in adults [28]. For this reason, we have chosen to integrate both types of measurement in the present study. The aim of the present article is therefore to analyze how children with ASD view a robot, using both an implicit measure (their social behavior towards the robot during the dictator game) and an explicit measure (adaptation of the Godspeed Questionnaire Series).

However, measures of anthropomorphism are not independent: prosocial behavior towards robots could be modulated by how they are perceived. Indeed, TD children who attribute the most mental characteristics to robots (e.g. the ability to experience sensations and think) would be those who show the most concern for their well-being [68]. In a moral dilemma task, TD adults who attribute mental states (emotions, intentions) to robots are less likely to sacrifice them to save humans [53]. Development type is then likely to impact children's prosocial behaviors towards a robot: individuals with ASD might perceive robots differently from TD individuals, altering interaction accordingly. However, given that to our knowledge there are no studies investigating the prosocial behaviors (helping and sharing) of children with ASD towards a robot, it is not possible at present to verify this link within this population. We can, however, turn our attention to the small number of studies that focus on the perception of robots by individuals with ASD.

Like TD children, children with ASD categorize robots as toys [58,80] and appear to interact with them in the same way as they do with humans [78]. Nevertheless, this perception is likely to vary from one individual to another: some children interact with the robot as with a social agent, while others treat it as an object [67], and the reactions observed during interaction can be diverse [70]. Factors related to the interaction situation (the robot's location, and the distance between it and the child) can also modulate the interaction [62].

TD children are prone to anthropomorphism, that is, they tend to attribute human characteristics to non-human entities, such as robots [21,26,44]. They may thus attribute desires and physiological states to robots [9], interpret their actions as goal-directed [46]), and seek to help them [45]. We had previously highlighted the ToM deficit in individuals with ASD, resulting in difficulties attributing mental states to others [5].

Anthropomorphism is a concept close to ToM, in that it involves attributing human characteristics to an agent. Some studies therefore suggest that the tendency towards anthropomorphism in this population would be impacted by ToM deficits [14]. This tendency may be reduced in children with ASD: They attribute fewer false beliefs (erroneous beliefs about reality) to a robot compared with TD children [80]. Conversely, other studies indicate that individuals with ASD may be more likely to anthropomorphize than TD individuals. First, they are less sensitive to the physical irregularities of non-human agents [40], and adolescents with ASD perceive an android robot (Actroid-F) as more human-like than typically developing adolescents [41]. Furthermore, the difference in perceived humanity between a synthetic voice and a human voice is more marked for TD adults than for adults with ASD, who report a similar impression of humanity for both voices [42]. Finally, ASD children are more likely to think that a robot can grow up and experience pain compared with TD children, although the latter tendency remains marginal [81]. These findings highlight differences in perception, with a tendency to differentiate less between humans and robots in individuals with ASD, but further research is needed to conclude on the issue.

Furthermore, some studies argue that individuals with ASD prefer robots to non-robotic toys or humans [33,60,69]. Individuals with ASD engage more in a task when faced with a robot rather than a human [38,39,65]. Children with ASD touch and look at a robot more than a human [13,17,33] and follow its gaze more than a human's [77]. This preference for interacting with robots could lead to increased sharing with this type of agent in individuals with ASD.

## 1.3   Children with ASD Would Share More with the Robot

TD children tend to prefer equal distributions between themselves and other recipients, whether humans [57] or robots [18]. Nevertheless, for children with ASD the results are more difficult to interpret. Two studies report a preference for equal distribution of resources in the same way as TD children [66,73], but another suggests conversely that ASD children give away more resources than they keep [57]. Moreover, while several studies have looked at TD children sharing with a robot [18,52], none to our knowledge has looked at ASD children sharing with a robot. Given the above arguments, we may wonder whether ASD children exhibit the same sharing behavior as TD children, whether towards a human or a robot. On the one hand, we aim to replicate the results obtained by [73] with a human experimenter and a physically present recipient, and on the other hand to determine whether the preference for the robot noted in the literature leads children with ASD to share more with a robot experimenter. The link between the attribution of human characteristics to the robot and prosocial behavior towards it will also be investigated.

In the human experimenter condition, we expect to observe similar sharing between ASD and TD children (H1a) since we use a similar methodology to that of [73]. Conversely, in the robot condition, we assume that ASD children will share more than TD children with the robot (H1b). TD children should share as much with a human as with a robot (H2a), while ASD children will share more with the robot than with the human (H2b). Finally, the higher the score on the positive items of the adapted Godspeed Questionnaire Series (i.e., the score on all questions except Q5 and Q6, as these questions deal with fear experienced during interaction), the higher the number of resources given will be (H3).

## 2 Method

### 2.1 Participants

The sample consisted of 38 children aged 9 to 13, including 10 verbal children with a diagnosis of ASD ($Mage = 10.95, SDage = 1.60$, 3 girls) and 28 typically developing children ($Mage = 10.29, SDage = 0.98$, 14 girls). All the children in the sample were educated in the same town (Eaubonne, 95600), the TD children in a regular school and the ASD children in a Montessori school. The diagnosis of ASD was confirmed by the school psychologist. Consent was obtained from legal guardians (and from the participants themselves) for all children included in the study, and the study followed the ethical principles outlined in the Helsinki Declaration of 1964. One child with ASD was excluded from the results due to difficulties in understanding the dictator game task.

The children were randomly divided into two groups: The first group interacted first with the robot and then with the human (Order 1: n = 18, $Mage = 10.8, SDage = 1.26$, 7 girls, 5 children with ASD), while the second group interacted first with the human and then with the robot (Order 2: n = 20, $Mage = 10.16, SDage = 1.06$, 10 girls, 5 children with ASD). The order of presentation of the agents is counterbalanced. The effect of order of presentation was analyzed later.

### 2.2 Materials

**NAO Robot.** We used a NAO robot created by Aldebaran Robotics (Aldebaran version 4 - "Evolution"). This humanoid robot is 58 cm tall, has a movable head, arms and legs, and is equipped with LEDs on its eyes, head and torso. The robot's programming is based on the Wizard of Oz method: the robot is semi-teleoperated from a computer using Choregraphe software, i.e. programs are prepared in advance, but are launched in real time by an experimenter during the run. This type of programming allows for greater flexibility in interaction with the children, who may have variable reaction times. In addition, the Wizard of Oz method implies that the experimenter programming the robot is not visible to the participant, in order to give the illusion that the robot is acting autonomously. In the present study, the experimenter programming the robot is hidden behind a folding screen (see Fig. 1).

**Mini Dictator Game.** The "mini dictator game" is a version of the dictator game specifically adapted for children with ASD and TD aged 3 to 6 [57], as this task is performed in a single round instead of several. In this study, we used this task but included a receiver physically present in the room (which could be either an adult or a robot). Two identical cardboard boxes were used as receptacles into which the child distributes the resources entrusted to him (box 1 corresponds to the resources he gives to his opponent and box 2 to the resources he keeps for himself).

While other authors presented resources in the form of stickers [73], we chose to use star-shaped cards, a form already used by [52]. Concerning the number of resources entrusted to the child for the distribution task, we chose to propose 5 stickers to share. Indeed, several studies point out that TD and ASD children tend to share equally between the two recipients (for example, when the authors entrusted the child with 10 stickers, the child tends to keep 5 for himself and give 5 away) [18,52,66]. Offering an odd number of stickers therefore forces the child to allocate more to one recipient. Consequently, 10 star cards were provided for each participant: 5 cards in the robot agent condition and 5 cards in the human agent condition (in case the latter decided to keep all the stars for himself with each of the agents).

**Godspeed Questionnaire Series.** The material used also involved an adaptation of the Godspeed Questionnaire Series (GQS), used in the literature to measure anthropomorphism [6]. This questionnaire measures the extent to which the child attributes human characteristics to a non-human agent. The GQS comprises 7 sub-scales: Anthropomorphism, Animacy, Likability, Perceived intelligence, Safety before interaction, Safety after interaction, Friendliness. We chose to select just one question per subscale for a total of 7 questions, in order to limit the length of the test since each child answered the questionnaire twice (once for the robot, once for the human) (see GQS adapted). For each question (each representing a subscale), children were asked to express their response on a Likert scale ranging from 1 to 5 (1: Strongly disagree; 2: Somewhat disagree; 3: Neither agree nor disagree; 4: Somewhat agree; 5: Strongly agree). We also used a colored emoticon scale to facilitate the child's responses to the questionnaire (see Emojis). This scale consisted of 5 emoticons, ranging from a very unhappy red emoticon to represent "No, totally disagree" (response rated 1) to a very smiley green emoticon to represent "Yes, totally agree" (response rated 5), with a neutral yellow emoticon in the middle of the scale. The child could respond verbally or simply point to one of the emoticons.

### 2.3 Procedure

The test was carried out individually, in a quiet room in the children's respective schools. It required the presence of two experimenters. The first experimenter (Experimenter 1, E1) accompanied the children to the testing room, and collected their responses to the questionnaire after the interaction, while the second experimenter (Experimenter 2, E2) set up the testing room, programmed the robot, and acted as the human agent in the human experimenter condition.

In the robot condition, E2 installed the robot in a sitting position on the table, facing the child's chair (at a distance of around 80 cm), and then hid behind a folding screen. In the human condition, E2 sits on a chair behind the table. In this way, when the child enters the room, accompanied by E1, the agent (robot or human) is already seated at the table (or on the table in the case of the robot). The boxes are also on the table, positioned on either side of the agent (see Fig. 1). E1 first asked the child to sit down on the chair provided, then introduced the agent. The procedure thus consisted of a priming phase, followed by a test phase. The experiment lasted about 20 min (5 min for the dictator game and 5 min for answering the GQS, for each condition). E1 leaves the room after the priming phase in both conditions.

**Priming Phase.** Our protocol was inspired by that of [52]. In this study, the authors asked the robot questions before starting the mini dictator game, in order to present the robot as having cognitive and emotional skills. We have used the same questions in the present study, in order to present both agents as having cognitive and emotional abilities. The questions asked are strictly identical in the robot and human conditions, and the answers provided are also the same.

E1 began by asking the agent: "Hello [name of the agent], can you count to 4?" The agent then replies, "Of course, I'll show you: 1, 2, 3, 4". This question is intended to suggest the robot's cognitive abilities. E1 then asked "What makes you happy?" The agent replies "I like playing with children and robots." This question is intended to suggest the robot's emotional abilities.

**Testing Phase.** E1 then explains the mini dictator game to the child: "I'm going to give you five cards, which you can share with Nao/Marion (name of E2) or keep for yourself. In any case, you must deal all the cards. This is Nao/Marion's box (showing box 1), when the cards are for him, you put them in. This is your box (showing box 2), when you want to keep them for yourself, you put them in. You choose how many you give him. I don't care what you do, I'm going to stand outside. Tell Nao/Marion when you've finished." E1 hands the cards to the child and leaves the room to allow the child to do the sharing, as in [52].

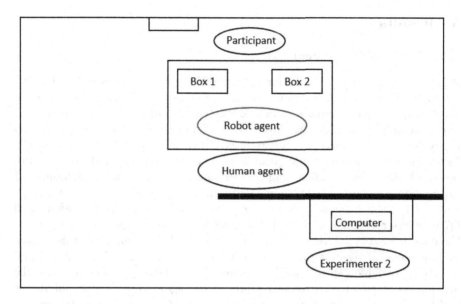

**Fig. 1.** Schematic diagram of the testing room. Experimenter 1 is not shown in the figure as they leave the room immediately after the priming phase. Experimenter 2 plays the role of the human agent in the human condition.

When the child says he's finished, the agent says "Thank you." (unless he hasn't received any cards). When she hears that the child has finished, E1 returns to the testing room, and notes the number of cards in each container.

E1 then asks the child to go to another room to answer a few questions: "Come with me, we will go to another room so you can explain." Once in the other room, E1 asks the child to sit down at a table opposite him. Before asking the child any questions, the experimenter presents the emoticon scale (see Emojis) and explains the meaning of each emoji (this instruction is also repeated on the second test). "Now I'm going to ask you some questions. To answer, you'll have to point to an emoticon. You can see that the emoticons go from least happy to most happy. You can show me the least happy to say no, and the most happy to say yes. Got it?" After checking that the child has understood (by repeating the instructions as many times as necessary), the experimenter asks the 7 questions adapted from the GQS (see Godspeed Questionnaire Series adapted) and collects the child's answers on a scoring sheet. Once the child has answered the questionnaire, E1 introduces the second test to the child: "Now we're going to start the game again with someone else." and takes him back to the first room. The same procedure is then followed with the second agent: priming phase, then test phase, consisting of the mini dictator game and the adapted GQS. At the end of the test, E1 thanks the child and takes him back to the classroom.

# 3   Results

## 3.1   Principal Component Analysis

All statistical analyses were performed using R 4.3.0 software (see R code) [59]. We first performed a Principal Component Analysis (PCA) of the responses to the Godspeed Questionnaire Series. Bartlett's test of sphericity showed that our data differed significantly from the null hypothesis, indicating a dependency between the variables ($\chi^2(91) = 113.70$, $p < .001$). PCA revealed a first factor including questions Q3 (likability), Q4 (intelligence) and Q7 (friendliness), which explained 26.90% of the variance; a second factor including Q1 (anthropomorphism) and Q2 (animacy), which explained 25.04% of the variance; and a third factor including Q5 (safety before), Q6 (safety after) and Q7 to a lesser extent, which explained 20.22% of the variance. Together, these three factors explain 72.15% of the variance (see PCA results). For this reason, in subsequent analyses we have grouped the answers to questions Q3, Q4 and Q7 into a variable called "Good personality", as these questions measure the attribution of positive personality traits to the agent (how likable, intelligent and friendly it is). Responses to questions Q1 and Q2, which assess the robot's appearance (human-like appearance and animacy), were combined in a variable called "Human-likeness". Similarly, answers to questions Q5 and Q6 (and Q7 to a lesser extent), which deal with perceived safety before and after interaction, will be combined in a variable called "Safety".

## 3.2   Generalized Linear Mixed Models

The variables Good personality, Human-likeness, Safety and Don (number of cards given by the participant) were corrected so that the data were strictly between 0 and 1 not included.

**Donation.** Because of the threshold and ceiling effects generated by responses ranging from 0 to 5, beta regression analyses were performed on the discrete variable donation (ranging from 0 to 5, and from 0 to 1 once corrected). The aim was (i) to test whether donation varies according to condition (human partner vs. robot), (ii) according to development type (ASD vs. TD), and (iii) according to questionnaire score.

A null model including no predictors was calculated (model A0).

At level 1, this null model was compared with several models including different predictors (experimenter, order, development type, gender, good personality, human-likeness, safety). Predictor quality was determined according to the Akaike Information Criterion (AIC): the predictor with the lowest AIC represents the predictor that best fits the data. We also took into account the Bayes Information Criterion (BIC) in our model selection, a fit index that penalizes complex models. As with the AIC, the lowest BIC represents the predictor that best fits the data. The model with the best predictor was retained for further analysis.

We first compared simple models with the null model, then combined models including several predictors (see Table 1). We retained the model with the lowest

AIC and BIC. Among the simple models, the model containing the Good personality factor (model A5) predicted donation better than all the other models, including the model containing no predictor, on both AIC and BIC [Model A5: AIC $= -14.264$, BIC $= -4.942$; Model A0: AIC $= -11.348$, BIC $= -4.356$, BFA5, A1 $= 1.34$]. Among the combined models, our aim was to see whether another model could generate better predictions of donation than Model A5. With regard to BIC, no model was better than the A5 model. However, the AIC gave the A11 model (also containing the interaction between condition and type of development in addition to the Good personality score) as slightly better [Model A11: AIC $= -15.235$, BIC $= 1.08$; Model A5: AIC $= -14.264$, BIC $= -4.942$; BFA11, A5 $= 0.744$]. According to the A11 model, there is no simple effect of experimenter or development type on giving. In the human experimenter condition, children with ASD shared as much as TD children. Moreover, TD participants shared as much with the robot as with the human. However, ASD children showed a tendency to share less with a robot than with a human (see Fig. 2). This result should be treated with caution, however, as the A11 model is not the best in terms of BIC. Like the A5 model, this model also indicates that the Good personality score significantly predicts giving: the higher the score, the higher the level of giving.

**Table 1.** Models predicting donation. *: Best model(s); BF x0: Bayes Factor which indicates how the given model (AX) compares to the null model (A0). The higher the number, the better the model.

| | AIC | BIC | BF x0 | |
|---|---|---|---|---|
| **Level 1** | | | | |
| **Simple models vs. random** | | | | |
| Model A0 (Null) | −11.348 | −4.356 | 1 | |
| Model A1 (Experimenter) | −10.502 | −1.179 | 0.204 | |
| Model A2 (Order) | −9.634 | −0.311 | 0.132 | |
| Model A3 (Development) | −10.233 | −0.91 | 0.179 | |
| Model A4 (Gender) | −11.514 | −2.191 | 0.339 | |
| Model A5 (Good Personality) | −14.264 | −4.942 | 1.34 | * |
| Model A6 (Human-likeness) | −9.71 | −0.388 | 0.138 | |
| Model A7 (Safety) | −11.708 | −2.385 | 0.373 | |
| **Combined models vs. random** | | | | |
| Model A8 (Experimenter + Development) | −9.448 | 2.206 | 0.038 | |
| Model A9 (Experimenter * Development) | −13.496 | 0.488 | 0.089 | |
| **Improving model A9** | | | | |
| Model A10 (Experimenter * Development * Gender) | −9.059 | 14.249 | 0.001 | |
| Model A11 (Experimenter * Development + Good Personality) | −15.235 | 1.08 | 0.744 | * |
| Model A12 (Experimenter * Development + Human-likeness) | −13.608 | 2.708 | 0.33 | |
| Model A13 (Experimenter * Development + Safety) | −14.251 | 2.064 | 0.455 | |
| **Level 2** | | | | |
| **Improving model A5** | | | | |
| Model A14 (Good Personality + Development) | −12.28 | −0.626 | 0.116 | |
| Model A15 (Good Personality + Experimenter) | −13.599 | −1.946 | 0.224 | |

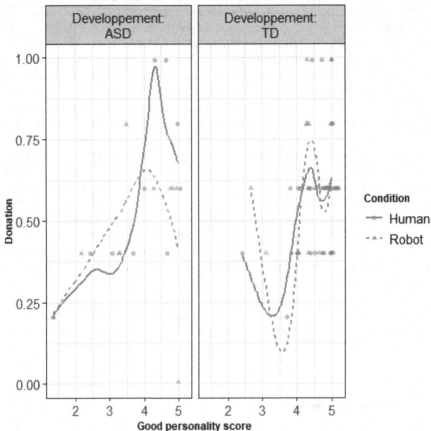

**Fig. 2.** Difference in donation by type of development and condition; ASD = Participants with Autism Spectrum Disorders, TD = Participants with Typical Development; Human = Human experimenter, Robot = Robot experimenter

**Post-hoc Analysis.** A beta regression analysis was performed on the ordinal variable questionnaire score (ranging from 1 to 5, and from 0 to 1 once corrected). We were specifically interested in the factors Good personality, Human-likeness and Safety, as well as the variation between questions Q5 (safety before) and Q6 (safety after). The aim was to assess whether the questionnaire score varied according to condition and type of development.

*Good Personality Predictors.* Model B2, which contains only development type as a predictor, best predicts the Good personality score [Model B2: AIC = −176.065, BIC = −166.742, Model B0: AIC = −172.17, BIC = −165.18; BFB2, B0 = 2.186] (see Table 2). Indeed, TD children assign higher Good personality scores than children with ASD.

**Table 2.** Models predicting Good personality. *: Best model; BF x0: Bayes Factor.

| Models | AIC | BIC | BF x0 | |
|---|---|---|---|---|
| Model B0 (null, Good personality reduced) | −172.17 | −165.18 | 1 | |
| Model B1 (Experimenter) | −170.618 | −161.295 | 0.144 | |
| Model B2 (Development) | −176.065 | −166.742 | 2.186 | * |
| Model B3 (Experimenter + Development) | −174.503 | −162.85 | 0.312 | |
| Model B4 (Experimenter * Development) | −172.733 | −158.789 | 0.041 | |
| Model B5 (Gender) | −170.499 | −161.176 | 0.135 | |
| Model B6 (Order) | −170.188 | −160.865 | 0.116 | |

*Human-Likeness Predictors.* Regarding the Human-likeness score, the AIC and BIC indices prefer different models (respectively, Model C4 and Model C1). Model C1, which contains only the experimenter factor as a predictor, is the one that, according to the BIC, best predicts the Human-likeness score [Model C1: AIC = −106.048, BIC = −96.725; Model C0: AIC = −75.849, BIC = −68.857, BFC1, C0 = 1125885] (see Table 3). Children assign higher Human-likeness scores to a human experimenter than to a Robot. However, the C4 model, which contains the interaction between development type and condition as a predictor, is the one that, according to the AIC, is the best at predicting the Human-likeness score. [Model C4: AIC = −106.484, BIC = −92.5; Model C1: AIC = −106.048, BIC = −96.725, BFC4,C1 = 0.121]. The difference in Human-likeness between human and robot would be smaller in ASD children than in NT children. However, the difference in AIC between model C1 and C4 is very small, indicating that this effect is probably small if it exists at all.

**Table 3.** Models predicting Human-likeness. *: Best model(s); BF x0: Bayes Factor.

| Models | AIC | BIC | BF x0 | |
|---|---|---|---|---|
| Model C0 (null, Human-likeness reduced) | −75.849 | −68.857 | 1 | |
| Model C1 (Experimenter) | −106.048 | −96.725 | 1125885 | * |
| Model C2 (Development) | −74.068 | −64.745 | 0.128 | |
| Model C3 (Experimenter + Development) | −104.079 | −92.425 | 131161.4 | |
| Model C4 (Experimenter * Development) | −106.484 | −92.5 | 136136.5 | * |
| Model C5 (Gender) | −73.863 | −64.611 | 0.115 | |
| Model C6 (Order) | −73.934 | −64.611 | 0.12 | |

*Safety Predictors.* Regarding the score at Safety, no model is better than the null model (see Table 4).

**Table 4.** Models predicting Safety. *: Best model; BF x0: Bayes Factor.

| Models | AIC | BIC | BF x0 | |
|---|---|---|---|---|
| Model D0 (null, Safety reduced) | −32.797 | −25.805 | 1 | * |
| Model D1 (Experimenter) | −31.959 | −22.236 | 0.205 | |
| Model D2 (Development) | −31.513 | −22.19 | 0.164 | |
| Model D3 (Experimenter + Development) | −30.692 | −19.038 | 0.034 | |
| Model D4 (Experimenter * Development) | −29.201 | −15.216 | 0.005 | |
| Model D5 (Gender) | −31.174 | −21.851 | 0.138 | |
| Model D6 (Order) | −31.615 | −22.292 | 0.173 | |

*Predictors of the Difference Between Q6 and Q5.* Concerning the difference between Q5 (safety before) and Q6 (safety after), the best model on AIC is the E2 model, which contains only the development type factor, while on BIC the null model is the best predictor [Model E2: AIC = 301.506, BIC = 310.829; Model E0: AIC = 302.834, BIC = 309.826, BFE2, E0 = 0.606]. ASD children would show slightly less difference between Q5 and Q6 than TD children, but this effect, if it exists, remains very small (See Table 5).

**Table 5.** Models predicting the difference between Q5 and Q6. *: Best model(s); BF x0: Bayes Factor.

| Models | AIC | BIC | BF x0 | |
|---|---|---|---|---|
| Model E0 (null, q65 reduced) | 302.834 | 309.826 | 1 | * |
| Model E1 (Experimenter) | 304.422 | 313.744 | 0.141 | |
| Model E2 (Developpement) | 301.506 | 310.829 | 0.606 | * |
| Model E3 (Experimenter + Developpement) | 303.109 | 314.763 | 0.085 | |
| Model E4 (Experimenter * Developpement) | 303.364 | 317.348 | 0.023 | |
| Model E5 (Gender) | 304.685 | 314.008 | 0.124 | |
| Model E6 (Order) | 304.829 | 314.152 | 0.115 | |

# 4  Discussion

## 4.1  Good Personality as the Best Predictor of Donation

Principal Component Analysis revealed three factors among the responses to the adapted GQS. The "Good personality" variable includes questions that measure the attribution of positive personality traits to the agent (how likable, intelligent and friendly it is, i.e. questions Q3, Q4 and Q7 respectively). The "Human-likeness" variable includes responses to questions evaluating the robot's appearance (human-like appearance and animacy, questions Q1 and Q2 respectively).

The "Safety" variable refers to responses to questions on perceived safety before and after interaction (Q5 and Q6 respectively, and Q7 to a lesser extent).

Statistical analyses of generalized linear mixed models indicate that the Good personality factor best predicts giving. Participants' tendency to share resources with the agent thus depends mainly on perceived agreeableness (Q3), perceived intelligence (Q4) and willingness to have the agent as a friend (Q7). In other words, the higher the score on these questions, the more participants shared with the agent, regardless of experimenter or type of development (ASD/TD). The attributions made to the experimenter therefore seem to determine the tendency to perform prosocial behavior towards they, which partially validates our initial hypothesis (H3). There is no simple effect of development type or experimenter on giving. Nevertheless, our analyses may suggest an interaction effect between development type and experimenter on giving. These results will be discussed in detail below.

We had assumed that children with ASD would share as much as TD children with the human agent (H1a). Our results are in line with our first hypothesis: in the human experimenter condition, there is no significant difference in sharing between TD children and children with ASD. This is consistent with previous studies suggesting that children with ASD show a similar tendency to sharing behaviors as TD children [66,73]. Importantly, these results were replicated despite the recipient being an adult in our methodology, unlike that of [73] where the recipient is a child. Our results, however, stand in contradiction with other studies, which highlighted less sharing in children with ASD [24,76], or conversely an increased tendency to share [57]. This discrepancy may first be explained by methodological differences, as these studies do not necessarily involve a recipient present in the testing room with the participant as we did in the present study (e.g. [57,76]). However, this is likely to impact the results [73] and does not provide an ecological measure reflecting the tendency to share in a real interaction situation. In addition, some studies include Chinese participants [76], while others include American participants [73]. Yet the culture (Western/Oriental) to which individuals belong can have an impact on their sharing behavior [31]. Finally, the contradictory results of these studies may reflect the wide inter-individual variability that exists within ASD [65]. In the present study, we did indeed find greater dispersion in giving and questionnaire responses in children with ASD compared with TD children.

We had also hypothesized that TD children would share as much with the robot experimenter as with the human (H2a), unlike children with ASD who would share more with the robot (H2b). In line with our initial hypothesis (H2a), TD children give as many resources to the robot as to the human, meaning that they show the same tendency towards prosocial behavior as with a human. This result suggests that TD children regard the robot as a social agent, and is consistent with the study showing that they behave similarly with a robot and a human [18]. However, for children with ASD, the results go in the opposite direction to our hypotheses (H1b and H2b). Contrary to our expectations, children with ASD tend to share less with the robot than with the human, and the magnitude of the difference in giving towards the robot and towards the human is greater among children with ASD than among TD children. This result can be interpreted in two ways. Firstly, children with ASD would not consider robots as social agents, or at least to a lesser extent than TD children, which would cause the observed decrease in prosocial behavior. We will come back to this point later in the post-hoc analysis of responses to the adapted GQS. Secondly, it is also possible that children with ASD see the robot as a social agent, but find it more difficult to generalize the prosocial behaviors acquired with humans to a non-human agent. Indeed, as children with ASD have had access to fewer social experiences than TD children due to lower social motivation [10], they would be less likely than TD children to apply social knowledge acquired during human-human interactions to a robot. This idea seems coherent, considering the social and generalization difficulties observed in ASD [8,75]. This lesser tendency to behave prosocially towards the robot also seems to call into question the idea of a preference for robots in children with ASD previously evoked in the literature [33]. Nevertheless, caution should be exercised in interpreting this result given the small sample size, and further studies will be needed to determine more precisely the interaction between development type and experimenter in sharing behaviors.

No effect of gender or order was found on giving. Boys gave as many resources as girls, and participants shared a similar number of resources regardless of the order in which they saw the agent (e.g. the robot agent first vs. second).

## 4.2 Predictors of Good Personality, Human-Likeness and Safety Factors

Post-hoc analyses of the questionnaire responses also provide some interesting insights.

Development type predicts Good Personality score (Q3, Q4 and Q7). More specifically, children with ASD score significantly lower than TD children on the attribution of positive personality traits: they judge the recipient to be significantly less pleasant, intelligent and friendly than TD children, whether the whether the experimenter was a human or a robot. This could be due to differences in the perception and interpretation of social cues in autism, including reduced connectivity and brain activity when processing social cues such as laughter, compared to typically developing individuals [47]. The increase of autistic traits is thus associated with a decrease in the attribution of positive social intention in laughing behavior. Furthermore, Good personality scores did not vary according to experimenter. This indicates that the condition (human vs. robot) has no impact on attributions of positive personality traits towards the experimenter. Children seem to enjoy the robot experimenter as much as the human experimenter, whatever their development type.

Human-likeness scores (Q1 and Q2) are significantly predicted by experimental condition. Children report significantly higher scores for the human experimenter than for the robot experimenter. This means that they judge the human to be significantly more human-like and more animated than the robot, whatever their development type. There was also a trendy interaction effect between development type and experimenter on the Human-likeness score: the difference in score between human and robot was slightly smaller in ASD children than in TD children, meaning that they tended to judge the robot as slightly more human-like than TD children, and to judge the human as slightly less human-like. This could indicate a difference in the perception of robots between TD children and those with ASD, but it could also be explained by categorization difficulties frequently noted in individuals with ASD [8]. This result remains uncertain, however, and requires further study to explore.

None of the factors included in the analysis predicted Safety scores (Q5 and Q6 and Q7 to a lesser extent). The development type and experimenter had no impact on children's perceived safety. This result indicates that children were no more afraid of being alone with the robot than they were of being alone with the human, whether they were TD or ASD. Children with ASD like TD children therefore did not find a robot more frightening than a human. This seems consistent with the literature showing the benefits brought by social robots, and particularly to individuals with ASD [12,13,17,33].

Finally, neither the gender of the participant nor the order of administration had any effect on Good Personality, Human-likeness and Safety scores.

The difference between the score reported for question Q5 and the score reported for question Q6 (corresponding respectively perceived to safety at the start of the interaction vs. at the end of the interaction) was also measured, by subtracting the score for Q5 from the score for Q6. The experimenter does not predict the difference in perceived safety between the beginning and end of the interaction, which means that the safety perceived by participants evolves over time in a similar way whether the experimenter is human or robot. Development type, however, showed little tendency to predict outcomes, with a slightly lower score difference between Q5 and Q6 in TD children than in children with ASD. The TD children therefore reported feeling less afraid at the end of the interaction than at the beginning, unlike children with ASD who provided similar responses to Q5 and Q6. These two questions were asked after the interaction, so it's possible that this trend is only observed in TD children, because children with ASD have more difficulty projecting themselves into the temporal frame. Indeed, children with ASD have a less precise understanding of temporal conjunctions such as "before" and "after' [56]. In other words, children with ASD would not show the same decrease in score between Q5 and Q6 as TD children only because they would be less able to differentiate between the fear felt at the beginning of the interaction and the fear felt at the end, leading them to answer the two questions more identically.

To summarize the predictors of questionnaire scores, the type of development seems to impact both the attribution of positive personality traits to agents (by decreasing the Good personality score in children with ASD whatever the condition) and the perceived resemblance to a human to a lesser extent (the difference in Human-likeness score between robot and human condition tends to be less marked in children with ASD). While the robot's perceived resemblance to a human does not seem to be related to prosocial behaviors towards this agent, the attribution of positive traits predicts sharing towards it. Consequently, the attribution of positive traits to the agent could be the mediating variable between children's development type and sharing behavior. Development type would impact sharing differently depending on the condition. Despite a lower attribution of positive personality traits to the agent in ASD children compared to TD children, they share similarly in the human experimenter condition. On the other hand, children with ASD give fewer resources in the robot experimenter condition, although they attribute as many positive traits to robots as to humans.

### 4.3 Comparison Between Explicit Measure (Questionnaire) and Implicit Measure (Donation) of Anthropomorphism

Having analyzed prosocial behavior towards the robot, which can be considered an implicit measure of anthropomorphism [72], and responses to adapted GQS, which refer to an explicit measure of anthropomorphism [6], we will compare these two types of measurement.

We have seen that the score of positive traits attributed to the agent is the factor that best predicts resource sharing. This indicates the existence of a positive link between the explicit measure of anthropomorphism and the implicit measure only on the items that designate the positive traits attributed to the agent (likability, intelligence and friendliness). Conversely, an agent's perceived anthropomorphism and animacy (Human-likeness score) and perceived safety before and after interaction (Safety score) do not explain the resource donation made to it, which is why our initial hypothesis (H3) is not fully validated.

Comparing explicit and implicit measures in children with ASD, they tend to share less with the robot than with the human, and to attribute fewer positive personality traits to it than TD children, but slightly more human physical characteristics. Therefore they seem less likely to anthropomorphize the robot than TD children on an implicit measure (amount of resource shared) and on part of the explicit measure, since they generally judge the agent less positively than TD children (answers to questions Q3, Q4 and Q7). Nevertheless, ASD children would anthropomorphize the robot more than TD children on the aspect of physical resemblance to humans. This incongruity between implicit and explicit measurement is consistent with that found in the literature in TD children [18], suggesting that prosocial behaviors toward an agent are not related to how the agent is perceived in terms of physical characteristics, but rather to how it is perceived in terms of mental characteristics. Future studies should focus on distinguishing these different aspects of explicit anthropomorphism, which will also allow us to better investigate the relationship between explicit and implicit measures of anthropomorphism.

Furthermore, in TD children, the tendency to anthropomorphize robots (which includes prosocial behaviors) can be modulated by various factors [20, 21], linked to the robot's design (e.g. its appearance), the context in which it is presented (e.g. the abilities lent to it by experimenters) [52] and the user himself (e.g.

his age) [19]. More studies involving the dictator game with a robot with children with ASD are needed to determine the conditions that determine the application of prosocial behaviors. Finally, it might be relevant to replicate this experiment with younger children, which might allow us to observe a more marked difference between the two experimenters. Indeed, children under the age of 7 tend to be less fair in their distribution of resources [27]. Further studies are therefore needed to investigate the differences in prosocial behavior towards a robot and anthropomorphism between children with ASD and typically developing children.

Some aspects of this study could be improved in the future. Firstly, the sample of children with ASD remains small, so it is essential to replicate this experiment with a larger sample. In addition, the adapted version of the GQS is proposed to the participant after the interaction has taken place, which may play on the response to the perceived safety question before the interaction (Q5) and therefore impact the interpretation of these results, particularly for participants with ASD who have more difficulty than TD children in representing different temporal cues [56].

## 5    Conclusion

The mini-dictator game with a humanoid robot as a recipient is an interesting experimental method for gaining a deeper understanding of the social-cognitive abilities of children with ASD, and in particular of prosocial behaviors. It also makes it possible to study anthropomorphism towards a robot in this population.

Children with ASD gave as many resources to the human experimenter as TD children, indicating preserved prosocial behaviors despite the social difficulties inherent in this disorder. However, children with ASD tended to share less with the robot than with the human, suggesting that children with ASD generalize less of their human social knowledge to robots than TD children. However, it remains to be clarified whether this lesser tendency is explicable by the generalization difficulties of children with ASD [8], or whether it is the consequence of a fundamentally different perception of robots. Indeed, children with ASD also attribute fewer positive personality traits than TD children, but this effect does not vary by experimenter and could be the consequence of an atypical perception of social cues in autism [47]. This divergence in perception could also explain the differences in perceived humanity between robot and human tending to be less marked in children with ASD.

With regard to anthropomorphism towards a robot, the results of the present study suggest that prosocial behaviors towards an agent are related to the mental characteristics attributed to it rather than to physical characteristics, regardless of the type of development. Attributions of positive traits to the agent (an explicit measure of anthropomorphism) predict sharing towards the agent, suggesting that they may be the mediating variable between children's development type and sharing behavior (an implicit measure). Nevertheless, further studies are needed to clarify the link between implicit and explicit measures of anthropomorphism, and the factors that modulate them, in TD children and children with ASD.

**Acknowledgements.** We would like to thank the Plurivalente school and Educa'son association in Eaubonne for their welcome and interest in this research project, as well as all the children who took part in the study. We would also like to thank Baptiste Jacquet for his help with the statistical analysis.

# References

1. American Psychiatric Association: Diagnostic and Statistical Manual of Mental Disorders. American Psychiatric Association, 5th edn (2013). https://doi.org/10.1176/appi.books.9780890425596. https://psychiatryonline.org/doi/book/10.1176/appi.books.9780890425596
2. Baratgin, J., Dubois-Sage, M., Jacquet, B., Stilgenbauer, J.L., Jamet, F.: Pragmatics in the false-belief task: let the robot ask the question! Front. Psychol. **11**, 593807 (2020). https://doi.org/10.3389/fpsyg.2020.593807
3. Baratgin, J., Jacquet, B., Dubois-Sage, M., Jamet, F.: "Mentor-child and naive-pupil-robot" paradigm to study children's cognitive and social development. In: Workshop: Interdisciplinary Research Methods for Child-Robot Relationship Formation, HRI-2021 (2021)
4. Baratgin, J., Jamet, F.: Le paradigme de "l'enfant mentor d'un robot ignorant et naïf" comme révélateur de competences cognitives et sociales précoces chez le jeune enfant. In: WACAI 2021. Centre National de la Recherche Scientifique [CNRS], Saint Pierre d'Oléron, France (2021). https://hal.archives-ouvertes.fr/hal-03377546
5. Baron-Cohen, S., Leslie, A.M., Frith, U.: Does the autistic child have a "theory of mind"? Cognition **21**(1), 37–46 (1985). https://doi.org/10.1016/0010-0277(85)90022-8. https://linkinghub.elsevier.com/retrieve/pii/0010027785900228
6. Bartneck, C., Kulić, D., Croft, E., Zoghbi, S.: Measurement instruments for the anthropomorphism, animacy, likeability, perceived intelligence, and perceived safety of robots. Int. J. Soc. Robot. **1**(1), 71–81 (2009). https://doi.org/10.1007/s12369-008-0001-3
7. Beaunay, B., Jacquet, B., Baratgin, J.: A selfish chatbot still does not win in the ultimatum game. In: Ahram, T., Taiar, R. (eds.) IHIET 2021, pp. 255–262. Springer, Cham (2022). https://doi.org/10.1007/978-3-030-85540-6_33
8. Brown, S.M., Bebko, J.M.: Generalization, overselectivity, and discrimination in the autism phenotype: a review. Res. Autism Spectr. Disord. **6**(2), 733–740 (2012). https://doi.org/10.1016/j.rasd.2011.10.012
9. Chernyak, N., Gary, H.E.: Children's cognitive and behavioral reactions to an autonomous versus controlled social robot dog. Early Educ. Dev. **27**(8), 1175–1189 (2016). https://doi.org/10.1080/10409289.2016.1158611
10. Chevallier, C., Kohls, G., Troiani, V., Brodkin, E.S., Schultz, R.T.: The social motivation theory of autism. Trends Cogn. Sci. **16**(4), 231–239 (2012). https://doi.org/10.1016/j.tics.2012.02.007
11. Chita-Tegmark, M.: Attention allocation in ASD: a review and meta-analysis of eye-tracking studies. Rev. J. Autism Dev. Disord. **3**(3), 209–223 (2016). https://doi.org/10.1007/s40489-016-0077-x

12. Conti, D., Trubia, G., Buono, S., Di Nuovo, S., Di Nuovo, A.: Evaluation of a robot-assisted therapy for children with autism and intellectual disability. In: Giuliani, M., Assaf, T., Giannaccini, M.E. (eds.) TAROS 2018. LNCS, vol. 10965, pp. 405–415. Springer, Cham (2018). https://doi.org/10.1007/978-3-319-96728-8_34

13. Costa, A., et al.: A comparison between a person and a robot in the attention, imitation, and repetitive and stereotypical behaviors of children with Autism Spectrum Disorder. In: Proceedings Workshop on Social Human-Robot Interaction of Human-Care Service Robots at HRI 2018 (2018)

14. Cullen, H., Kanai, R., Bahrami, B., Rees, G.: Individual differences in anthropomorphic attributions and human brain structure. Soc. Cogn. Affect. Neurosci. 9(9), 1276–1280 (2014). https://doi.org/10.1093/scan/nst109

15. Dacey, M.: Anthropomorphism as cognitive bias. Philos. Sci. 84(5), 1152–1164 (2017). https://doi.org/10.1086/694039. https://www.cambridge.org/core/product/identifier/S0031824800010278/type/journal_article

16. Dacey, M., Coane, J.H.: Implicit measures of anthropomorphism: affective priming and recognition of apparent animal emotions. Front. Psychol. 14 (2023). https://www.frontiersin.org/articles/10.3389/fpsyg.2023.1149444

17. David, D.O., Costescu, C.A., Matu, S., Szentagotai, A., Dobrean, A.: Effects of a robot-enhanced intervention for children with ASD on teaching turn-taking skills. J. Educ. Comput. Res. 58(1), 29–62 (2020). https://doi.org/10.1177/0735633119830344. http://journals.sagepub.com/doi/10.1177/0735633119830344

18. Di Dio, C., et al.: It does not matter who you are: fairness in pre-schoolers interacting with human and robotic partners. Int. J. Soc. Robot. 12(5), 1045–1059 (2020). https://doi.org/10.1007/s12369-019-00528-9

19. Di Dio, C., et al.: Shall i trust you? From child-robot interaction to trusting relationships. Front. Psychol. 11, 469 (2020). https://doi.org/10.3389/fpsyg.2020.00469

20. Dubois-Sage, M., Jacquet, B., Jamet, F., Baratgin, J.: The mentor-child paradigm for individuals with autism spectrum disorders. In: CONCATENATE Social Robots Personalisation - International Conference on Human Robot Interaction (HRI) 2023. Association for Computing Machinery (2023). https://doi.org/10.48550/arXiv.2312.08161

21. Dubois-Sage, M., Jacquet, B., Jamet, F., Baratgin, J.: We do not anthropomorphize a robot based only on its cover: context matters too! Appl. Sci. 13(15) (2023). https://doi.org/10.3390/app13158743. https://www.mdpi.com/2076-3417/13/15/8743

22. Dubois-Sage, M., Jacquet, B., Jamet, F., Baratgin, J.: People with autism spectrum disorder could interact more easily with a robot than with a human: reasons and limits. Behav. Sci. (Basel, Switzerland) 14(2), 131 (2024). https://doi.org/10.3390/bs14020131

23. Dunfield, K.A.: A construct divided: prosocial behavior as helping, sharing, and comforting subtypes. Front. Psychol. 5, 958 (2014). https://www.frontiersin.org/articles/10.3389/fpsyg.2014.00958

24. Dunfield, K.A., Best, L.J., Kelley, E.A., Kuhlmeier, V.A.: Motivating moral behavior: helping, sharing, and comforting in young children with autism spectrum disorder. Front. Psychol. 10, 25 (2019). https://doi.org/10.3389/fpsyg.2019.00025. https://www.frontiersin.org/article/10.3389/fpsyg.2019.00025/full

25. Dunfield, K.A., Kuhlmeier, V.A.: Classifying prosocial behavior: children's responses to instrumental need, emotional distress, and material desire. Child Dev. 84(5), 1766–1776 (2013). https://doi.org/10.1111/cdev.12075

26. Epley, N., Waytz, A., Cacioppo, J.T.: On seeing human: a three-factor theory of anthropomorphism. Psychol. Rev. **114**(4), 864–886 (2007). https://doi.org/10.1037/0033-295X.114.4.864. http://doi.apa.org/getdoi.cfm?doi=10.1037/0033-295X.114.4.864

27. Fehr, E., Bernhard, H., Rockenbach, B.: Egalitarianism in young children. Nature **454**(7208), 1079–1083 (2008). https://doi.org/10.1038/nature07155

28. Fussell, S.R., Kiesler, S., Setlock, L.D., Yew, V.: How people anthropomorphize robots. In: Proceedings of the 3rd International Conference on Human Robot Interaction - HRI 2008, Amsterdam, The Netherlands, p. 145. ACM Press (2008). https://doi.org/10.1145/1349822.1349842. http://portal.acm.org/citation.cfm?doid=1349822.1349842

29. Galang, C.M., Obhi, S.S.: Automatic imitation does not predict levels of prosocial behaviour in a modified dictator game. Acta Psychologica **204**, 103022 (2020). https://doi.org/10.1016/j.actpsy.2020.103022. https://www.sciencedirect.com/science/article/pii/S0001691819304640

30. Hartley, C., Fisher, S.: Do children with autism spectrum disorder share fairly and reciprocally? J. Autism Dev. Disord. **48**(8), 2714–2726 (2018). https://doi.org/10.1007/s10803-018-3528-7

31. Henrich, J., et al.: "Economic man" in cross-cultural perspective: behavioral experiments in 15 small-scale societies. Behav. Brain Sci. **28**(6), 795–815; discussion 815–855 (2005). https://doi.org/10.1017/S0140525X05000142

32. Ibbotson, P.: Little dictators: a developmental meta-analysis of prosocial behavior. Curr. Anthropol. **55**(6), 814–821 (2014). https://doi.org/10.1086/679254

33. Šimleša, S., Stošić, J., Bilić, I., Cepanec, M.: Imitation, focus of attention and social behaviours of children with autism spectrum disorder in interaction with robots. Interact. Stud. Soc. Behav. Commun. Biol. Artif. Syst. **23**(1), 1–20 (2022). https://doi.org/10.1075/is.21037.sim. http://www.jbe-platform.com/content/journals/10.1075/is.21037.sim

34. Jameel, L., Vyas, K., Bellesi, G., Cassell, D., Channon, S.: Great expectations: the role of rules in guiding pro-social behaviour in groups with high versus low autistic traits. J. Autism Dev. Disord. **45**(8), 2311–2322 (2015). https://doi.org/10.1007/s10803-015-2393-x

35. Jamet, F., Masson, O., Jacquet, B., Stilgenbauer, J.L., Baratgin, J.: Learning by teaching with humanoid robot: a new powerful experimental tool to improve children's learning ability. J. Robot. **2018** (2018). https://doi.org/10.1155/2018/4578762. https://www.hindawi.com/journals/jr/2018/4578762/cta/

36. Johnson, S.C.: Detecting agents. Philos. Trans. Roy. Soc. London Ser. B Biol. Sci. **358**(1431), 549–559 (2003). https://doi.org/10.1098/rstb.2002.1237. https://royalsocietypublishing.org/doi/10.1098/rstb.2002.1237

37. Kahn, P.H., et al.: "Robovie, you'll have to go into the closet now": children's social and moral relationships with a humanoid robot. Dev. Psychol. **48**(2), 303–314 (2012). https://doi.org/10.1037/a0027033

38. Kumazaki, H., et al.: Job interview training targeting nonverbal communication using an android robot for individuals with autism spectrum disorder. Autism Int. J. Res. Pract. **23**(6), 1586–1595 (2019). https://doi.org/10.1177/1362361319827134

39. Kumazaki, H., et al.: Optimal robot for intervention for individuals with autism spectrum disorders. Psychiatry Clin. Neurosci. **74**(11), 581–586 (2020). https://doi.org/10.1111/pcn.13132. https://onlinelibrary.wiley.com/doi/abs/10.1111/pcn.13132

40. Kumazaki, H., et al.: A pilot study for robot appearance preferences among high-functioning individuals with autism spectrum disorder: implications for therapeutic use. PLoS ONE **12**, e0186581 (2017). https://doi.org/10.1371/journal.pone.0186581

41. Kumazaki, H., et al.: Impressions of humanness for android robot may represent an endophenotype for autism spectrum disorders. J. Autism Dev. Disord. **48**(2), 632–634 (2018). https://doi.org/10.1007/s10803-017-3365-0

42. Kuriki, S., Tamura, Y., Igarashi, M., Kato, N., Nakano, T.: Similar impressions of humanness for human and artificial singing voices in autism spectrum disorders. Cognition **153**, 1–5 (2016). https://doi.org/10.1016/j.cognition.2016.04.004

43. Liebal, K., Colombi, C., Rogers, S.J., Warneken, F., Tomasello, M.: Helping and cooperation in children with autism. J. Autism Dev. Disord. **38**(2), 224–238 (2008). https://doi.org/10.1007/s10803-007-0381-5

44. Manzi, F., et al.: A robot is not worth another: exploring children's mental state attribution to different humanoid robots. Front. Psychol. **11**, 2011 (2020). https://doi.org/10.3389/fpsyg.2020.02011. https://www.ncbi.nlm.nih.gov/pmc/articles/PMC7554578/

45. Martin, D.U., MacIntyre, M.I., Perry, C., Clift, G., Pedell, S., Kaufman, J.: Young children's indiscriminate helping behavior toward a humanoid robot. Front. Psychol. **11** (2020). https://doi.org/10.3389/fpsyg.2020.00239

46. Martin, D.U., Perry, C., MacIntyre, M.I., Varcoe, L., Pedell, S., Kaufman, J.: Investigating the nature of children's altruism using a social humanoid robot. Comput. Hum. Behav. **104**, 106149 (2020). https://doi.org/10.1016/j.chb.2019.09.025. https://www.sciencedirect.com/science/article/pii/S0747563219303590

47. Martinelli, A., Hoffmann, E., Brück, C., Kreifelts, B., Ethofer, T., Wildgruber, D.: Neurobiological correlates and attenuated positive social intention attribution during laughter perception associated with degree of autistic traits. J. Neural Transm. **130**(4), 585–596 (2023). https://doi.org/10.1007/s00702-023-02599-5

48. Masson, O., Baratgin, J., Jamet, F.: Nao robot and the "endowment effect". In: 2015 IEEE International Workshop on Advanced Robotics and its Social Impacts (ARSO), Lyon, France, pp. 1–6 (2015). https://doi.org/10.1109/ARSO.2015.7428203

49. Masson, O., Baratgin, J., Jamet, F.: Nao robot as experimenter: social cues emitter and neutralizer to bring new results in experimental psychology. In: Proceedings of the International Conference on Information and Digital Technologies, IDT 2017, pp. 256–264 (2017). https://doi.org/10.1109/DT.2017.8024306

50. Masson, O., Baratgin, J., Jamet, F.: Nao robot, transmitter of social cues: what impacts? In: Benferhat, S., Tabia, K., Ali, M. (eds.) IEA/AIE 2017. LNCS, vol. 10350, pp. 559–568. Springer, Cham (2017). https://doi.org/10.1007/978-3-319-60042-0_62

51. Masson, O., Baratgin, J., Jamet, F., Ruggieri, F., Filatova, D.: Use a robot to serve experimental psychology: some examples of methods with children and adults. In: International Conference on Information and Digital Technologies (IDT-2016), Rzeszow, Poland, pp. 190–197 (2016). https://doi.org/10.1109/DT.2016.7557172

52. Nijssen, S.R.R., Müller, B.C.N., Bosse, T., Paulus, M.: You, robot? The role of anthropomorphic emotion attributions in children's sharing with a robot. Int. J. Child-Comput. Interact. **30**, 100319 (2021). https://doi.org/10.1016/j.ijcci.2021.100319. https://www.sciencedirect.com/science/article/pii/S2212868921000465

53. Nijssen, S.R.R., Müller, B.C.N., Baaren, R.B.v., Paulus, M.: Saving the robot or the human? Robots who feel deserve moral care. Soc. Cogn. **37**(1), 41–S2

(2019). https://doi.org/10.1521/soco.2019.37.1.41. https://guilfordjournals.com/doi/10.1521/soco.2019.37.1.41

54. O'Connor, R.A.G., Stockmann, L., Rieffe, C.: Spontaneous helping behavior of autistic and non-autistic (Pre-)adolescents: a matter of motivation? Autism Res. **12**(12), 1796–1804 (2019). https://doi.org/10.1002/aur.2182. https://onlinelibrary.wiley.com/doi/abs/10.1002/aur.2182

55. World Health Organization: International Statistical Classification of Diseases and Related Health Problems. World Health Organization, 11th edn (2019). https://icd.who.int/browse11.%20Licensed%20under%20Creative%20Commons%20Attribution-NoDerivatives%203.0%C2%A0IGO%20licence%20(CC%C2%A0BY-ND%C2%A03.0%C2%A0IGO)

56. Overweg, J., Hartman, C.A., Hendriks, P.: Temporarily out of order: temporal perspective taking in language in children with autism spectrum disorder. Front. Psychol. **9** (2018). https://www.frontiersin.org/articles/10.3389/fpsyg.2018.01663

57. Paulus, M., Rosal-Grifoll, B.: Helping and sharing in preschool children with autism. Exp. Brain Res. **235**(7), 2081–2088 (2017). https://doi.org/10.1007/s00221-017-4947-y

58. Peca, A., Simut, R., Pintea, S., Pop, C., Vanderborght, B.: How do typically developing children and children with autism perceive different social robots? Comput. Hum. Behav. **41**, 268–277 (2014). https://doi.org/10.1016/j.chb.2014.09.035

59. R Core Team: R: A language and environment for statistical computing. R Foundation for Statistical Computing, Vienna, Austria (2023). https://www.R-project.org/

60. Robins, B., Dautenhahn, K., Dubowski, J.: Does appearance matter in the interaction of children with autism with a humanoid robot? Interact. Stud. Soc. Behav. Commun. Biol. Artif. Syst. **7**(3), 479–512 (2006). https://doi.org/10.1075/is.7.3.16rob. http://www.jbe-platform.com/content/journals/10.1075/is.7.3.16rob

61. Ryan-Enright, T., O'Connor, R., Bramham, J., Taylor, L.K.: A systematic review of autistic children's prosocial behaviour. Res. Autism Spectr. Disord. **98**, 102023 (2022). https://doi.org/10.1016/j.rasd.2022.102023. https://www.sciencedirect.com/science/article/pii/S1750946722001106

62. Saadatzi, M.N., Pennington, R.C., Welch, K.C., Graham, J.H.: Small-group technology-assisted instruction: virtual teacher and robot peer for individuals with autism spectrum disorder. J. Autism Dev. Disord. **48**(11), 3816–3830 (2018). https://doi.org/10.1007/s10803-018-3654-2

63. Sandoval, E.B., Brandstetter, J., Obaid, M., Bartneck, C.: Reciprocity in human-robot interaction: a quantitative approach through the prisoner's dilemma and the ultimatum game. Int. J. Soc. Robot. **8**(2), 303–317 (2016). https://doi.org/10.1007/s12369-015-0323-x

64. Scassellati, B., Admoni, H., Matarić, M.: Robots for use in autism research. Annu. Rev. Biomed. Eng. **14**, 275–294 (2012). https://doi.org/10.1146/annurev-bioeng-071811-150036

65. Schadenberg, B.R., Reidsma, D., Heylen, D.K.J., Evers, V.: Differences in spontaneous interactions of autistic children in an interaction with an adult and humanoid robot. Front. Robot. AI **7** (2020). https://www.frontiersin.org/articles/10.3389/frobt.2020.00028

66. Schmitz, E.A., Banerjee, R., Pouw, L.B., Stockmann, L., Rieffe, C.: Better to be equal? Challenges to equality for cognitively able children with autism spectrum disorders in a social decision game. Autism Int. J. Res. Pract. **19**(2), 178–186 (2015). https://doi.org/10.1177/1362361313516547

67. Short, E.S., Deng, E.C., Feil-Seifer, D., Matarić, M.J.: Understanding agency in interactions between children with autism and socially assistive robots. J. Hum.-Robot Interact. **6**(3), 21–47 (2017). https://doi.org/10.5898/JHRI.6.3.Short

68. Sommer, K., Nielsen, M., Draheim, M., Redshaw, J., Vanman, E., Wilks, M.: Children's perceptions of the moral worth of live agents, robots, and inanimate objects. J. Exp. Child Psychol. **187**, 104656 (2019). https://doi.org/10.1016/j.jecp.2019.06.009

69. Syriopoulou-Delli, C.K., Gkiolnta, E.: Review of assistive technology in the training of children with autism spectrum disorders. Int. J. Dev. Disabil. **68**(2), 73–85 (2022). https://doi.org/10.1080/20473869.2019.1706333. https://www.ncbi.nlm.nih.gov/pmc/articles/PMC8928843/

70. Telisheva, Z., Amirova, A., Rakhymbayeva, N., Zhanatkyzy, A., Sandygulova, A.: The quantitative case-by-case analyses of the socio-emotional outcomes of children with ASD in robot-assisted autism therapy. Multimodal Technol. Interact. **6**, 46 (2022). https://doi.org/10.3390/mti6060046

71. Thellman, S., Giagtzidou, A., Silvervarg, A., Ziemke, T.: An implicit, nonverbal measure of belief attribution to robots. In: Companion of the 2020 ACM/IEEE International Conference on Human-Robot Interaction, Cambridge, United Kingdom, pp. 473–475. ACM (2020). https://doi.org/10.1145/3371382.3378346. https://dl.acm.org/doi/10.1145/3371382.3378346

72. Torta, E., van Dijk, E., Ruijten, P.A.M., Cuijpers, R.H.: The ultimatum game as measurement tool for anthropomorphism in human-robot interaction. In: Herrmann, G., Pearson, M.J., Lenz, A., Bremner, P., Spiers, A., Leonards, U. (eds.) ICSR 2013. LNCS, vol. 8239, pp. 209–217. Springer, Cham (2013). https://doi.org/10.1007/978-3-319-02675-6_21

73. Townsend, L., Robeson, A., Vonk, J., Rohrbeck, K.: Autism does not dictate children's lack of sharing in a prosocial choice test. J. Autism Dev. Disord. **51**(6), 2029–2035 (2021). https://doi.org/10.1007/s10803-020-04691-1

74. Ulber, J., Tomasello, M.: Young children's prosocial responses toward peers and adults in two social contexts. J. Exp. Child Psychol. **198**, 104888 (2020). https://doi.org/10.1016/j.jecp.2020.104888. https://www.sciencedirect.com/science/article/pii/S0022096519303194

75. Valeri, G., Speranza, M.: Modèles neuropsychologiques dans l'autisme et les troubles envahissants du développement. Développements **1**(1), 34–48 (2009). https://doi.org/10.3917/devel.001.0034. https://www.cairn.info/revue-developpements-2009-1-page-34.htm

76. Wang, X., et al.: Empathy, theory of mind, and prosocial behaviors in autistic children. Front. Psychiatry **13** (2022). https://www.frontiersin.org/article/10.3389/fpsyt.2022.844578

77. Wiese, E., Müller, H.J., Wykowska, A.: Using a gaze-cueing paradigm to examine social cognitive mechanisms of individuals with autism observing robot and human faces. In: Beetz, M., Johnston, B., Williams, M.A. (eds.) ICSR 2014. LNCS, vol. 8755, pp. 370–379. Springer, Cham (2014). https://doi.org/10.1007/978-3-319-11973-1_38

78. Wood, L.J., Dautenhahn, K., Rainer, A., Robins, B., Lehmann, H., Syrdal, D.S.: Robot-mediated interviews - how effective is a humanoid robot as a tool for interviewing young children? PLoS ONE **8**(3), e59448 (2013). https://doi.org/10.1371/journal.pone.0059448

79. Wykowska, A., Chaminade, T., Cheng, G.: Embodied artificial agents for understanding human social cognition. Philos. Trans. Roy. Soc. B Biol. Sci.

**371**(1693), 20150375 (2016). https://doi.org/10.1098/rstb.2015.0375. https://royalsocietypublishing.org/doi/10.1098/rstb.2015.0375

80. Zhang, Y., et al.: Theory of robot mind: false belief attribution to social robots in children with and without autism. Front. Psychology **10**, 1732 (2019). https://doi.org/10.3389/fpsyg.2019.01732. https://www.frontiersin.org/article/10.3389/fpsyg.2019.01732/full

81. Zhang, Y., et al.: Could social robots facilitate children with autism spectrum disorders in learning distrust and deception? Comput. Hum. Behav. **98**, 140–149 (2019). https://doi.org/10.1016/j.chb.2019.04.008. https://www.sciencedirect.com/science/article/pii/S0747563219301487

82. Zhao, X., Li, X., Song, Y., Shi, W.: Autistic traits and prosocial behaviour in the general population: test of the mediating effects of trait empathy and state empathic concern. J. Autism Dev. Disord. **49**(10), 3925–3938 (2019). https://doi.org/10.1007/s10803-018-3745-0

83. Zheng, Z., Young, E.M., Swanson, A.R., Weitlauf, A.S., Warren, Z.E., Sarkar, N.: Robot-mediated imitation skill training for children with autism. IEEE Trans. Neural Syst. Rehabil. Eng. **24**(6), 682–691 (2016). https://doi.org/10.1109/TNSRE.2015.2475724. https://ieeexplore.ieee.org/document/7239626/

# An Exploratory Literature Review of Robots and Their Interaction as Assistive Technology for Persons with Disabilities: Focus on Promoting Activity and Participation

Kai Seino(✉) iD

National Rehabilitation Center for Persons With Disabilities, Namiki 4-1, Tokorozawa, Japan
seino-kai@rehab.go.jp

**Abstract.** According to the United Nations, approximately 15% of the world's population has some form of disability. Although the development of service robots to support those people with disabilities is increasing, there has not been enough comprehensive discussion on the research and effectiveness of robots as assistive technology for people with disabilities. This study conducted an exploratory literature review to determine the current status of previous research on robots to assist people with disabilities and to provide a basis for research on robots to assist people with disabilities. A literature review was conducted using PubMed as the database to identify current research trends. The search was conducted on August 1, 2021, using the keywords "disabled" and "robotics" and 49 articles were identified. After an in-depth screening process, 43 studies were deemed eligible for further analysis. As a result, the previous studies covered a variety of robot applications, including robot-assisted gait training, robot-assisted play activities, and robot-assisted task-oriented upper limb skills training. Physical disabilities accounted for the majority of the disability types in the previous study (81.4%), of which 94.3% were targeted for movement disorders. Therefore, the results of this study suggest that research on robots for the disabled mainly targets movement disorders such as limb inconvenience caused by physical disabilities. Regarding the purpose of using robots to overcome these motor impairments, most of the research was on support and training for upper and lower limb movements. Effects on participation or activity were found in 54.5% of the articles dealing with movement disorders. It was suggested that robots can support activity and participation for movement disorders of people with disabilities and have some effectiveness. On the other hand, 11.6% of the previous studies dealing with cognitive aspects of disability were less common than those dealing with motor aspects. In conclusion, robots hold great promise in improving the quality of life of people with disabilities, especially in promoting activity and participation.

**Keywords:** Persons with Disabilities · Rehabilitation · Human-Robot Interaction · Movement disorder

J. Baratgin et al. (Eds.): HAR 2023, LNCS 14522, pp. 323–344, 2024.
https://doi.org/10.1007/978-3-031-55245-8_21

# 1 Introduction

## 1.1 Background

According to the United Nations, approximately one billion people, accounting for approximately 15% of the global population, have some form of physical, mental, or sensory disability. Recently, support robots have been increasingly introduced and applied into the field of rehabilitation to support persons with disabilities (PWD). Rehabilitation here refers to medical, welfare, educational, employment, and other support for children and PWD.

For example, Hirokawa et al. [1, 2] proposed robotic interventions for children with developmental disabilities such as autism spectrum disorder (ASD) to improve their social skills and promote their communication abilities. They developed a robot manipulation interface that supporters can use for this purpose, along with a method for quantitatively measuring face-to-face behavior and facial expressions during intervention activities. The study also verified the effectiveness of the robot interface. Moreover, Jain et al. [3] developed a machine-learning algorithm that automatically instructs a robot regarding the time when therapeutic interventions must be encouraged in ASD. Humanoid robots, with mechanical features that are highly compatible with children with ASD and physical and social characteristics similar to those of humans, are expected to play a crucial role in interventions for children with ASD.

In addition, "Orihime", an avatar robot developed by Ory Laboratory, can be operated through remote control by persons with severe impairments, such as amyotrophic lateral sclerosis or spinal injurieor even patients with spinal injuries. For example, the robot can perform the tasks of a waiter in a café, and the robot supports work and communication tasks [4]. Thus, for PWD, using robot applications is expected to aid in medical treatment, skill acquisition, and communication. Although the development of various robots is progressing, the opportunities and issues associated with using robots for PWD have not yet been fully discussed.

## 1.2 Activities and Participation of Persons with Disabilities

The international definition of disability is the World Health Organization's (WHO) International Classification of Functioning, Disability, and Health (ICF) [5]. The ICF is the WHO framework for measuring health and disability at both individual and population levels. The ICF consists of two divisions, each with two components. The first division is Life Functioning and Disability. This consists of Body Functions and Body Structures and Activities Participation. The second division is Environmental and Personal Factors, not "Browse Restoration". This consists of Environmental Factors and Personal Factors. Each factor can be described in both positive and negative terms, and each factor interacts with the others. The terms are defined as follows: Activities are the performance by an individual of a task or action. In the negative, Activity Limitations refer to the difficulties that arise when an individual performs an activity. Participation refers to involvement in life situations. In the negative, it is Participation Restrictions, which refer to the difficulties experienced when an individual is involved in some life situation.

In the ICF, the term "persons with disabilities" is defined within a comprehensive framework that goes beyond a mere medical perspective. The ICF defines persons with disabilities as individuals who experience significant impairments, limitations in functioning, and restrictions in their participation in various life situations due to an interaction between their health condition and contextual factors. These factors include both environmental factors, such as societal attitudes and physical barriers, and personal factors like age, gender, and coping strategies [5].

Robots have the potential to assist or improve the Body Functions and Body Structures of people with disabilities and promote Activities and Participation through their functions and effects. However, there has not been a comprehensive examination of how robots can promote participation and activity among persons with disabilities.

### 1.3  Purpose

Therefore, this study aims to do the following in order to contribute to the basic data for research on assistive robots for PWD. To review the current status of previous research on robots that assist PWD, and to discuss the future prospects of such research.

## 2  Method: Literature Review

In this study, we conducted a literature review to identify scientific research relevant to robots for PWD. We examined papers written in English to understand international research trends. In this study, first, it was to target abstracts, after which the full text was reviewed. We have provided a step-by-step description of the process involved in the literature review.

### 2.1  Search Process

For document retrieval, PubMed, a medical database published by the National Center for Biotechnology Information in the National Library of Medicine, was used. The search was conducted on August 1, 2021.

### 2.2  Search Terms

The titles and abstracts of the search were examined. The keywords were set to "Persons with disability" and "Robot". Consequently, 49 papers were extracted.

### 2.3  Eligibility Criteria

Based on the objectives of this review, we used a two-stage screening process. For initial screening, the title and abstract were included. Secondary screening of the complete text was performed. Exclusion criteria were used in all steps of this review.

Papers were selected for inclusion if they met three specific criteria. First, studies were classified as articles or academic works; thus, popular press articles were excluded.

The second criterion was that the titles or abstracts retrieved must have explicitly mentioned both the term "robot" and the term "disability". The final criterion used for including studies in this review required not only that the study met all of the aforementioned eligibility requirements but also that the robot in the study provided support for PWD.

## 2.4 Screening Procedure

The titles and abstracts of 49 previously identified unique entries were screened manually. The screening procedure identified 43 eligible studies. After that, the full texts of the studieswere selected for full-text screening, which involved reading each of the selected papers in detail to determine their suitability based on all previously listed criteria. Finally, this screening process identified 43 eligible studies. Figure 1 visually represents this review process and the associated counts.

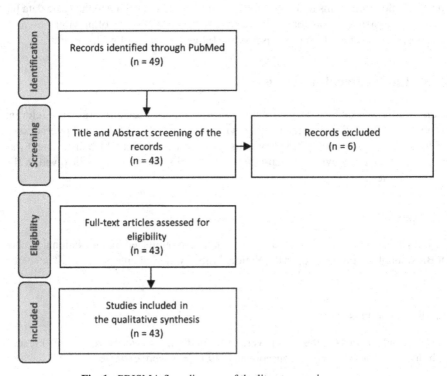

**Fig. 1.** PRISMA flow diagram of the literature review process

## 3  Results 1: All Disabilities and Summary

### 3.1  Summary of the Research

A summary of the extracted papers is given in Tables 1 and 2.

**Table 1.** Extracted papers (In order from the most recent).

|   | Year | Authors | Title | Outlet | |
|---|------|---------|-------|--------|---|
| 1 | 2021 | Gimigliano F, et al. | Robot-assisted arm therapy in neurological health conditions: rationale and methodology for the evidence synthesis in the CICERONE Italian Consensus Conference | Eur J Phys Rehabil Med | [6] |
| 2 | 2020 | Aliaj K, et al. | Replicating dynamic humerus motion using an industrial robot | PLoS One | [7] |
| 3 | 2020 | Buitrago JA, et al. | A motor learning therapeutic intervention for a child with cerebral palsy through a social assistive robot | Disabil Rehabil Assist Technol | [8] |
| 4 | 2020 | Kostrubiec V and Kruck J | Collaborative research project: Developing and testing a robot-assisted intervention for children with autism | Front Robot AI | [9] |
| 5 | 2019 | Mohammadi M, et al. | A high-resolution tongue-based joystick to enable robot control for individuals with severe disabilities | IEEE Int Conf Rehabil Robot | [10] |
| 6 | 2019 | Sattelmayer M, et al. | Over-ground walking or robot-assisted gait training in people with multiple sclerosis: does the effect depend on baseline walking speed and disease related disabilities? A systematic review and meta-regression | BMC Neurol | [11] |
| 7 | 2019 | Zhang J, et al. | An EEG/EMG/EOG-based multimodal human-machine interface to real-time control of a soft robot hand | Front Neurorobot | [12] |
| 8 | 2018 | Ricklin S, et al. | Dual-task training of children with neuromotor disorders during robot-assisted gait therapy: prerequisites of patients and influence on leg muscle activity | J Neuroeng Rehabil | [13] |
| 9 | 2018 | Sakamaki I, et al. | Preliminary testing by adults of a haptics-assisted robot platform designed for children with physical impairments to access play | Assist Technol | [14] |

*(continued)*

**Table 1.**  (*continued*)

| | Year | Authors | Title | Outlet | |
|---|---|---|---|---|---|
| 10 | 2017 | Guo C, et al. | Iterative learning impedance for lower limb rehabilitation robot | J Healthc Eng | [15] |
| 11 | 2017 | Straudi S, et al. | The effectiveness of Robot-Assisted Gait Training versus conventional therapy on mobility in severely disabled progressIve MultiplE sclerosis patients (RAGTIME): study protocol for a randomized controlled trial | Trials | [16] |
| 12 | 2017 | Tidoni E, et al. | The role of audio-visual feedback in a thought-based control of a humanoid robot: A BCI study in healthy and spinal cord injured people | IEEE Trans Neural Syst Rehabil Eng | [17] |
| 13 | 2017 | van den Heuvel RJF, et al. | Robot ZORA in rehabilitation and special education for children with severe physical disabilities: a pilot study | Int J Rehabil Res | [18] |
| 14 | 2016 | Ríos-Rincón AM, et al. | Playfulness in children with limited motor abilities when using a robot | Phys Occup Ther Pediatr | [19] |
| 15 | 2015 | Ferm UM, et al | Participation and enjoyment in play with a robot between children with cerebral palsy who use AAC and their peers | Augment Altern Commun | [20] |
| 16 | 2015 | Feys P, et al. | Robot-supported upper limb training in a virtual learning environment: a pilot randomized controlled trial in persons with MS | J Neuroeng Rehabil | [21] |
| 17 | 2015 | Peri E, et al. | An ecological evaluation of the metabolic benefits due to robot-assisted gait training | Annu Int Conf IEEE Eng Med Biol Soc | [22] |
| 18 | 2015 | Phelan SK, et al. | What is it like to walk with the help of a robot? Children's perspectives on robotic gait training technology | Disabil Rehabil | [23] |
| 19 | 2015 | Vanmulken DA, et al. | Robot-assisted task-oriented upper extremity skill training in cervical spinal cord injury: a feasibility study | SpinalCord | [24] |
| 20 | 2015 | Yamaguchi J, et al. | Measuring benefits of telepresence robot for individuals with motor impairments | Stud Health Technol Inform | [25] |
| 21 | 2014 | Encarnação P, et al. | Using virtual robot-mediated play activities to assess cognitive skills | Disabil Rehabil Assist Technol | [26] |

**Table 1.** (*continued*)

| | Year | Authors | Title | Outlet | |
|---|---|---|---|---|---|
| 22 | 2013 | Pearson Y and Borenstein J | The intervention of robot caregivers and the cultivation of children's capability to play | Sci Eng Ethics | [27] |
| 23 | 2013 | Lebec O, et al. | High level functions for the intuitive use of an assistive robot | IEEE Int Conf Rehabil Robot | [28] |
| 24 | 2012 | Fasoli SE, et al. | New horizons for robot-assisted therapy in pediatrics | Am J Phys Med Rehabil | [29] |
| 25 | 2012 | Reinkensmeyer DJ, et al. | Comparison of 3D, Assist-as-Needed Robotic Arm/Hand Movement Training Provided with Pneu-WREX to Conventional Table Top Therapy Following Chronic Stroke | Am J Phys Med Rehabil | [30] |
| 26 | 2012 | Swinnen E, et al. | Treadmill training in multiple sclerosis: can body weight support or robot assistance provide added value? A systematic review | Mult Scler Int | [31] |
| 27 | 2011 | Carrera I, et al. | ROAD: domestic assistant and rehabilitation robot | Med Biol Eng Comput | [32] |
| 28 | 2011 | Jardón A, et al. | Usability assessment of ASIBOT: a portable robot to aid patients with spinal cord injury | Disabil Rehabil Assist Technol | [33] |
| 29 | 2011 | Tonin L, et al. | Brain-controlled telepresence robot by motor-disabled people | Annu Int Conf IEEE Eng Med Biol Soc | [34] |
| 30 | 2009 | Micera S, et al. | On the control of a robot hand by extracting neural signals from the PNS: Preliminary results from a human implantation | Annu Int Conf IEEE Eng Med Biol Soc | [35] |
| 31 | 2008 | Kulyukin V, et al. | Robot-assisted shopping for the visually impaired: proof-of-concept design and feasibility evaluation | Assist Technol | [36] |
| 32 | 2007 | Billard A, et al. | Building Robota, a mini-humanoid robot for the rehabilitation of children with autism | Assist Technol | [37] |
| 33 | 2007 | Kamnik R and Bajd T | Human voluntary activity integration in the control of a standing-up rehabilitation robot: A simulation study | Med Eng Phys | [38] |
| 34 | 2005 | Parsons B, et al. | The Middlesex University rehabilitation robot | J Med Eng Technol | [39] |
| 35 | 2004 | Brisben AJ, et al. | Design evolution of an interactive robot for therapy | Telemed J E Health | [40] |

(*continued*)

**Table 1.** (*continued*)

|  | Year | Authors | Title | Outlet |  |
|---|---|---|---|---|---|
| 36 | 2003 | Hesse S, et al. | Robot-assisted arm trainer for the passive and active practice of bilateral forearm and wrist movements in hemiparetic subjects | Arch Phys Med Rehabil | [41] |
| 37 | 2001 | Bai O, et al. | Compensation of hand movement for patients by assistant force: Relationship between human hand movement and robot arm motion | IEEE Trans Neural Syst Rehabil Eng | [42] |
| 38 | 2001 | Driessen BJ, et al. | MANUS—a wheelchair-mounted rehabilitation robot | Proc Inst Mech Eng H | [43] |
| 39 | 2001 | Hoppenot P and Colle E | Localization and control of a rehabilitation mobile robot by close human-machine cooperation | IEEE Trans Neural Syst Rehabil Eng | [44] |
| 40 | 1998 | Chen S, et al. | Performance statistics of a head-operated force-reflecting rehabilitation robot system | IEEE Trans Rehabil Eng | [45] |
| 41 | 1997 | Morvan JS, et al. | Technical aids for the physically handicapped: a psychological study of the master robot | Int J Rehabil Res | [46] |
| 42 | 1994 | Lancioni GE, et al. | Promoting ambulation and object manipulation in persons with multiple handicaps through the use of a robot | Percept Mot Skills | [47] |
| 43 | 1988 | Harwin WS, et al. | A robot workstation for use in education of the physically handicapped | IEEE Trans Biomed Eng | [48] |

**Table 2.** Summary of research (In order from the most recent).

|  | Disability and disease | Research design | Subject/sample size | The purpose of research on robots | Robot's details | Result | Relation to Activity | Relation to Participation | Promotional effect on activity and participation |  |
|---|---|---|---|---|---|---|---|---|---|---|
| 1 | Physical disability (movement disorder: MD) | The methodology for evidence integration is explained | – | Robot-assisted Arm Therapy (RAT) | Upper limb robotic | – | ○ |  |  | [6] |
| 2 | Physical disability (MD) | Development | – | The duplicate of the upper limbs by a robot of operation | A 6-axis industrial robotic manipulator | This investigation demonstrates the ability to replicate human motion robotically | ○ |  |  | [7] |
| 3 | Physical disability (MD)/cerebral palsy | Case Reports | 1 child with cerebral palsy | Support of the motor learning of a walk | A social aid robot (details -- unknown) | Humanoid robots can aid therapy for kids with movement disorders. Our method offers a structured approach for cerebral palsy treatment | ○ |  | ○ | [8] |

(*continued*)

**Table 2.** (*continued*)

| | Disability and disease | Research design | Subject/sample size | The purpose of research on robots | Robot's details | Result | Relation to Activity | Relation to Participation | Promotional effect on activity and participation | |
|---|---|---|---|---|---|---|---|---|---|---|
| 4 | Developmental disease/autism | Controlled Trial | Children with autism spectrum disorder (ASD) (N = 20) | Support in the study of social skill | A white, spherical prototype with a smile. It delivers rewards, displays colors, spins, and provides cues | Results showed that sensory rewards provided by the robot elicited more positive reactions than verbal praise from humans | | ○ | ○ | [9] |
| 5 | Physical disability (MD) | Development | 2 persons | A robot controller | An inductive tongue computer interface (ITCI) for robots | The results show that the contact on the touchpads can be localized by almost 1 mm accuracy | ○ | | | [10] |
| 6 | Physical disability (MD)/multiple sclerosis (MS) | Systematic review/meta-regression | 9 studies were included in the review | Comparison of the effect of robot-assisted gait training (RAGT) | – | The pooled estimates of the effects regarding performance over short- and long-distance tests were small and nonsignificant | ○ | | | [11] |
| 7 | Physical disability (MD)/apoplexy | Development | 6 healthy persons | A robot's control | Soft robot hand | The system increased the number of possible control commands to the soft robot in its customary expressive way | ○ | | | [12] |
| 8 | Physical disability (MD) | Quasi-experimental study | 21 children with neurological gait disorders | Gait training support | The walk decoration of a robot drive (the Lokomat: Hocoma AG, Volketswil, Switzerland) | Several performance measures could differentiate well between patients who walked with physiological versus compensatory movements while performing the DT exergame | ○ | | ○ | [13] |
| 9 | Physical disability (MD) | Development | 10 healthy persons and one cerebral palsy patient | A robot controller | A robot system with a haptic interface | For participants with physical disabilities, two of the three tasks were slower with the proposed system | ○ | | ○ | [14] |
| 10 | Physical disability (MD)/apoplexy | Development | Healthy person | A robot's control | The robot system mainly comprises an omnidirectional mobile chassis (OMC) and body weight support (BWS) system | Assistance was provided with a proposed iterative learning method | ○ | | | [15] |
| 11 | Physical disability (MD)/MS | The protocol of randomized controlled trial | 98 patients with multiple sclerosis (49 persons per group) | Gait training support (robot support gait training (RAGT)) | Robot-driven walking orthosis (Lokomat: Hocoma AG, Volketswil, Switzerland) | – | | ○ | | [16] |
| 12 | Physical disability (MD)/spinal cord injury | Development | 1 person with spinal cord injury | A robot controller | Humanoid robot | Suggests that control of an external device may be improved | ○ | | ○ | [17] |
| 13 | Physical disability (MD) | Intervention study | 17 children with a physical disability | Support of play | ZORA is a 58cm humanoid robot with seven senses: movement, touch, hearing, speech, vision, connection, and thought | The results of this study show a positive contribution of ZORA in achieving therapy and educational goals | ○ | ○ | ○ | [18] |
| 14 | Physical disability (MD)/cerebral palsy | Details – unknown (intervention study) | 4 children and their mothers | Movement support, support of play | Lego robot | While playing with the robot, the level of all the children's playfulness increased sharply. Comparison of a baseline phase and a follow-up phase showed an improved level of playfulness | ○ | ○ | ○ | [19] |
| 15 | A communication obstacle/cerebral palsy | Details -- unknown (intervention study) | Unknown | Support of play | The developed robot talks or moves to promote the pleasant independent play | Unknown | ○ | ○ | | [20] |

(*continued*)

**Table 2.** (*continued*)

| | Disability and disease | Research design | Subject/sample size | The purpose of research on robots | Robot's details | Result | Relation to Activity | Relation to Participation | Promotional effect on activity and participation | |
|---|---|---|---|---|---|---|---|---|---|---|
| 16 | Physical disability (MD)/MS | Randomized controlled trial | 17 persons with multiple sclerosis | Support of upper-limb training | The Haptic-Master robot (MOOG, Netherlands) provides support and tactile feedback in a virtual learning setting | The moving task was performed in a brief time. In the intervention group and contrast group, the clinically measured value did not show a significant change | ○ | | | [21] |
| 17 | Physical disability (MD)/cerebral palsy | Controlled Trial | 4 children with cerebral palsy | Walk support | Unknown | The metabolic advantage of the medical treatment that may improve a disabled child's movement efficiency was suggested | ○ | | ○ | [22] |
| 18 | Physical disability (MD)/cerebral palsy | Case Reports | Unknown | Gait training support | The walk decoration of a robot drive (the Lokomat: Hocoma AG, Volketswil, Switzerland) | Four themes related to children's expectations and experiences using Lokomat were identified | ○ | | ○ | [23] |
| 19 | Physical disability (MD) | Case Reports | 5 cervical cord injury patients | Support of upper-limb training | Unknown | Therapists report that working with the HM is easy to learn and easy to perform. Usability of the HM may be improved | ○ | | ○ | [24] |
| 20 | Physical disability (MD) | Development | Unknown | Support of friendship by remoteness | Telepresence robot | Unknown | | ○ | | [25] |
| 21 | Unknown | Unknown | Unknown | Support of play | Virtual robot (details -- unknown) | Interaction and communication aspects revealed favoring the virtual robot compared to physical robots | | | | [26] |
| 22 | Physical disability/cognitive disorder | Literature review | Children with disabilities | Development of the child's ability to play | Robot to take care of children | Appropriate robotic caregiver intervention can contribute to the realization of the ability to play. Robotic caregiver intervention can reduce the burden of caregiving | ○ | ○ | ○ | [27] |
| 23 | Physical disability (MD) | Development | 4 healthy individuals and 29 patients with quadriplegia (muscular dystrophy, spinal injury, SMA, MS, ALS, cerebral palsy, rheumatoid arthritis, post-polio, locked-in syndrome, and other motor paralysis) | A robot's control | The robot combines a RobuLAB10 mobile platform from Robosoft and a 6-DOF JACO arm by KINOVA mounted on it | Technology assessment was affirmative | ○ | | ○ | [28] |
| 24 | Physical disability (MD) | Review | None | Robot support of the upper limbs and the leg | None | – | ○ | | | [29] |
| 25 | Physical disability (MD)/apoplexy | Randomized controlled trial | 26 apoplexy patients (13 each in the robot and target groups) | To verify the effectiveness of robot-assisted upper limb movement training | Actuating arm support for 3-dimensional movement | After training, arm motor skills in the robot group After 3 months, sensory scores improved and motor skills improved | ○ | | ○ | [30] |
| 26 | Physical disability (MD)/MS | Systematic review | 5 true and 3 pre-experimental studies included, covering 161 MS patients | Reviewing the effectiveness of treadmill training (TT), body-supported TT (BWSTT), and robot-assisted TT (RATT) in those with MS, emphasizing gait outcomes | – | There is a limited number of published papers related to TT in persons with MS, concluding that TT, BWSTT, and RATT improve walking speed and endurance | ○ | | ○ | [31] |
| 27 | Physical disability (MD) | Review | None | Support of movement a walk | None | – | ○ | | | [32] |
| 28 | Physical disability (MD)/spinal cord injury | Development | 6 persons with spinal cord injury | Support/complement of a motor function | A portable manipulator, ASIBOT, is used, developed by the Robotics Lab at University Carlos III of Madrid | Unknown | ○ | | | [33] |

(*continued*)

**Table 2.** *(continued)*

| | Disability and disease | Research design | Subject/sample size | The purpose of research on robots | Robot's details | Result | Relation to Activity | Relation to Participation | Promotional effect on activity and participation | |
|---|---|---|---|---|---|---|---|---|---|---|
| 29 | Physical disability (MD) | Development | Users with mental disabilities controlling a telepresence robot (N = 2) and healthy users (N = 4) | A robot controller | Control for brain-controlled telepresence robots | The user with obstacles attained the same performance as a healthy user | ○ | | ○ | [34] |
| 30 | Physical disability (MD) | Development | Unknown | A robot controller | A smart device for a robot hand | Both extractions of movement information and recovery of a sensory function are possible | ○ | | ○ | [35] |
| 31 | Physical disability (visual impairment) | Development | Shoppers (n = 10) with visual impairment | Support of shopping (a shopper is led to a particular place, and information necessary for searching a product is offered) | The device of a robot shopping cart | Shoppers could search for products independently and reliably in supermarkets | ○ | ○ | ○ | [36] |
| 32 | Developmental disease/autism | Development | An experimental platform for studies with autistic children by two | Support of acquisition of the concerted action by an imitation robot | Multiple degrees of freedom doll-shaped humanoid robots | The results support the view of adopting non-invasive interfaces for interacting with the robot | | ○ | ○ | [37] |
| 33 | Unknown | Development | Unknown | Support of stand-up training | Unknown | The subject's non-performance was confirmed | | | | [38] |
| 34 | Physical disability (MD) | Development | Spinal cord injury patient | A robot controller | The manipulator of wheelchair loading | They observed that it was easy to use but was approximately 50% lower than comparable systems before design modifications were incorporated | ○ | | ○ | [39] |
| 35 | Unknown | Development | Unknown | Medical treatment, education, support of play | An interactive robot for therapy (details -- unknown) | Unknown | | | | [40] |
| 36 | Physical disability (MD)/apoplexy | Controlled trial | 12 patients of hemiplegia | Support of training of the upper limbs | Robot arm | Reduction in spasticity, ease of finger health, and pain mitigation were observed. There were improvements in the articular motor function of a wrist and a finger | ○ | | ○ | [41] |
| 37 | Physical disability (mo(MD)/cerebellar dysfunction, Parkinson's disease | Development | Unknown | A robot controller | Robot arm | Motion of a patient's hand was improved | ○ | | ○ | [42] |
| 38 | Physical disability (MD) | Development | Unknown | Support/complement of a motor function | The manipulator of wheelchair loading | Unknown | ○ | | | [43] |
| 39 | Unknown | Development | Unknown | A robot controller | Unknown | Unknown | | | | [44] |
| 40 | Physical disability (MD) | Development | Development | Unknown | A pair of master–slave robots | – | | | | [45] |
| 41 | Physical disability (MD) | Development (Interview) | Movement disorder 28 people | Assessment of psychological aspects | arm-type robot | There were concerns and threats when using the robot | ○ | | | [46] |
| 42 | Multiple handicaps (MD, visual impairment, intellectual disability) | Development | 2 persons with multiple handicaps (movement disorder, visual impairment, intellectual disability) | Walk support etc. | Unknown | The data showed that both subjects successfully learned to use the robot, transporting, and putting away objects, and achieved independent ambulation times of over 22 and 20 min | ○ | | ○ | [47] |
| 43 | Physical disability (MD) | Development (Experiment) | Children with physical disabilities | Support for education | Workstations using robots and vision systems | The child could perform the task by using the robot | ○ | | ○ | [48] |

## 3.2  Publication Outlets

The literature search identified 43 papers published on robot rehabilitation for PWD that met the criteria. These publications were primarily from five conferences and 38 journals. A breakdown of publications by type is shown in Fig. 2. There were no dominant publication venues, with 17 of the 43 studies published in unique venues. Of the studies that were published in the same locations, three were published in "Assistive Technology", "Disability and Rehabilitation: Assistive Technology", "IEEE Transactions on Neural Systems and Rehabilitation Engineering", and "Annual International Conference of the IEEE Engineering in Medicine and Biology Society". Two each were published in the "International Journal of Rehabilitation Research", the "Journal of Neuro Engineering and Rehabilitation", and the "International Conference on Rehabilitation Robotics".

The oldest paper was published in 1988. Since the 2010s, the number of studies tended to increase compared to the number in the 2000s. A breakdown of annual publications is presented in Fig. 3.

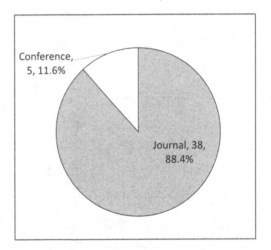

**Fig. 2.**  Publications by type

## 3.3  The Target Disability

Among the 43 identified articles, the research objectives and the disorders and diseases covered were as follows Thirty-five studies (one was co-listed with cognitive impairment) (81.4% of the total) were related to physical disabilities. Of these, 33 (94.3% of the physical disabilities) were related to movement disorders. Motor impairments will be treated independently in the next section. Only one study was based on visual impairment. Target disabilities in the peer-reviewed articles were as follows: two (4.7%) studies focused on developmental disabilities, two of which were related to autism. One (2.3%) study focused on communication disorders and another (2.3%) on multiple disabilities (motor, visual, and intellectual disabilities). Four (9.3%) could not be identified due to

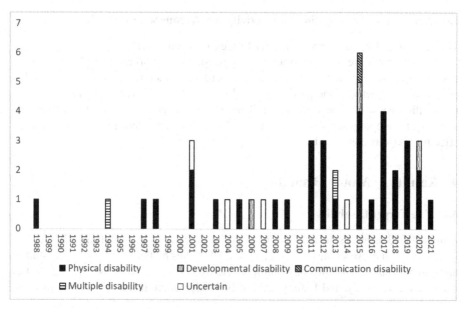

**Fig. 3.** Publications by year and disability

disability classification. No clear trend could be identified for the year of publication or disability.

### 3.4 Research Design

This section analyzes the studies that were conducted on non-movement disorders. The research designs used were as follows. One study for developmental disabilities was a controlled trial. The other three studies (one each for developmental disabilities, visual impairment, and multiple disabilities) were developmental. The other study was of unknown research design.

### 3.5 Purpose of the Robot Used or Investigated

The robots in two of the studies targeting developmental disabilities were interactive. The purpose of using the robots was to train social skills in one study and to learn cooperative behavior in the other study. The robot in one study of communication disorders was an interactive robot. The purpose of the use of the robot was play training. The robot in one study with a visual impairment was a robotic shopping cart. The purpose of the robot was to assist with shopping. The robot in one study with an unknown target audience was virtual, but the details and purpose of use were unknown. The other study's robot and purpose were unknown.

### 3.6    Relation of Participation and Activity for Persons with Disabilities

The two studies on developmental disabilities were on social skills and cooperative movement, which are associated with participation. They also confirmed its effectiveness. One study targeting communication disorders is a study of play support and is associated with activity and participation. However, the effectiveness of this study was not verified because it was not tested. The one case targeting visual impairment was a study of shopping support, which was related to activity and participation and its effectiveness was confirmed.

## 4    Results 2: Motor Disability

### 4.1    The Target Disability

Among the 43 articles selected, 33 (76.7%) were on movement disabilities. The disabilities covered in those studies were as follows. 19 studies had no disease listed, 4 studies had cerebral palsy, 4 studies had multiple sclerosis, 4 studies had apoplexy, 2 studies had spinal cord injury, and 1 study had cerebellar dysfunction and Parkinson's disease.

### 4.2    Research Design

About the types of study designs used in the papers, we will first mention the medical studies in order of their level of evidence. One study used systematic review and meta-analysis, one systematic review, two randomized controlled trials, two intervention studies, two controlled trials, two quasi-experimental studies, three case reports, and two qualitative reviews. There were also 17 non-medical engineering studies on development. One of these used interviews and the other used experiments as methodology. The other two were protocols for randomized controlled trials, and one was a methodological study of evidence synthesis.

Here is a summary of the systematic reviews with the highest level of evidence. Sattelmayer et al. (2019) aimed to evaluate the effectiveness of robot-assisted gait training (RAGT) in comparison to conventional over-ground walking training (CGT) for patients with multiple sclerosis (MS). The research focused on studies involving adult MS patients that compared RAGT with non-robot-assisted gait training. Only randomized controlled studies were considered. Five databases were searched. The review included nine studies. The findings of the research indicated that there was no significant difference between RAGT and CGT in terms of walking ability, whether measured over short or long distances.

Swinnen et al. (2012) objective was to critically analyze the existing literature on the effectiveness of treadmill training (TT), body-weight-supported TT (BWSTT), and robot-assisted TT (RATT) for individuals with multiple sclerosis (MS), with a particular focus on gait-related outcome measurements. The researchers conducted extensive searches across electronic databases, including four databases. They also examined the reference lists of articles and narrative reviews. The inclusion criteria were set to encompass pre-, quasi-, and true-experimental studies that involved adult MS patients participating in TT, BWSTT, or RATT2. From their search, eight studies met the criteria for inclusion in the review. Of these, five were true-experimental studies, while three were pre-experimental. In total, 161 MS patients participated in the interventions, which varied in duration and frequency, ranging from 6 to 42 sessions, conducted 2 to 5 times a week, over a span of 3 to 21 weeks. The findings from these studies indicated significant improvements in walking speed and endurance. There were also noted improvements in step length, double-support time, and scores on the Expanded Disability Status Scale. In conclusion, published studies on TT for MS patients suggest potential benefits of TT, BWSTT, and RATT in enhancing walking speed and endurance, it remains inconclusive as to which type of TT is the most effective.

### 4.3 Purpose of the Robot Used or Investigated

The robots addressed in the movement disorder studies were as follows. 11 were interfaces for robots, 7 were robots for upper limb assistance, 6 were robots for gait assistance, 2 were humanoid robots, and 1 was a remote control robot. Others were either indistinguishable or not described. The humanoid robots included one for the development of a robot controller and one for a play support intervention study.

The purposes of the robots were as follows 10 cases were for the development of robot controllers and robot control methods, 9 cases for walking support and training, 8 cases for upper limb support and training, 4 cases for movement support, 1 case for play support, 1 case for friendship support, 1 case for psychological evaluation, and 1 case for educational support.

### 4.4 Relation of Participation and Activity for Persons with Disabilities

The 33 extracted studies of robots for movement disorders were related to the participation or activity of people with disabilities and their effect on them as follows. Thirty-one studies (93.9% of the motor disability articles) were considered to be related to activity. In the remaining two cases, the details of the robot were unknown and could not be determined.

Next, three articles (9.1%) were considered to be associated with participation. The reason why they were considered to be related to participation is that two of them were intended to support play and one was intended to support friendship, not only assisting physical functions but also supporting activities to interact with people and society.

Here is a summary of three papers on participation, which can be considered a more complex and social activity. Rios A et al. (2016) delves into the impact of the robotic intervention on the playfulness of children with cerebral palsy (CP). Recognizing that children with motor impairments often have limited opportunities for free play, the research aimed to explore the potential benefits of an adapted Lego robot during play sessions. The method employed a partially non-concurrent multiple baseline design, involving four children with CP and their mothers. Over the course of the study, the children were provided with an adapted Lego robot, referred to as the "robot", during 15-min free play sessions with their mothers. These sessions took place twice a week for approximately 14 weeks. The results were promising. All participating children exhibited a significant increase in their levels of playfulness while interacting with the robot. Furthermore, mothers reported an improvement in their perceptions of their child's play performance and expressed greater satisfaction with their child's play during the intervention phase.

Renée J.F. van den Heuvela et al. (2017) aimed to explore the potential of interventions based on the ZORA robot in rehabilitation and special education for children with severe physical disabilities. Conducted over 2.5 months across two centers, the research involved children with developmental ages ranging from 2 to 8 years. These children participated in six sessions with the ZORA robot, either individually or in groups. ZORA is a humanoid robot, originally produced as the NAO robot by Softbank Robotics. It is equipped with seven senses for natural interaction, including movement, feeling, hearing, speaking, seeing, connecting, and thinking. The robot can be pre-programmed to perform various activities, such as dancing or interacting with users. As a result, the sessions with ZORA were perceived as playful, and the robot made a positive contribution towards achieving therapy and educational goals. Three primary domains were identified as the most promising for ZORA's application: movement skills, communication skills, and cognitive skills. Furthermore, the robot was found to enhance motivation, concentration, initiative-taking, and attention span in children.

Yamaguchi et al. (2015) aimed to investigate the effect of using a telepresence robot in the marketplace by individuals with motor impairments. The telepresence robot could be an alternative means to attend social activities such as going to school and work for people who have difficulty going out because of motor impairments. Three participants were involved with trial use for attending university courses for a month and the initial results are shown.

Next, 18 (54.5%) affected participation or activity. These effects were those that showed some significant difference or change in the study design. Those study designs included one systematic review, one randomized controlled trial, two intervention studies, two controlled trials, one quasi-experimental study, three case studies, and eight development studies. Although the level of evidence of effectiveness differed, a certain level of effectiveness was verified. In addition, if the research design was designed to verify effectiveness in engineering research development, effectiveness could be confirmed even if the research was not a medical study.

## 5  Consideration

### 5.1  Summary of Results

The review conducted for this study provided a comprehensive analysis of robotic-assisted interventions and developments for PWD. A screening procedure identified 43 eligible studies. The studies covered a variety of robot applications, including robot-assisted gait training, robot-assisted play activities, and robot-assisted task-oriented upper extremity skills training.

The particular field of the published studies was divided into two or more fields, roughly classified into engineering systems, rehabilitation study systems, and rehabilitation engineering systems of the two boundary domains. From the perspective of rehabilitation for PWD, the research has focused on evaluating the robot for the required task of the studied rehabilitation system. Most of the analyzed robots are at the research and development stages. That is, current rehabilitation robots have not been widely implemented in practice. After confirming the effectiveness of these robots, they might be used for various practical purposes.

## 5.2   Disability and Disease, and the Research Purpose

Physical disability accounted for the majority of the disability types in the previous study (81.4%), of which 94.3% were for movement disorders. Therefore, the results of this study suggest that research on robots for the disabled mainly targets movement disorders such as limb inconvenience caused by physical disabilities. Regarding the purpose of using robots to overcome such motor impairments, most of them provided support and training for upper and lower limb movements. The diseases reported as contributing factors to movement disorders were cerebral palsy, multiple sclerosis, stroke, spinal cord injury, cerebellar dysfunction, and Parkinson's disease.

## 5.3   The Research Design of Robots to Assist with Movement Disorders

Two systematic reviews (one including a meta-analysis) of robotics for motor impairments were reported with a high level of medical evidence [11, 31]. Both focused on assisted walking in patients with multiple sclerosis; Sattelmayer et al. (2019) [11] compared the effectiveness of robot-assisted gait training (RAGT) to conventional ground gait training (CGT) but found no significant differences between RAGT and CGT concerning walking ability Swinnen et al. (2012) [31] critically analyzed the effectiveness of treadmill training (TT), body weight bearing TT (BWSTT), and robot-assisted TT (RATT). As a result, published studies on TT for MS patients suggested the potential benefits of TT, BWSTT, and RATT in improving walking speed and endurance. Thus, the suggested effectiveness of RATT indicates that robotic assistance may be effective in the treatment of motor disability in patients with multiple sclerosis.

## 5.4   Robots to Assist with Movement Disorders

Robots being studied to assist with motor disabilities were reported to include robotic interfaces, robots for upper limb support, robots for gait support, humanoid robots, and remotely controlled robots. The research objectives of those robots were development, movement support including walking and upper limb, play and friendship support, psychological evaluation, and educational support. Since the function of a robot does not always correspond to the purpose of research using the robot, it is necessary to focus on the purpose of research using the robot in terms of support for PWD.

## 5.5   Robot-Mediated Interactions

According to the ICF [5], disability is caused by the interaction between people and the environment. Accordingly, the robot for PWD mainly works with a person or environment. Based on this, the idea is as follows: The robot for supporting the motor function of a person with a physical disability (movement disorder) is related to the interaction between the person and robot. Therefore, it provides support, with the robot conducting the input and output tasks of the human and working in the environment. Furthermore, robots that support the learning of persons with developmental diseases are related to the interaction between humans and robots. However, this type of robot reacts like a fake human. That is, it functions as an interaction between a fake human being as a robot and a human being. Accordingly, it is expected that forms of the general view and function for which a robot is used and the devices themselves will differ greatly.

## 5.6   Cognitive Robotics and Helping People with Disabilities

As many of the previous studies focused on helping with movement disorders, most robots for assisting PWD have been designed to supplement the functions of physical disabilities. Therefore, the robots are expected to be controlled by the user to perform the necessary tasks. In addition, since the robots will be used in medical and welfare settings, the safety of the robot's behavior is a top priority. Therefore, little attention has been paid to the intelligent behavior of robots and robot autonomy guided by artificial intelligence and deep learning. Similarly, evolutionary robotics through adaptation to the environment has not been studied. Furthermore, the functions that are expected of PWD are those that are inherently human, and functions that are expected to be compensated by robots are those of arms and legs.

On the other hand, fewer previous studies (5 (11.6%)) dealt with the cognitive aspects of disability than with the motor function aspects. As for the cognitive aspect, robots that talk and move are used to support learning and play for children with developmental disabilities. There is a possibility that cognitive modeling, knowledge representation, and cognitive architecture can be used to develop such robots. For robots designed for PWD, the specific environment and aspects of the human–robot interaction are important. It is expected that more research will be done on the functions and utilization of the robot itself, taking into account the expected and cognitive behaviors, as well as the effects of the user, environmental constraints, and technical and psychological aspects.

In a subsequent study in the period covered by this study, Papadopoulou MT, et al. (2022) [49] conducted a special education intervention using a humanoid social robot with children with learning disabilities. Results of controlled trials showed that the group using the robot was comparable to human educators, with an advantage only in phonological awareness exercises. Research on robots to assist with cognitive aspects is developing and expanding.

## 5.7   Relation of Participation and Activity for Persons with Disabilities

Of the studies of robots for motor disability, 93.9% were associated with activity, 9.1% were associated with participation. Those associated with activity were those that supported physical function. On the other hand, those associated with participation were studied to support activities to interact with people and society: play and friendship support.

Next, effects on participation or activity were found in 54.5% of the articles dealing with motor disability. Overall, although the level of evidence is limited by the study design, robots can support activity and participation for movement disorders in people with disabilities and have some effectiveness, as shown in the systematic review by Sattelmayer et al. (2019) on the effectiveness of robots for walking support In terms of effectiveness, robots will continue to be used in the future. In terms of effectiveness, we believe that the accumulation of randomized controlled trials focusing on specific diseases and motor functions will make it possible to demonstrate the effectiveness of robots for other activities and participation in the future.

### 5.8  Limitations and Suggestion for Future Research

This study is an exploratory review of all disability types. Therefore, a broad concept of search terms was used. Future research focusing on specific disabilities, diseases, or purposes of assistance could provide a basis for developing more practical robots. In addition to the development of robots, medical and rehabilitation research to evaluate the effectiveness of robot-assisted supports and interventions is also important in the study of robots to assist people with disabilities.

In addition, to the previous studies, the development studies included several healthy individuals on the test list for user evaluation in these studies. The engineering, medical, and rehabilitation research methods differed greatly. However, evaluation and evidence from people with disabilities are necessary for future development.

### 5.9  Conclusion

Considering the percentage of the world's population with disabilities, the evolution and application of robot technology could be an extremely important means of improving the quality of life of these individuals. In particular, the use of robots in the areas of daily operational support and rehabilitation is expected to promote the independence of people with disabilities. Robot-assisted interventions can be a promising means of supplementing or improving the functional abilities of people with disabilities, thereby promoting their activity and participation and improving their quality of life. Diverse applications, ranging from gait training to cognitive support, suggest the potential of robots in rehabilitation and support. However, more rigorous research is essential to ascertain the best practices and long-term impact of these interventions.

## References

1.  Hirokawa, M., Funahashi, A., Suzuki, K.: A doll-type interface for real-time humanoid tele-operation in robot-assisted activity: a case study. In: Proceedings of the 2014 ACM/IEEE International Conference on Human-Robot Interaction, pp. 174–175 (2014). https://doi.org/10.1145/2559636.2563680
2.  Hirokawa, M., Funahashi, A., Itoh, Y., Suzuki, K.: Design of affective robot-assisted activity for children with autism spectrum disorders. In: Proceedings of the 23rd IEEE International Symposium on Robot and Human Interactive Communication, pp. 365–370 (2014). https://doi.org/10.1109/ROMAN.2014.6926280
3.  Jain, S., Thiagarajan, B., Shi, A., Clabaughand, C., Matarić, M.J.: Modeling engagement in long-term, in-home socially assistive robot interventions for children with autism spectrum disorders. Sci. Robot. 5(39), eaaz3791 (2020). https://doi.org/10.1126/scirobotics.aaz3791

4. Shizume, C.: Enabling technology at robot cafe redefines work for disability. Zenbird (2019). https://zenbird.media/enabling-technology-at-robot-cafe-redefines-work-for-disability/. Accessed 01 Aug 2021

5. WHO: International Classification of Functioning, Disability and Health. World Health Organization, Geneva (2002)

6. Gimigliano F, et al.: Robot-assisted arm therapy in neurological health conditions: rationale and methodology for the evidence synthesis in the CICERONE Italian Consensus Conference. Eur. J. Phys. Rehabil. Med. (2021). https://doi.org/10.23736/S1973-9087.21.07011-8. Italian consensus conference on robotics in neurorehabilitation (CICERONE)

7. Aliaj, K., et al.: Replicating dynamic humerus motion using an industrial robot. PLoS One 15(11), e0242005 (2020). https://doi.org/10.1371/journal.pone.0242005

8. Buitrago, J.A., Bolaños, A.M., Caicedo, B.E.: A motor learning therapeutic intervention for a child with cerebral palsy through a social assistive robot. Disabil. Rehabil. Assist. Technol. 15(3), 357–362 (2020). https://doi.org/10.1080/17483107.2019.1578999

9. Kostrubiec, V., Kruck, J.: Collaborative research project: developing and testing a robot-assisted intervention for children with autism. Front. Rob. AI 7, 37 (2020). https://doi.org/10.3389/frobt.2020.00037

10. Mohammadi, M., Knoche, H., Gaihede, M., Bentsen, B., Andreasen Struijk, L.N.S.: A high-resolution tongue-based joystick to enable robot control for individuals with severe disabilities. In: 16th IEEE International Conference on Rehabilitation Robotics, pp. 1043–1048 (2019). https://doi.org/10.1109/ICORR.2019.8779434

11. Sattelmayer, M., Chevalley, O., Steuri, R., Hilfiker, R.: Over-ground walking or robot-assisted gait training in people with multiple sclerosis: does the effect depend on baseline walking speed and disease related disabilities? A systematic review and meta-regression. BMC Neurol. 19(1), 93 (2019). https://doi.org/10.1186/s12883-019-1321-7

12. Zhang, J., Wang, B., Zhang, C., Xiao, Y., Wang, M.Y.: An EEG/EMG/EOG-based multimodal human-machine interface to real-time control of a soft robot hand. Front. Neurorobot. 29(13), 7 (2019). https://doi.org/10.3389/fnbot.2019.00007

13. Ricklin, S., Meyer-Heim, A., van Hedel, H.J.A.: Dual-task training of children with neuromotor disorders during robot-assisted gait therapy: prerequisites of patients and influence on leg muscle activity. J. Neuroeng. Rehabil. 15(1), 82 (2018). https://doi.org/10.1186/s12984-018-0426-3

14. Sakamaki, I., et al.: Preliminary testing by adults of a haptics-assisted robot platform designed for children with physical impairments to access play. Assist. Technol. 30(5), 242–250 (2018). https://doi.org/10.1080/10400435.2017.1318974

15. Guo, C., Guo, S., Ji, J., Xi, F.: Iterative learning impedance for lower limb rehabilitation robot. J. Healthc. Eng. 2017, 6732459 (2017). https://doi.org/10.1155/2017/6732459

16. Straudi, S., et al.: The effectiveness of robot-assisted gait training versus conventional therapy on mobility in severely disabled progressive MultiplE sclerosis patients (RAGTIME): study protocol for a randomized controlled trial. Trials 18(1), 88 (2017). https://doi.org/10.1186/s13063-017-1838-2

17. Tidoni, E., Gergondet, P., Fusco, G., Kheddar, A., Aglioti, S.M.: The role of audio-visual feedback in a thought-based control of a humanoid robot: a BCI study in healthy and spinal cord injured people. IEEE Trans. Neural Syst. Rehabil. Eng. 25(6), 772–781 (2017). https://doi.org/10.1109/TNSRE.2016.2597863

18. van den Heuvel, R.J.F., Lexis, M.A.S., de Witte, L.P.: Robot ZORA in rehabilitation and special education for children with severe physical disabilities: a pilot study. Int. J. Rehabil. Res. 40(4), 353–359 (2017). https://doi.org/10.1097/MRR.0000000000000248

19. Ríos-Rincón, A.M., Adams, K., Magill-Evans, J., Cook, A.: Playfulness in children with limited motor abilities when using a robot. Phys. Occup. Ther. Pediatr. 36(3), 232–246 (2016). https://doi.org/10.3109/01942638.2015.1076559

20. Ferm, U.M., Claesson, B.K., Ottesjö, C., Ericsson, S.: Participation and enjoyment in play with a robot between children with cerebral palsy who use AAC and their peers. Augment. Altern. Commun. **31**(2), 108–123 (2015). https://doi.org/10.3109/07434618.2015.1029141

21. Feys, P., et al.: Robot-supported upper limb training in a virtual learning environment: a pilot randomized controlled trial in persons with MS. J. Neuroeng. Rehabil. **12**, 60 (2015). https://doi.org/10.1186/s12984-015-0043-3

22. Peri, E., et al.: An ecological evaluation of the metabolic benefits due to robot-assisted gait training. In: 37th Annual International Conference of the IEEE Engineering in Medicine and Biology Society, pp. 3590–3593 (2015). https://doi.org/10.1109/EMBC.2015.7319169

23. Phelan, S.K., Gibson, B.E., Wright, F.V.: What is it like to walk with the help of a robot? Children's perspectives on robotic gait training technology. Disabil. Rehabil. **37**(24), 2272–2281 (2015). https://doi.org/10.3109/09638288.2015.1019648

24. Vanmulken, D.A., Spooren, A.I., Bongers, H.M., Seelen, H.A.: Robot-assisted task-oriented upper extremity skill training in cervical spinal cord injury: a feasibility study. Spinal Cord. **53**(7), 547–551 (2015). https://doi.org/10.1038/sc.2014.250

25. Yamaguchi, J., Parone, C., Di Federico, D., Beomonte Zobel, P., Felzani, G.: Measuring benefits of telepresence robot for individuals with motor impairments. In: Sik-Lányi, C., Hoogerwerf, E.J., Miesenberger, K., Cudd, P. (eds.) Studies in Health Technology and Informatics, Volume 2017: Assistive Technology, pp. 703–709. IOS Press (2015). https://doi.org/10.3233/978-1-61499-566-1-703

26. Encarnação, P., Alvarez, L., Rios, A., Maya, C., Adams, K., Cook, A.: Using virtual robot-mediated play activities to assess cognitive skills. Disabil. Rehabil. Assist. Technol. **9**(3), 231–241 (2014). https://doi.org/10.3109/17483107.2013.782577

27. Pearson, Y., Borenstein, J.: The intervention of robot caregivers and the cultivation of children's capability to play. Sci. Eng. Ethics **19**(1), 123–137 (2013). https://doi.org/10.1007/s11948-011-9309-8

28. Lebec, O., et al.: High level functions for the intuitive use of an assistive robot. In: 2013 IEEE International Conference on Rehabilitation Robotics, pp. 1–6 (2013). https://doi.org/10.1109/ICORR.2013.6650374

29. Fasoli, S.E., Ladenheim, B., Mast, J., Krebs, H.I.: New horizons for robot-assisted therapy in pediatrics. Am. J. Phys. Med. Rehabil. **91**(11 Suppl 3), S280-289 (2012). https://doi.org/10.1097/PHM.0b013e31826bcff4

30. Reinkensmeyer, D.J., Wolbrecht, E.T., Chan, V., Chou, C., Cramer, S.C., Bobrow, J.E.: Comparison of 3D, assist-as-needed robotic arm/hand movement training provided with Pneu-WREX to conventional table top therapy following chronic stroke. Am. J. Phys. Med. Rehabil. **91**(11 Suppl 3), S232–S241 (2012). https://doi.org/10.1097/PHM.0b013e31826bce79

31. Swinnen, E., Beckwée, D., Pinte, D., Meeusen, R., Baeyens, J.P., Kerckhofs, E.: Treadmill training in multiple sclerosis: can body weight support or robot assistance provide added value? A systematic review. Mult. Scler. Int. **2012**, 240274 (2012). https://doi.org/10.1155/2012/240274

32. Carrera, I., Moreno, H.A., Saltarén, R., Pérez, C., Puglisi, L., Garcia, C.: ROAD: domestic assistant and rehabilitation robot. Med. Biol. Eng. Comput. **49**(10), 1201 (2011). https://doi.org/10.1007/s11517-011-0805-4

33. Jardón, A., Gil, Á.M., de la Peña, A.I., Monje, C.A., Balaguer, C.: Usability assessment of ASIBOT: a portable robot to aid patients with spinal cord injury. Disabil. Rehabil. Assist. Technol. **6**(4), 320–330 (2011). https://doi.org/10.3109/17483107.2010.528144

34. Tonin, L., Carlson, T., Leeb, R., del R Millán, J.: Brain-controlled telepresence robot by motor-disabled people. In: 2011 Annual International Conference of the IEEE Engineering in Medicine and Biology Society, pp. 4227–4230 (2011). https://doi.org/10.1109/IEMBS.2011.6091049

35. Micera, S., et al.: On the control of a robot hand by extracting neural signals from the PNS: preliminary results from a human implantation. In: 2009 Annual International Conference of the IEEE Engineering in Medicine and Biology Society, pp. 4586–4589 (2009). https://doi.org/10.1109/IEMBS.2009.5332764

36. Kulyukin, V., Gharpure, C., Coster, D.: Robot-assisted shopping for the visually impaired: proof-of-concept design and feasibility evaluation. Assist. Technol. **20**(2), 86–98 (2008). https://doi.org/10.1080/10400435.2008.10131935

37. Billard, A., Robins, B., Nadel, J., Dautenhahn, K.: Building Robota, a mini-humanoid robot for the rehabilitation of children with autism. Assist. Technol. **19**(1), 37–49 (2007). https://doi.org/10.1080/10400435.2007.10131864

38. Kamnik, R., Bajd, T.: Human voluntary activity integration in the control of a standing-up rehabilitation robot: a simulation study. Med. Eng. Phys. **29**(9), 1019–1029 (2007). https://doi.org/10.1016/j.medengphy.2006.09.012

39. Parsons, B., White, A., Prior, S., Warner, P.: The Middlesex university rehabilitation robot. J. Med. Eng. Technol. **29**(4), 151–162 (2005). https://doi.org/10.1080/03091900412331298898

40. Brisben, A.J., Lockerd, A.D., Lathan, C.: Design evolution of an interactive robot for therapy. Telemed. J. E Health **10**(2), 252–259 (2004). https://doi.org/10.1089/tmj.2004.10.252

41. Hesse, S., Schulte-Tigges, G., Konrad, M., Bardeleben, A., Werner, C.: Robot-assisted arm trainer for the passive and active practice of bilateral forearm and wrist movements in hemiparetic subjects. Arch. Phys. Med. Rehabil. **84**(6), 915–920 (2003). https://doi.org/10.1016/S0003-9993(02)04954-7

42. Bai, O., Nakamura, M., Shibasaki, H.: Compensation of hand movement for patients by assistant force: relationship between human hand movement and robot arm motion. IEEE Trans. Neural Syst. Rehabil. Eng. **9**(3), 302–307 (2001). https://doi.org/10.1109/7333.948459

43. Driessen, B.J., Evers, H.G., van Woerden, J.A.: MANUS–a wheelchair-mounted rehabilitation robot. Proc. Inst. Mech. Eng. H **215**(3), 285–290 (2001). https://doi.org/10.1243/0954411011535876

44. Hoppenot, P., Colle, E.: Localization and control of a rehabilitation mobile robot by close human-machine cooperation. IEEE Trans. Neural Syst. Rehabil. Eng. **9**(2), 181–190 (2001). https://doi.org/10.1109/7333.928578

45. Chen, S., Rahman, T., Harwin, W.: Performance statistics of a head-operated force-reflecting rehabilitation robot system. IEEE Trans. Rehabil. Eng. **6**(4), 406–414 (1998). https://doi.org/10.1109/86.736155

46. Morvan, J.S., Guichard, J.P., Torossian, V.: Technical aids for the physically handicapped: a psychological study of the master robot. Int. J. Rehabil. Res. **20**(2), 193–197 (1997). https://doi.org/10.1097/00004356-199706000-00009

47. Lancioni, G.E., Oliva, D., Signorino, M.: Promoting ambulation and object manipulation in persons with multiple handicaps through the use of a robot. Percept. Mot. Skills **79**(2), 843–848 (1994). https://doi.org/10.2466/pms

48. Harwin, W.S., Ginige, A., Jackson, R.D.: A robot workstation for use in education of the physically handicapped. IEEE Trans. Biomed. Eng. **35**(2), 127–131 (1988). https://doi.org/10.1109/10.1350

49. Papadopoulou, M.T., et al.: Efficacy of a robot-assisted intervention in improving learning performance of elementary school children with specific learning disorders. Child. (Basel) **9**(8), 1155 (2022). https://doi.org/10.3390/children9081155

# The Learning Model for Data-Driven Decision Making of Collaborating Enterprises

Charles El-Nouty[1]([✉])(iD) and Darya Filatova[2](iD)

[1] LAGA, Université Sorbonne Paris Nord, 93430 Villetaneuse, France
`elnouty@math.univ-paris13.fr`
[2] RDS, Nancy, France
`daria_filatova@interia.pl`

**Abstract.** Determining the principles for building an intelligent management system aimed at making optimal decisions becomes an urgent need to achieve sustainable development of cooperating enterprises and requires the search for appropriate management models based on data dynamics. The purpose of the work was to indicate the determinants of sustainable development of collaborating enterprises used to determine the management model. The application of systems analysis and multi-agent simulation principles to human-system interactions has indicated a way to construct the data-driven decision making reinforcement learning model. The results obtained indicate that such management eliminates subjectivity in decision-making, leading to the stabilization of the economic development of cooperating enterprises, and hence to sustainable development.

**Keywords:** Human-system interaction · Data-driven decision making · Learning model

## 1 Introduction

Progressive globalization and rapid development of artificial intelligence methods require the search for new effective methods of enterprise management. Business entities taking independent actions based on a variety of data and analyzed using machine learning methods to reduce costs or increase profitability are often not enough. Currently, one of the ways to increase own strength and achieve success in the market is to establish cooperation with other enterprises as data-driven collaborative human-AI decision making [3]. While there is enormous economic and business potential generated by integrating business analytics with artificial intelligence, in some cases such a solution can exacerbate potential dangers in the management of a firm. This is due to the fact that the optimal solutions offered by AI methods are local in nature, i.e. characteristic of operational management. Strategic management requires a global solution, assuming a certain development model [21].

J. Baratgin et al. (Eds.): HAR 2023, LNCS 14522, pp. 345–356, 2024.
https://doi.org/10.1007/978-3-031-55245-8_22

Before talking about decision-making with respect strategic development, it is necessary to find out exactly how enterprises cooperate [19]. This will determine ecosystem-based management and the choice of information and the construction of predictive and prescriptive models. In general, the cooperation of enterprises can occur at the level of concentration or cooperation. The second type of the cooperation is one of the main forms of enterprise integration. This approach makes possible the increase the competitiveness of divisions while maintaining their legal and economic independence. Cooperative associations can be created at various levels of formalization and legal legitimacy. They can also take various forms, such as strategic alliances, franchising or agreements. The latter, due to the direction of cooperation, can take the form of horizontal cooperation carried out between competitors, or vertical cooperation linking entities operating at different levels of trade, representing different stages of production or distribution, i.e. enterprises that do not compete with each other [23, 24].

The type of the cooperation is very important. It is related to value system and, thus, to vision of strategical development of each business unit. With respect to final customers we can distinct hierarchy-based, ecosystem-based, and market-based value system [11]. Abstracting from any data and methods of their processing, the analysis of the legal framework for the creation and operation of vertical agreements has shown that vertical agreements with ecosystem-based value system are currently one of the most beneficial forms of cooperation between enterprises [9, 18]. Their main advantage lies in the possibility of better coordination among the entities involved, resulting in increased economic efficiency of the activities carried out ([19]). The European Commission adheres to a similar position, arguing that these agreements, in particular, can help reduce transaction and distribution costs, as well as optimize the level of sales and investments of the cooperating parties [26].

In this paper, further attention will be focused only on companies interacting vertically (i.e., "groups"). It should be noted that the cooperation begun will satisfy all parties to the agreement and bring them the expected benefits in the long term, only if the interests and sustainable development of all parties involved are taken into account [18–20]. Most often, sustainable development is defined as "development that meets the current needs of society without compromising the development opportunities of future generations".

This definition is so broad that it does not allow the introduction of strict methods of managing a complex economic structure. One can understand the desire to get "better" business analytics by recording all transactions and collecting as much information as possible about markets, competitors, etc. However, due to these large and diverse data sets it becomes difficult to obtain accurate descriptive, predictive and prescriptive models. Therefore to focus on decision making, it is necessary to specify the conditions for such development, taking into account the increasing complexity, uncertainty and exacerbating market volatility [16]. In order to address these points, the paper develops a learning model that integrates system analysis of ecosystem and optimal control. The model adopts multi-agent interaction principle and provides a concrete approach for data-driven human-AI collaboration.

The rest of the paper is organized in the following manner. In Sect. 2, we discuss principles of ecosystem-based cooperation and propose definitions of ecosystem and its sustainable development. Section 3 contains formal representation of the ecosystem dynamics, possible scenario of cooperation as well as a learning model. Finally, in Sect. 4, we present the concluding remarks concerning the implementation of the learning model for data-driven decision making.

## 2    Principles of Ecosystem-Based Cooperation

Intuitively, an ecosystem can be defined as a system of elements that make up an ordered whole or some kind of internal environment [2]. In one of the formal definitions the ecosystem is presented as

*"a community of organizations, institutions, and individuals that impact the enterprise and the enterprise's customers and supplies"* (see [27], p. 1325).

Obviously, enterprises have competitors whose behavior is difficult to predict, so the ecosystem itself is surrounded by an uncertain external environment. Since the conditions of the contract do not change during cooperation, and if they do, then according to certain rules, the ecosystem must have properties whose evolution can be described quite accurately. From the foregoing, it can be seen that the ecosystem is a special environment for which the enterprise must observe and to which it must respond. Such a reaction-action affects the dynamic capabilities of the ecosystem and, consequently, the ability of the ecosystem to create sustainable competitive advantages.

Another definition of the ecosystem directly refers to a collaboration agreement among business partners. According to (see [1], p. 42)

*"the ecosystem is defined by the alignment structure of the multilateral set of partners that need to interact in order for a focal value proposition to materialize".*

The alignment structure as a kind of mutual agreement among the members regarding their roles during cooperation. The natural desire for constant alignment of the structure can be called a sustainable ecosystem strategy.

We agreed to consider enterprises with vertical cooperation. This results in technological modularity, which means separability along the production chain. Companies bound by a contract are not in competition, their cooperation is a symbiosis. There is no hierarchy between partners, they complement each other. There are two main types of complementarity: unique and supermodular [1]. A unique complementarity occurs when product $A$ is needed to produce product $B$ but not vice versa, or when product $A$ and product $B$ are needed at the same time (in both cases $A$ and $B$ are considered as nongeneric complements). A supermodular complementarity can be described as "the better product $A$ is, the more valuable product $B$ is." Taking together, *the ecosystem is a group of firms with non-generic complementarity and non-hierarchical structure.*

The ecosystem must have all of the following properties, namely [4–6, 29]:

- integrity and emergence (an ecosystem is a complex entity. The interaction of enterprises, the exchange of information between them, the joint assessment of the behavior of competitors and product markets develop new properties of competitiveness that individual enterprises do not possess),
- structure (each enterprise has its own organizational structure, its elements perform different functions, for example, the production department is engaged in the production of goods, the sales department is engaged in the sale of finished products, etc. The interaction of enterprises complements the functions of the departments of each enterprise, depending on the conditions of cooperation.),
- functionality (manifestation of competitiveness under uncertain environmental influences to achieve business goals),
- development (irreversible, purposeful, natural change in the internal environment of the enterprise, the result of which is new qualities such as the way of personnel management, production management, relations with suppliers),
- self-regulation (the desire to control and to maintain the internal potential of the enterprise, despite adverse changes in the external environment, pursuing long-term business goals),
- resilience (each company in the group must continue to fulfill its mission despite the difficulties that arise, i.e. resist environmental stresses - competition in product markets caused by changing market conditions, political conditions, changes in legislation. In addition, the group must respond to such disturbances and recover from them, while maintaining dynamic stability in product markets),
- adaptation (the ability to change production and market both short-term and long-term strategies in order to maintain or improve existing positions in the product market and acquire new properties in an uncertain competitive environment),
- reliability (the ability of the enterprise to function under stated conditions for example under selected business goals).

Analyzing the above properties, we can identify the determinants of ecosystem-based sustainable development of the group:

- self-regulation,
- resilience,
- adaptation,
- reliability.

These properties testify to the possibility of achieving the business goals not only of the group, but also of each of its member, as well as long-term cooperation, regardless of the harmful actions of competitors [7,11]. Therefore, under the ecosystem-based sustainable development of the group, we mean

*a self-regulating and adaptive desire to stay on the market in a competitive environment, which ensures reliable cooperation, the result of which is development aimed at achieving business goals.*

# 3    Modeling of Ecosystem-Based Sustainable Development

## 3.1    Ecosystem Dynamics

Before to introduce the model of the ecosystem-based sustainable development, we discuss dynamics of ecosystem, which includes [11,13,25,30]:

- coordination (type of modular structure [17]),
- collaboration (tactics and governance [14]),
- value creation (competitive aspect [15,22,28]),
- governance and regulation (strategic aspects [8]).

The technological modularity is the endogenous factor in the ecosystem genesis [12]. To obtain the optimal structure of ecosystem one has to underlie the type of complementarity and to fix the objective with respect to customers of the final products ([10]). In other words, it is necessary to distinct when

- the ecosystem tries to stay in homeostasis (internal environment of entities related to production),
- the ecosystem reacts on the external environment (product market or consumption).

The complementarity of partners can be related to production as well as to consumption, defining for basic structures of ecosystem, namely:

- type 1 – unique complementarity in production and unique complementarity in consumption,
- type 2 – supermodular complementarity in production and unique complementarity in consumption,
- type 3 – unique complementarity in production and supermodular complementarity in consumption,
- type 4 – supermodular complementarity in production and supermodular complementarity in consumption.

To maintain homeostasis, each enterprise determines the vision of its development, chooses a value system and sets key objectives with respect to its production capacities (see Fig. 1). Imposing relationships with partners also requires maintaining the homeostasis of the enterprise. The dynamics of partnership-coordinating activities depends on the type of complementarity and brings with

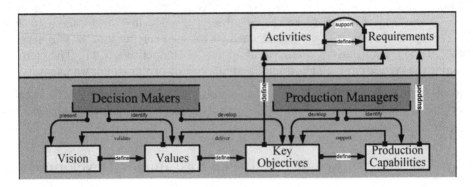

**Fig. 1.** An internal environment of the entity: strategic perspective

it various advantages expressed in the effects and impacts of the group-level cooperation (see Table 1).

The effects of group coordination can be measured by setting metrics such as KPI or OEE. Observation of these indicators makes it possible to track the dynamics of the ecosystem, and therefore to establish the conditions for sustainable development. We also need to recognize that the type of complementarity chosen will influence the complexity of the learning model. In the next subsection we will consider the ecosystem of *type 1* just to show principles of learning model formulation.

## 3.2   Scenario

Let us consider a certain product market (environment) in which two types of enterprises act (agents). We will refer to these enterprises as two-types agents, namely: the first-type-producers (Vendors), offering semi-finished products – $A_1$, and the second-type-producers (Buyers), offering finished products – $A_2$ – assembled from semi-finished products. Vendors and Buyers can cooperate under the bilateral vertical agreement in order to develop some joint strategy for managing the sale of their products in markets with some guaranteed profit. They also can establish the duration of this cooperation, guaranteeing the necessary funds and capacities to ensure the production of semi-finished products and finished products. Buyer can purchase the semi-product directly on the market at the current price, and Vendor can put up his semi-product for sale at the market price. This situation does not guarantee either the desired sales volume or guaranteed profit. It also does not require the signing of an agreement. Therefore, the bilateral cooperation is beneficial, if Vendor and Buyer are settled the price of the semi-product to ensures partial or its complete consumption with respect to the production capacities of both parties.

This concept can be presented as the following scenario. Vendor is interested in cooperating with Buyer until it ensures partial or complete consumption of the semi-finished products produced by it. For such cooperation to be beneficial

**Table 1.** Advantages, effects, and impacts of group level coordination with respect to complementarity principle and nature of environment

| Structure of ecosystem | Advantage | Effect | Impact |
|---|---|---|---|
| type 1 | production of compatible components | increase utility by join or separate consumption | increase reachability of key objectives |
| type 2 | greater utility than separate consumption | quality improvement production cost reduction availability increase | increase reachability of key objectives improve production capabilities |
| type 3 | production of compatible components | increase return by join consumption | increase reachability of key objectives |
| type 4 | greater return than separate consumption | quality improvement production cost reduction availability increase | increase reachability of key objectives improve production capabilities |

Vendor must offer a resale price below the market price, otherwise products will be sold directly on the market to the other buyers, which may result in a decrease in revenues from the sale. Buyer can buy the semi-finished products from Vendor or on the market. This means that the demand for these products exceeds the production capacity of Vendor. If Vendor refuses to sell the required quantity of the semi-finished products, Buyer must purchase it from the other vendors of the market. The income of Buyer depends on the terms of purchase of the semi-finished products, on the demand for the finished product and its selling price. The vendor-buyer interactions take place during the cooperation time or instance. Once cooperation is finished Vendor and Buyer can choose other business partners to continue their activities toward business goal.

### 3.3   Constraint Learning Model

The main principles of learning model can be seen as "agent – environment automated interaction". That is to say, being in some state the agent selects some optimal action such that to maximize the total amount of reward (or discounted utility) over long-term horizon under uncertain environmental conditions (decision process as training and inference).

Let $[t_0, t_1] \subset \mathbb{R}_+$ be the time interval defining the duration of cooperation and let both entities have the necessary means and powers to enable production. Thus, the production volume - $X_j$, $j \in \{1, 2\}$ - of the product $A_j$ depends only on the parameter vector defining the production capacity $\theta^{(j)} = [\theta_1^{(j)}, \theta_2^{(j)}]$, where $\theta_1^{(j)} : [t_0, t_1] \rightarrow \mathbb{R}$ is the production growth factor, $\theta_2^{(j)} : [t_0, t_1] \rightarrow \mathbb{R}$ indicates the maximum production capacity relative to the product $A_j$. Its dynamics is

$$dX_j(t) = f_j(\theta^{(j)}, X_j), \tag{1}$$

with known initial conditions $X_j(t_0) = x_{j0}$, such that $x_{j0} \leq \theta_2^{(j)}$, $f_j$ is the convex function of the class $C^2$.

We suppose that production volume $X_j$ depends also on the quality factor $\gamma_j \in (0,1)$ of the produced goods. This allows to rewrite (1)

$$dX_j^{\gamma_j}(t) = f_j(\theta^{(j)}, X_j^{\gamma_j}), \tag{2}$$

keeping initial conditions and define "equilibrium $X^*$ of production" with respect to the quality factor and the production growth factor (see Fig. 2 and Fig. 3).

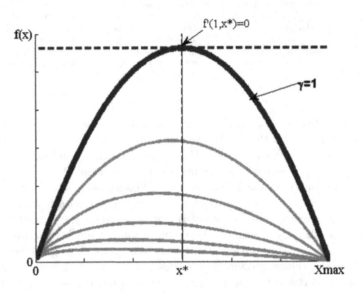

**Fig. 2.** The production volume dynamics: stability and quality issues

The market price of $A_j$ is determined by the function $p_j^M : [t_0, t_1] \to \mathbb{R}_+$. The total quantity of $A_j$ available in the market is presented as $X_j^M : [t_0, t_1] \to \mathbb{R}_+$ with some demand $d_j = \zeta(p_j^M, X_j^M)$, where $\zeta : R_+^2 \to \mathbb{R}_+$ is the decreasing convex function of the class $C^2$. The agreement selling price of $A_1$ depends on the demand and the total quantity of the product available in the market, i.e. $p_1^s = h_1^s(d_1, X_1^M)$ where $h_1^s : R_+^2 \to \mathbb{R}_+$ is the decreasing convex function of the class $C^2$. In addition, according some selected management strategy the total production $X_j$ should not exceed demand $d_j$ and should not generate losses, ensuring that production costs are covered.

For each enterprise, income $\pi_j$ is defined as the difference between sales revenue – $S_j$ – and the sum of variable costs and fixed costs – $C_j$, namely:

$$\pi_j(X_j^{\gamma_j}(t, \theta^{(j)}), \mathbf{u}_j(t)) = S_j(t) - C_j(t), \, t \in [t_0, t_1], \, j \in \{1, 2\}, \tag{3}$$

where

– for $\mathbf{u}_1 = [u_{11}, u_{12}]$ such that $u_{11} \geq 0$, $u_{12} \geq 0$, $0 \leq u_{11} + u_{12} \leq 1$

$$S_1(t) = (u_{11}(t)p_1^s(t) + u_{12}(t)p_1^M(t))X_1^{\gamma_1}(t, \theta^{(j)}) \tag{4}$$

and

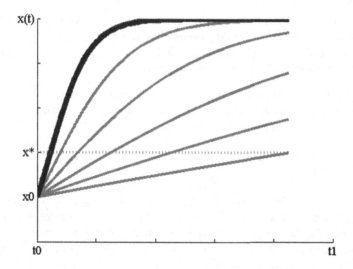

**Fig. 3.** The production volume dynamics: quality and adaptation issues

– for $\mathbf{u}_2 = [u_{21}]$ such that $0 \leq u_{21} \leq 1$

$$S_2(t) = u_{21}(t)p_2^M(t)X_2^{\gamma_2}(t,\theta^{(j)}). \tag{5}$$

The total income of the group during the cooperation period is

$$\Pi(\Delta) = \sum_{j=1}^{2} \int_{t_0}^{t_1} e^{\kappa_j t} \pi_j \left( X_j^{\gamma_j}(t,\theta^{(j)}), \mathbf{u}_j(t) \right) dt, \tag{6}$$

where $\kappa_j$ is some selected discount value.

The objective of learning is to find both market and production policies which maximizing the income, i.e.

$$\mathcal{J}(X^\gamma, \theta, \mathbf{u}) = \max_{\theta, \mathbf{u}} \sum_{j=1}^{2} \int_{t_0}^{t_1} e^{\kappa_j t} \pi_j \left( X_j^{\gamma_j}(t,\theta^{(j)}), \mathbf{u}_j(t) \right) dt, \tag{7}$$

for data-driven decisions on the production volume as

– adaptation to the product market, i.e. production of $A_1$ and $A_2$ does not dominate the market (antitrust regulations regarding market segments $X_j^{(0)}$, $j \in \{1,2\}$):

$$X_j(t) \leq X_j^{(0)}, \forall t, t \in [t_0, t_1]; \tag{8}$$

– adaptation and resistance to competition (antitrust regulations), i.e. the quantity of $A_1$ and $A_2$ does not exceed market demand during cooperation period:

$$(u_{11}(t) + u_{12}(t)) X_1^{\gamma_1}(t) \leq d^{(1)}(t), \tag{9}$$

$$u_{21} X_2^{\gamma_2}(t) \leq d^{(2)}(t) \ \forall t, t \in [t_0, t_1];$$

– sustainability of the production program, i.e. for the selected model of production dynamics (2), stability areas should be determined and a solution to the problem (7) should be sought only in the feasible areas, which could involve setting additional constraints on (7).

For the remaining three ecosystem types, a learning model can also be defined using the same reasoning. Particular attention must be paid when selecting the goal function, which will relate to utility or return.

## 4     Conclusion

In order to stay on the market, enterprises have to cooperate. Very often, such cooperation creates ecosystems in which partners complement each other in the value chain. This brings partners mutual benefits, expressed in increased competitiveness, profitability, return. The sustainable development of cooperation requires data-driven decision-making under conditions of market uncertainty. Learned from the data, model that will be used while decision-making should comprehensively cover the problems that partners face and give recommendations on how to act in the situation that arises.

By defining an ecosystem as a group of firms with non-generic complementarity and a non-hierarchical structure, we pointed out some properties of ecosystems that allow to control the sustainable development of business partners. An analysis of the properties of ecosystem dynamics and ways of complementarity of partners made it possible to identify four main types of ecosystems. The learning model was created for one type of ecosystems. The main benefits of this model are reduction of subjectivity in the decision-making process as well as easy implementation of AI-methods such as reinforcement learning or machine learning.

**Acknowledgments.** We thank two anonymous reviewers for their very useful and relevant comments.

## References

1. Adner, R.: Ecosystem as structure: an actionable construct for strategy. J. Manag. **43**(1), 39–58 (2017)
2. Croom, S., Vidal, N., Spetic, W., Marshall, D., McCarthy, L.: Impact of social sustainability orientation and supply chain practices on operational performance. Int. J. Oper. Prod. Manag. **38**(12), 2344–2366 (2018)
3. Davenport, T.H.: From analytics to artificial intelligence. J. Bus. Analytics **1**(2), 73–80 (2018)
4. Eckstein, D., Goellner, M., Blome, C., Henke, M.: The performance impact of supply chain agility and supply chain adaptability: the moderating effect of product complexity. Int. J. Prod. Res. **53**(10), 3028–3046 (2015)

5. Filatova, D., El-Nouty, C. and Fedorenko, R. V.: Some theoretical backgrounds for reinforcement learning model of supply chain management under stochastic demand. In: 2021 International Conference on Information and Digital Technologies (IDT), pp. 24–30, Zilina, Slovakia, IEEE (2021)
6. Filatova D., El-Nouty, C.: Production process balancing: a two-level optimization approach. In: International Conference on Information and Digital Technologies (IDT), pp. 133–141, Zilina, Slovakia, IEEE (2019)
7. Haki, K., Blaschke, M., Aier, S., Winter, R., Tilson, D.: Dynamic capabilities for transitioning from product platform ecosystem to innovation platform ecosystem. Eur. J. Inf. Syst. (2022) https://doi.org/10.1080/0960085X.2022.2136542
8. Gatignon, A., Capron, L.: The firm as an architect of polycentric governance: building open institutional infrastructure in emerging markets. Strateg. Manag. J. **44**, 48–85 (2023)
9. Helfat, C.E., Campo-Rembado, M.A.: Integrative capabilities, vertical integration, and innovation over successive technology lifecycles. Organ. Sci. **27**(2), 249–264 (2015)
10. Helfat, C.E., Raubitschek, R.S.: Dynamic and integrative capabilities for profiting from innovation in digital platform-based ecosystems. Res. Policy **47**(8), 1391–1399 (2018)
11. Jacobides, M.G., Cennamo, C., Gawer, A.: Towards a theory of ecosystems. Strateg. Manag. J. **39**(8), 2255–2276 (2018)
12. Jacobides, M.G., MacDuffie, J.P., Tae, C.J.: Agency, structure, and the dominance of OEMs: change and stability in the automotive sector. Strateg. Manag. J. **37**(9), 1942–1967 (2016)
13. Ji, H., Zou, H., Liu, B.: Research on dynamic optimization and coordination strategy of value co-creation in digital innovation ecosystems. Sustainability **15**, 7616 (2023)
14. Kapoor, R.: Collaborating with complementors: what do firms do? Adv. Strateg. Manag. **30**, 3–25 (2013)
15. Kapoor, R., Agarwal, S.: Sustaining superior performance in business ecosystems: evidence from application software developers in the iOS and Android smartphone ecosystems. Organ. Sci. **28**(3), 531–551 (2017)
16. Kapoor, R., Furr, N.R.: Complementarities and competition: unpacking the drivers of entrants' technology choices in the solar photovoltaic industry. Strateg. Manag. J. **36**(3), 416–436 (2015)
17. Kapoor, R., Lee, J.M.: Coordinating and competing in ecosystems: how organizational forms shape new technology investments. Strateg. Manag. J. **34**(3), 274–296 (2013)
18. Li, P., Tan, D., Wang, G., Wei, H., Wu, J.: Retailer's vertical integration strategies under different business modes. Eur. J. Oper. Res. **294**(3), 965–975 (2021)
19. Liu, X.: Vertical integration and innovation. Int. J. Ind. Organ. **47**, 88–120 (2016)
20. Mikalef, P., Gupta, M.: Artificial intelligence capability: conceptualization, measurement calibration, and empirical study on its impact on organizational creativity and firm performance. Inf. Manag. **58**(3), 103434 (2021)
21. Rana, N.P., Chatterjee, S., Dwivedi, Y.K., Akter, S.: Understanding dark side of artificial intelligence (AI) integrated business analytics: assessing firm's operational inefficiency and competitiveness. Eur. J. Inf. Syst. **31**(3), 364–387 (2022)
22. Ritala, P., Agouridas, V., Assimakopoulos, D.: Value creation and capture mechanisms in innovation ecosystems: a comparative case study. Int. J. Technol. Manage. **63**, 244–267 (2013)

23. Sarker, S., Sarker, S., Sahaym, A., Bjorn-Andersen, N.: Exploring value cocreation in relationships between an ERP vendor and its partners: a revelatory case study. MIS Q. **36**(1), 317–338 (2012)
24. Schreieck, M., Wiesche, M., Krcmar, H.: Capabilities for value co-creation and value capture in emergent platform ecosystems: a longitudinal case study of SAP's cloud platform. J. Inf. Technol. **36**(4), 365–390 (2021)
25. Sultana, N., Turkina, E.: Collaboration for sustainable innovation ecosystem: the role of intermediaries. Sustainability **15**, 7754 (2023)
26. Tangeras, T.P., Tag, J.: International network competition under national regulation. Int. J. Ind. Organ. **47**, 152–185 (2016)
27. Teece, D.J.: Explicating dynamic capabilities: the nature and microfoundations of (sustainable) enterprise performance. Strateg. Manag. J. **28**(13), 1319–1350 (2007)
28. Uzzi, B.: Social structure and competition in interfirm networks: the paradox of embeddedness. Adm. Sci. Q. **42**(1), 35–67 (1997)
29. Wareham, J., Fox, P. B., Cano Giner, J. L.: Technology ecosystem governance. Organ. Sci. **25**(4), 1195–1215 (2014)
30. Zhang, Y., Liu, B., Sui, R.: Evaluation and driving determinants of the coordination between ecosystem service supply and demand: a case study in Shanxi province. Appl. Sci. **13**, 9262 (2023)

# Applied Reasoning

# Designing Automated Systems for Learning Analysis

Sophie Charles[1,2]([✉]) [ID] and Alain Jaillet[1] [ID]

[1] CY Cergy Paris Université, 95000 Cergy, France
sophie.charles@cyu.fr
[2] ISAE-Supméca, 93400 Saint-Ouen, France

**Abstract.** In the context of the e-FRAN program, our research focused on engineering freshmen's learning processes of 3-D modeling in relation to their spatial skills. One of the approaches developed consisted in identifying the students' sequences of micro operations using a double observation of their actions by recording their on-screen activity, and filming their gestures and postures. The objective consisted in determining the logical sequences of both their bodily attitudes, which are related to their manipulation of the software, and their conceptual approaches, which are linked to the results they obtained by interacting with the software. This preliminary characterization aimed to establish the possible recurring patterns in human behavior in the context studied, to then determine the robustness of the codings with regard to the software functionalities, to finally identify characteristic sequences of operations. The later phase consisted in feeding a parser with the encoded traces left by the subjects' interactions with the software interface, using a heuristic library of actions. The objective was to identify sequences indicative of efficient, or non-efficient, conceptual processes in the task performance. The encoded operations logs revealed recurring sequences, which were then processed statistically to identify prototypes of rationality of intentions. The encoded sequences were submitted to a further manual encoding to identify strategies that arose at the students' first encounter with the modeling interface.

**Keywords:** Learning Analysis · Activity Characterization · Limited-Intelligence Systems

Before Computer Aided Design (CAD) programs were invented, object design was carried out through 2-D drawings (Poitou 1984, p. 468). The advent of CAD in the 1980s was sometimes met with suspicion as the 2-D views the software generated required a mental conversion into 3-D objects, a skill which drawers considered specific to their profession (Poitou 1984, p. 476). It seems this artificial rationality was considered insufficient compared to a human rationality. Nevertheless, the drawing process was time-consuming and prone to errors, and 3-D modelers became ubiquitous in the professional world, and consequently in the institutions that train engineering students to mechanical design (Peng et al. 2012, p. 9). These software programs produce dynamic and trustworthy complex representations of objects, making "manufacturing more time and cost-efficient"

(Brown 2009, p. 54). In the end, an artificial rationality took over from a human rationality. However, these programs are not self-sufficient and require humans to operate them, reintroducing human rationality into the 3-D design process, although this human rationality needs to adapt to interface with the artificial rationality (Bertoline, Hartman, et al. 2009, p. 641). Our initial research, which takes root in the EXAPP_3D project, financed by the e-FRAN program[1], was concerned specifically with three kinds of rationalities: teaching impact, student behaviors and software operation. It aimed at investigating students' spatial ability, backgrounds and learning opportunities, and behaviors when using a 3-D modeler. Its purpose was to identify whether some empirical rationalism developed through the subjects' experience could model spatial skills, and to study their behaviors in response to the artificial rationality of the functioning of a modeling software, which is based on 3-D representations. To define our subjects' characteristics, we submitted them to a battery of spatial tests and questionnaires exploring their learning opportunities. The students were also asked to complete a modeling task, so that actions, recurrences and modeling characteristics could be identified and recorded. This first approach is now further developed in the present study with a questioning prompted by the initialization of limited-intelligence systems. These need to be programmed to characterize what they observe, to determine what is noticeable in what they observe, so that they can predict what is to come when they detect recurrences. In the context of learning, a limited-intelligence system could observe a student's actions and behavior, in order to identify the sequences of operations generated, so that it detects those which produce an action that contributes to the task completion. This is not dissimilar to the activity theory which aims at characterizing this process by identifying its three stages: operations, actions, and activity (Savoyant 2010, p. 94). The aim of limited-intelligence systems is to first learn subjects' operation patterns, then the action patterns they develop to complete a task. The system needs to be conceived so that it first identifies relevant traits, then operation sequences that form actions, and finally shortcomings and obstacles met by a subject, using performance criteria such as task completion time and the number of actions carried out. Before conceiving a limited-intelligence system that captures, retrieves, and analyses data according to selection criteria, it is necessary to structure a rationality that is likely to be reproduced artificially. This allows for the description of the steps of what could be a process of automation and autonomation of the criteria to be considered. This paper presents the projection of the rationalization of an observation, that is to say the determination of relevant traits that form a sequence, i.e. a succession of encoded observations. How can notable sequences be determined? After several attempts at statistical sequence analysis, as developed by Dubus (2000), the heuristic library approached was adopted. It consists in a collection of expected sequences based on the rationalities defined by the authors of the modeling platform as good practice. In other words, there are ideal chains of operations that enable the completion of a target action, given a specific software program. Once built, this library of actions is used by the limited rationality of a computer program, that aims at establishing a link between the subjects' observed rationality, and the rationality determined by the functioning of the 3-D modeler. To validate this process, a great number of sequences would be needed.

[1] Espace de formation, de recherche et d'animation numérique (e-FRAN) projects are supported by the French Ministère de l'enseignement supérieur, de la recherche et de l'innovation.

This would require a colossal amount of work. To assess the feasibility of this approach, limited numbers of subjects were used for preliminary characterizations.

# 1   How to Determine Learning Processes in 3-D Modeling?

Nowadays, designers use 3-D modelers, Computer Aided Design (CAD) software, to produce dynamic and trustworthy complex representations of objects, making "manufacturing more time and cost-efficient" (Brown 2009, p. 54). Similarly, students enrolled in modeling courses are taught how to design 3-D objects, using CAD software (Johnson & Diwakaran 2011, p. 22.305.2). 3-D modeling courses have a twofold objective: they aim at teaching students how 3-D modelers work and the functions they offer, and efficient strategies that make the most of parametric modeling (Chester 2007, p. 23; Rynne & Gaughran 2007, p. 57). Commands are specific to a modeler, whereas strategies can be used in any modeler (Hamade et al. 2005, p. 306). Unlike learning software commands, developing efficient strategies is difficult as there are several ways of designing an object (Bertoline, Wiebe, et al. 2009, p. 416). The difficulty lies in developing strategies which are time-efficient and limit the number of errors (Bhavnani et al. 2001, p. 230). This strategic knowledge is considered to be characteristic of CAD expertise (Bhavnani et al. 1993, p. 327).

To assess students' modeling performance, instructors can use criteria related to the accuracy of the model produced, and the modeling strategy developed (Lang et al. 1991, p. 260; Steinhauer 2012, p. 47). The accuracy can be evaluated thanks to geometrical, dimensional and completion criteria (Branoff & Dobelis 2012, p. 25.548.6). This can be completed with the number of errors made (Lang et al. 1991, p. 260) and the robustness of the model (Steinhauer 2012, p. 47). The assessment of the strategy can include the number and the variety of commands used (Lang et al. 1991, p. 260), the approach, the structure and the creativity demonstrated (Steinhauer 2012, p. 47), the number of expert strategies used (Chester 2007, p. 30) and the task completion time (Hamade et al. 2007, p. 645; Johnson & Diwakaran 2011, p. 22.305.3).

Most of the information needed to evaluate these criteria can be found in the students' models and in their feature trees, which show the final order of the sketches and the features used to produce the model (Lieu & Sorby 2009, pp. 6–49). As these observations account for the final version of the design, they need to be completed with the observation of the students' modeling activity if one is interested in determining how the students achieved these results, and whether the students employed expert strategies (Chester 2007, p. 30).

## 1.1   Observing Modeling Activity

Various protocols have been employed to investigate 3-D modeling strategies: Lang et al. (1991) filmed the screens, the keyboards and the data tablets their students used to model a part using CADD4X (Computervision 1983) (p. 261), whereas Hartman (2005) opted for a think-aloud protocol with the experts he recruited for his study (p. 8), and Chester (2007) screen-recorded his students' modeling activity (p. 29).

## 1.2  Coding Modeling Activity

These observations provided raw data that were processed to investigate the modeling strategies adopted by the subjects. Different approaches were developed: Lang et al. (1991) first determined a list of unit tasks needed to complete the part used in the experiment, and organized them in what the authors considered their efficient ordering. They then encoded the students' recordings according to this ideal sequence (p. 262). This coding meant that only part of the activity was processed, as errors for instance were omitted. Hartman's (2005) method consisted in analyzing the transcripts resulting from the talk-aloud protocol to determine his subjects' modeling processes: he took note of the geometry and the functions they used, the sketch planes they selected and the elements they copied (i.e. geometry, complex features) (p. 8). Chester (2007) first identified expert strategies in expert performance (p. 30). He then used this reference to determine the number of expert strategies his students employed.

## 1.3  Sequencing Modeling Activity

Because of the authors' various research concerns, their raw data processing methods differed not only in their approach, but also in the final data they yielded. Lang et al. (1991) generated sequences of actions, albeit limited to the previously identified unit tasks, and the transitions between these tasks. They used their efficiently ordered list to illustrate the students' modeling activity and highlight the students' transitions between the unit tasks the authors had determined, as illustrated in Fig. 1. This enabled them to characterize the students' sequences and identify two general strategies. On the other hand, Hartman (2005) focused on identifying common language and methods in all his subjects, to model a common procedure in expert modeling practice (p. 11). As for Chester (2007), he concentrated on the occurrence of expert strategies employed by his

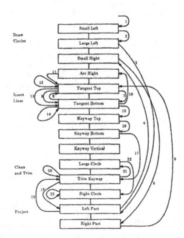

**Fig. 1.** Example of a modeling sequence highlighting transitions between the unit tasks. From "Extracting and Using Procedural Knowledge in a CAD Task" by G.T. Lang, R. E. Eberts, M. G. Gabel, and M. M. Barash 1991, *IEE Transactions on Engineering Management*, *38*, p. 264.

students, to assess the impact of his modeling course, designed to improve students' strategic knowledge (p. 30).

## 2 Experiment

### 2.1 Research Context

The increasing role played by digital tools in education encouraged the French government to investigate its impact on learning, and sponsor research programs addressing this issue. EXAPP_3D, an e-FRAN project aimed at better understanding how multi-purpose 3-D modeling software was used by learners at different levels of schooling, provided the opportunity to investigate spatial ability and its possible inferences as a necessary ability in 3-D modeling (Charles 2023). The present study focuses on the development of a semi-automated method designed to detect and characterize sequences of operations and actions constituent of engineering freshmen' modeling activity.

### 2.2 Participants

The sample consisted of 129 freshmen in a French engineering school, aged between 18 and 23, mean 19.9. There were $N_F = 26$ [20%] women and $N_H = 103$ [80%] men. 73 [57%] students had been exposed to Technology and Industrial Science courses prior to joining the school, whereas 56 [43%] came from courses deprived of technological content. 113 [88%] students had some experience with 3-D modelers, when 16 [12%] did not. The compliance of our data collection protocol with ethical regulations was validated by the school's data protection officer. All data was anonymised.

### 2.3 Method

To investigate students' potential learning of 3-D modeling, a test-retest protocol was adopted to identify differences, if any, in the modeling performance of first-year engineering students, before and after taking a 10-h modeling course using the 3-D modeler CATIA (Dassault Systèmes 2012). The students were asked to complete the same modeling task at the beginning and at the end of the first term, using the cloud-native product development platform Onshape (Hirschtick et al. 2014). In the first data collection, the students were asked to follow a tutorial to learn how to use the platform, then to create a 3-D model, using three views, one of which included dimensions, as illustrated in Fig. 2. In the second data collection, the students were asked to complete the same modeling task, after being given the opportunity to follow the same tutorial as in the first data collection.

In order to analyze the students' modeling activity, we recorded the students' on-screen activity using the screen recorder Camtasia (Techsmith 2019), and filmed the upper part of their bodies with a webcam. These two observations were combined to gain information on the students' on-screen activity, the response generated by the modeler to the students' actions, the students' visual focus and verbal expressions. Both recordings were synchronized so that the encoding of the two observations could produce information about the sequences of student-modeler interactions. Task completion

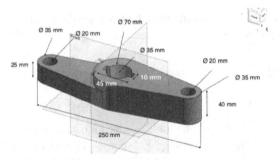

**Fig. 2.** View with dimensions used in the modeling experiment. From *"Habileté spatiale et stratégies de modélisation 3D"* by S. Charles 2023, p. 109.

times were determined from the time the students began modeling, to the time they proceeded to save their recordings.

The encoding process consisted in five steps: 1) detection of observable variables, 2) creation and validation of an encoding grid, 3) creation and stabilization of a library of actions, 4) automated treatment of the encoded sequence of operations with the action library, 5) characterization of the students' modeling activity.

**Detection of Observable Variables.** We first detected variables related to software manipulation and exploration in the screen recordings of five students: production of elements of the 2-D profile, selection of functions, creation of geometrical entities, selection of features, use of visualizing tools, erratic mouse movements. We also took note of software responses: display of geometrical elements and feature application, messages of error. In the videos of the upper-part of the body, we identified several variables which were not visible in the screen recordings: use of the keyboard, visual focus, and verbal expressions.

**Creation and Validation of an Encoding Grid.** This breakdown of the students' activity was organized in an encoding grid, which was submitted to two external coders with a sample video to assess its reliability. The intercoder agreement was 90%. Following a discussion with the intercoders, the grid was revised and completed with the most recurrent functions, geometrical constraints and visualizing tools observed in the videos of five other students. This produced the final encoding grid, presented in Table 1, which was used to code the students' videos with the video analysis software V-note (version 2.7.2). Each video was viewed three times: one to code the eye focus and the verbal expressions, one to describe the mouse movements, and one to identify the functions selected. The encoding of the videos followed an alphabetical order. We realized after encoding four videos that the time-consuming characteristic of this data treatment and decided to limit the encoding to the videos of the four students that had already been encoded and to those of 24 students, who had been chosen based on their participation in all the experiments of our initial research: the whole experimental protocol included a battery of spatial tests, questionnaires investigating the students' spatial strategies, extra-curricular activities and previous practice of modeling software, and the modeling experiment described in this paper. This data collection occurred at the beginning of the year and at the end of the first term. As participation was optional, the number of participants varied according to the experiment and data collection. The detail of the selecting criteria for the 24 students can be found in our dissertation (Charles 2023, p. 111). This restricted sample was composed of $N_F = 4$ [14%] women and $N_H = 24$ [86%] men. 17 [61%] students had been exposed to Technology and Industrial Science courses prior to joining the school, whereas 11 [39%] came from courses deprived of technological content. 27 [96%] students had some experience with 3-D modelers, when 1 [4%] did not.

**Table 1.** Encoding grid.

| Object | Code | Indicator | Description | Example |
|---|---|---|---|---|
| Keyboard | DIM | Dimension | Dimension typed | 35 mm |
| | ESC | Function disabling | ESC key | Can be seen on screen/webcam video, or heard |
| | ENTER | Dimension validation | Enter key | Can be seen on screen or webcam video, or heard |
| Mouse movement | DT | Erratic mouse movement | The subject moves the curser without a precise direction | Erratic or slow mouse movement on screen |
| | DTM | Function search | The subject moves the curser across a menu to look for a function | Erratic or slow mouse movement on menu bar |
| | DE | Element movement | The subject moves a sketch element with the mouse | Movement of a sketch element |
| | PS | Rotation | The subject moves the model with the mouse to visualize it in 3D | Movement of the model |
| | P | Mouse over | The subject moves the mouse over an object to reveal an information bubble | Information bubbles pop up when the mouse hovers over hidden menus |
| Mouse click on profile | CPD | Production of a sketch element | The subject clicks on:<br>– a position in the coordinate system where they want to create an element,<br>– an element they want to apply a function to,<br>– an element they want to dimension | – a point in the coordinate system,<br>– a segment, a circle,<br>– a radius, a length |

*(continued)*

**Table 1.** (*continued*)

| Object | Code | Indicator | Description | Example |
|---|---|---|---|---|
| Mouse click to select an icon | CPC, CPR | Function selection | The subject clicks on a function | Center point circle, Center point rectangle |
| | CPA | Production of a sketch element: other | The subject clicks on: <br>– the green check or the red cross, <br>– an area (plane, surface of the profile) to affect a feature | – green check, red cross <br>– Front plane |
| Modeler response | ME | Error message | The modeler indicates the subject's action does not comply with modeling rules | Red message alert, concerned sketch elements and constraint icons turn red |
| | A | Display of a sketch element or constraint | The modeler displays a new element/constraint | A circle appears, a circle and a segment become tangent |
| Eye focus | SCREEN | Visual focus on the screen | The subject looks at the screen | The subject looks straight ahead |
| | INSTR | Visual focus on the instructions | The subject reads the instructions | The subject looks towards the table |
| | KEYB | Visual focus on the keyboard | The subject looks at the keyboard | The subject looks at the keyboard |
| | NEIGH | Visual focus on the student sitting next to the subject | The subject looks at a person sitting next to them | The subject looks at their neighbor/ neighbor's screen |
| Verbal expressions | AIE | External information | The subject hears information related to the experiment in a conversation with their neighbor/researcher, between 2 students, a student and the researcher | S1: "Ah, there you go! <br>S2: You've done it? <br>S1: Equal on these 2 segments, and then it sorts the dimension." |

*Note.* Adapted from "*Habileté spatiale et stratégies de modélisation 3D*" by S. Charles 2023, p. 408

**Creation and Stabilization of a Library of Actions and Automated Treatment of the Encoded Sequences of Operations with the Action Library.** The coding of the videos produced a sequence of operations, which constitute actions, which compose an individual's activity (Savoyant 2010, p. 94). For example, the sequence of operations KEYBOARD DIMENSION ENTER SCREEN was encoded with the action DIMENSION. A library of 227 potential actions was created from the standard procedures described in the online glossary and tutorials provided by the platform (16 actions), and from the actions observed in the students' videos: although the platform tutorials describe a way of completing an operation or an action, there can be more than one way for doing so. For example, it is possible to exit a function by clicking on its icon, by pressing the Escape key or by selecting a new function. This library was injected in a Python program designed to detect the actions of the library in the students' coded sequences, list them in their chronological order and calculate their occurrences. The program was designed to ignore the operations referring to the categories visual focus, mouse movement, error messages, model movement and verbal expressions, so that it could identify the actions of the library. Table 2 illustrates the encoded sequence of a student who produced two successive circles and a segment.

This processing covered under 90% of the encoded operations: 89.29% of the operations of the first data collection and 85,58% of the operations of the second data collection were processed into actions. To increase the covering rate, the actions which included the operations referring to the categories visual focus (SCREEN, INSTR, KEYB, NEIGH), mouse movement (DT, DTM, DE, P), model movement (PS) and verbal expressions (AIE) of the encoding grid were designed to ignore the previously mentioned operations when they occurred in the middle of an action, as we could not predict all the variations in the actions which included these operations: for example, if the student looked at the instructions before entering a dimension, the program ignored the operation INSTR and encoded the action DIMENSION. In addition, a new manual coding was applied to the processed sequences. 25 new actions were added to the library, some actions were merged, and some errors due to the first manual coding of the videos were corrected: the three parallel original manual encodings produced sequences of successive operations although they occurred at the same time. This could prevent the Python program from detecting the actions, if the order of the operations was not identical to the one described in the library. The sequences were processed a second time, using the revised library reduced to 145 active actions, to calculate the action occurrences. To reduce the number of actions in the library which covered a variety difficult to process statistically, more actions were merged (for example, three actions which had been previously identified differently because they had been exited in different ways), to produce a library of 126 actions.

**Characterization of the Students' Modeling Activity.** The sequences produced were processed in Hector (Dubus 2000), a sequence analysis program which automatically classifies sequences in what the program considers homogenous categories (p. 33). Our objective was to identify groups of students who had followed similar trajectories in their modeling activity. Unfortunately, the low number of student videos we had encoded produced too great a variety of sequences for the program to identify fine enough similarities.

While we encoded the videos, we noticed that some students repeated some elements of the object, for example because they had problems identifying an appropriate surface to extrude. This repetition was an opportunity to observe whether learning had taken place during the first modeling task, that is to say before teacher intervention. One of these elements is the central key groove of the model, illustrated in Fig. 2. It is made of a rectangle which is concentric with the central circle. The portions of the videos dedicated to the modeling of the groove were identified each time a rectangle was

**Table 2.** Extract of an encoded sequence

| Timestamp | Coded operation (manual) | Coded action (automated) |
|---|---|---|
| 00:02:13 | CPC | CIRCLE*2 |
| 00:02:14 | INSTR | |
| 00:02:18 | CPD | |
| 00:02:18 | SCREEN | |
| 00:02:19 | INSTR | |
| 00:02:20 | SCREEN | |
| 00:02:21 | CPD | |
| 00:02:21 | A | |
| 00:02:21 | INSTR | |
| 00:02:22 | SCREEN | |
| 00:02:23 | CPD | |
| 00:02:24 | CPD | |
| 00:02:24 | A | |
| 00:02:24 | INSTR | |
| 00:02:36 | SCREEN | |
| 00:02:36 | DTM | |
| 00:02:38 | CPC | |
| 00:02:40 | LINE | SEGMENT*1 |
| 00:02:41 | CPD | |
| 00:02:44 | CPD | |
| 00:02:44 | A | |
| 00:02:45 | KEYBOARD | |
| 00:02:46 | SCREEN | |
| 00:02:46 | ESC | |

*Note.* The abbreviations for the coded operations are described in Table 1

created. The number of rectangles produced, whether kept or deleted, was calculated for each student. With a view to identifying potential differences in the sequences dedicated to the production of the central groove, we selected the videos of the four students who created at least four rectangular grooves. This sample was composed of $N_F = 2$ [50%] women and $N_H = 2$ [50%] men. 3 [75%] students had been exposed to Technology and Industrial Science courses prior to joining the school and all of them had some experience with 3-D modelers.

The coded groove-modeling sequences were divided into smaller sequences when a series of actions led to the production of a rectangle. Each rectangle sequence was analyzed to complete the original coding with the elements of the rectangle produced, the elements of the rectangle acted upon, the type of rectangle created (unique, composed of two horizontal rectangles, composed of two vertical rectangles, composed of two semi-rectangles as illustrated in Table 3) and its rank in the sequence when more than one rectangle of the same nature were produced, and the dimensions entered.

**Table 3.** Types of rectangle used for the key groove

| Unique rectangle | 2 horizontal rectangles | 2 vertical rectangles | 2 semi-rectangles |
|---|---|---|---|

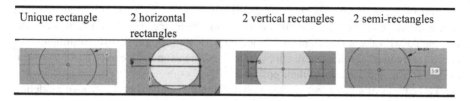

Table 4 illustrates the encoded sequence of a student who produced a fourth unique rectangle, and then dimensioned its length and its width.

**Table 4.** Extract of an encoded groove-modeling sequence

| Timestamp | Code | Action | Element acted upon | Element/constraint produced | Dimension in mm |
|---|---|---|---|---|---|
| 00:20:53 | CPR | Center Point Rectangle | – | – | – |
| 00:20:55 | CPD | | – | – | – |
| 00:20:58 | CPD | | – | – | – |
| 00:20:59 | A | | – | RU4 | – |
| 00:21:01 | DIM | Dimension*2 | – | – | – |
| 00:21:02 | CPD | | Upper segment | – | – |
| 00:21:03 | DIMENSION | | – | – | 45 |
| 00:21:05 | ENTER | | – | – | – |
| 00:21:05 | A | | – | Length | – |
| 00:21:06 | CPD | | Left segment | – | – |
| 00:21:08 | DIMENSION | | – | – | 10 |
| 00:21:09 | ENTER | | – | – | – |
| 00:21:09 | A | | – | Width | – |

*Legend.* DIM = function used to dimension an element; DIMENSION = value entered as a dimension; UR = Unique rectangle

# 3 Characterization of the Students' Modeling Performance

## 3.1 Groove-Modeling Sequences: Modeling Performance Before the Modeling Course ($N = 4$)

This encoding was further enriched with the nature of the changes made from a rectangle to the following one (function, geometry, software manipulation), the nature of the problem encountered (manipulation, geometry), and the condition of the rectangle at the end of the sequence (kept, deleted). These data were coded and organized in a chronological order, producing groove-modeling strategies. This characterization was completed with the assessment of the accuracy of the design, for which dimensional criteria were used: 2 marks were allocated to the dimensions of the length and the width of the groove, and 2 marks to their positioning.

Three types of strategy were identified through the analysis of the groove-modeling sequences. The first one is characterized by a succession of rectangles, which were all created and suppressed but for the final rectangle. Each rectangle was produced differently, using different functions, geometry or manipulation of the function. This strategy was named the trial-and-error method, as the student tried and failed until he produced a result which he deemed acceptable to keep. This student's sequence is described in Table 5.

**Table 5.** Example of a groove-modeling sequence: the trial-and-error approach

| Rectangle | Function used | Change made | Problem encountered | Condition | Accuracy |
|---|---|---|---|---|---|
| Semi-rectangles transformed into UR1 | Line | – | Manipulation | Deleted | 0/4 |
| UR2 | Line | Geometry | Manipulation | Deleted | 2/4 |
| UR3 & 4 | Center point rectangle | Function | Manipulation | Deleted | 0/4 |
| UR5 | Center point rectangle | Manipulation | – | Kept | 4/4 |

*Legend.* UR = Unique rectangle

The second strategy is characterized by the repetition of a modeling approach, building up on the first rectangle sequence: the students produced a rectangle which they deleted as they met a geometrical problem. They used a similar approach, with the addition of more efficient geometry in the production of their second rectangle. They later repeated this strategy when they produced two additional rectangles in further sketches. This strategy was named the upgrading method as the students modified a strategy which they had previously developed, then repeated once stabilized. The third strategy is characterized by the development of a strategy which is fairly accurate but time-consuming. The student later attempted different approaches but abandoned them due to failure, and reverted to her first approach. This strategy was named the return approach and is detailed in Table 6.

**Table 6.** Example of a groove-modeling sequence: the return approach

| Rectangle | Function used | Change made | Problem encountered | Condition | Accuracy |
|---|---|---|---|---|---|
| UR1 | Corner rectangle | – | Manipulation | Deleted | 0/4 |
| LG1 & RR1 | Corner rectangle | Geometry Function | – | Kept | 3/4 |
| UR2 transformed into LG2 & RR2 | Corner rectangle | – | Geometry | Kept | 3/4 |
| LG3 | Corner rectangle | - | - | Kept | 3/4 |

*Legend.* UR = Unique rectangle, LR = Left vertical rectangle; RR = Right vertical rectangle

### 3.2 Modeling Performance: Before and After Taking the Modeling Course ($N = 28$)

Statistical analysis using IBM SPSS Statistics (version 28) was performed to compare task completion times and the number of actions produced by the restricted sample ($N = 28$) before and after the modeling course. The normality of the distribution of the completion times, the number of actions and the number of erratic house movements were tested using a Shapiro-Wilk test (Shapiro & Wilk 1965): none of them followed a Gaussian distribution. We therefore opted for the non-parametric sign test to compare these variables before and after the CAD course. This revealed a significant ($p = 0.005$) change in the completion time and a non-significant ($p = 0.089$) result regarding the number of actions. The box plots, illustrated in Fig. 3, show that students completed the modeling task significantly quicker at the end of the term. The box plots in Fig. 4 seem to indicate that they produced fewer actions at the end of the term, although the decrease was not significant. A further sign test was performed to investigate the evolution of the number of erratic mouse movements: it revealed a significant ($p = 0.031$) change, which the box plots, illustrated in Fig. 5, showed to be an increase in the number of erratic mouse movements for the second data collection.

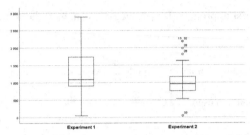

**Fig. 3.** Distribution of the completion time expressed in seconds for the first and the second data collections ($N = 28$) *Note.* From *"Habileté spatiale et stratégies de modélisation 3D"* by S. Charles 2023, p. 488.

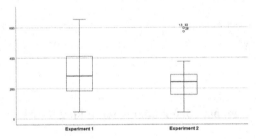

**Fig. 4.** Distribution of the number of actions for the first and the second data collections ($N = 28$) *Note.* From *"Habileté spatiale et stratégies de modélisation 3D"* by S. Charles 2023, p. 488.

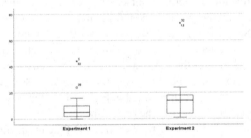

**Fig. 5.** Distribution of the number of erratic mouse movements for the first and the second data collections ($N = 28$) *Note.* From *"Habileté spatiale et stratégies de modélisation 3D"* by S. Charles 2023, p. 488.

## 4   Discussion and Conclusion

The approach described in this paper, whose original purpose was not to investigate the potential contribution of the links between rationalities to the structure of a limited-intelligence system, highlights two elements which contribute to this issue.

The first point concerns the key role of data collection in all investigations. For now, an artificial program with a rationality limited to counting, can detect, encode and order operations in sequences. For now, the artificial phase does not suffice. There is still need for a human rationality, which determines the occurrence of an observable variable in a syncretic approximation, that is unique to human beings, and that gives them a considerable head start on limited-intelligence systems. In all likelihood, the financing of a complete environment, which analyses gestures and postures, could in turn determine rational natures in observable variables by inferring logs and gestures, without the need for a human observer. These systems exist for simple functions. For 3-D modelers, investments are out of reach financially, but achievable humanly.

The second point has to do with the artificial rationality imposed by the software program. Even though the heuristic library presented in this study was developed by a human being, who deduced the relevant sequences of operations form the ideal manipulation of the software and the subjects' modeling activity, the automated proto-treatment detects them, including trials and errors, and poor performance. This constitutes the main challenge of an experimental work on limited-intelligence system initialization: how to substitute human rationality, which can conceive operation sequences detection and action-characterizing libraries, for an artificial rationality that self-elaborates the sequence of observable variables worthy of notice, as well as their worth in terms of actions. We mentioned our attempt to develop a study based on sequence analysis, that is to say to determine statistically the probability for sequences of operations, which form actions, to occur. This requires a large corpus of not only 28, but maybe hundreds or thousands of subjects. Other statistical approaches could also be used to attempt to determine human rationalities with the sole resort to artificial rationality treatments.

# References

Bertoline, G.R., Hartman, N., Adamo-Villani, N.: Computer-aided design, computer-aided engineering, and visualization. In: Nof, S.H. (eds.) Springer Handbook of Automation. SHB, pp. 639–652. Springer, Heidelberg (2009). https://doi.org/10.1007/978-3-540-78831-7_37

Bertoline, G.R., Wiebe, E.N., Hartman, N.W., Ross, W.A.: Technical Graphics Communications, 4th edn. McGraw-Hill Higher Education (2009)

Bhavnani, S.K., Garrett, J.H., Shaw, D.S.: Leading indicators of CAD experience. In: CAAD Futures 1993 Proceedings of the Fifth International Conference on Computer-Aided Architectural Design Futures, pp. 313–334 (1993)

Bhavnani, S.K., Reif, F., John, B.E.: Beyond command knowledge: Identifying and teaching strategic knowledge for using complex computer applications. In: Proceedings of the SIGCHI Conference on Human Factors in Computing Systems, pp. 229–236 (2001). https://doi.org/10.1145/365024.365107

Branoff, T., Dobelis, M.: Engineering graphics literacy: measuring students' ability to model objects from assembly drawing information. In: Proceedings of the 66th Midyear Conference of the Engineering Design Graphics Division of the American Society for Engineering Education, vol. 41 (2012). https://doi.org/10.18260/1-2--21306

Brown, P.: Do computers aid the design process after all? Intersect: The Stanford J. Sci. Technol. Soc. 2(1), 52–66 (2009)

Charles, S.: Habileté spatiale et stratégies de modélisation 3D [Doctoral dissertation, CY Cergy Paris Université] (2023).https://hal.science/tel-04097396

Chester, I.: Teaching for CAD expertise. Int. J. Technol. Des. Educ. 17(1), 23–35 (2007). https://doi.org/10.1007/s10798-006-9015-z

Computervision: CADDS 4X [Computer software] (1983)

Dassault Systèmes: CATIA (V6 2013x) [Computer software]. Dassault Systèmes (2012). https://www.3ds.com/fr/produits-et-services/catia/

Dubus, A.: Une Méthode d'Analyse des Séquences. Bulletin de Méthodologie Sociologique **65**(1), 33–57 (2000). https://doi.org/10.1177/07591063000650010

Hamade, R.F., Artail, H.A., Jaber, M.Y.: Learning theory as applied to mechanical CAD training of novices. Int. J. Hum.-Comput. Interact. **19**(3), 305–322 (2005). https://doi.org/10.1207/s15327590ijhc1903_2

Hamade, R.F., Artail, H.A., Jaber, M.Y.: Evaluating the learning process of mechanical CAD students. Comput. Educ. **49**(3), 640–661 (2007). https://doi.org/10.1016/j.compedu.2005.11.009

Hartman, N.W.: Defining expertise in the use of constraint-based CAD tools by examining practicing professionals. Eng. Des. Graph. J. **69**(1) (2005). Article 1

Hirschtick, J., McEleney, J., Li, T., Corcoran, D., Lauer, M., Harris, S.: Onshape (Version 103) [Computer software]. Onshape (2014). https://www.onshape.com/

Johnson, M.D., Diwakaran, R.P.: CAD model creation and alteration: a comparison between students and practicing engineers. In: Proceedings of the 2011 ASEE Annual Conference & Exposition 22.305.1–22.305.12 (2011). https://peer.asee.org/cad-model-creation-and-alteration-a-comparison-between-students-and-practicing-engineers

Lang, G.T., Eberts, R.E., Gabel, M.G., Barash, M.M.: Extracting and using procedural knowledge in a CAD task. IEEE Trans. Eng. Manag. **38**(3), 257–268 (1991). https://doi.org/10.1109/17.83758

Lieu, D.K., Sorby, S.A.: Visualization, Modeling, and Graphics for Engineering Design, 1st edn. Cengage Learning, Delmar (2009)

Peng, X., McGary, P., Johnson, M., Yalvac, B., Ozturk, E.: Assessing novice CAD model creation and alteration. Comput.-Aided Des. Appl. PACE(2), 9–19 (2012)

Poitou, J.-P.: L'évolution des qualifications et des savoir-faire dans les bureaux d'études face à la conception assistée par ordinateur. Sociologie Du Travail **26**(4), 468–481 (1984)

Rynne, A., Gaughran, W.F.: Cognitive modeling strategies for optimum design intent in parametric modeling (PM). Comput. Educ. J. **18**(3), 55–68 (2007)

Savoyant, A.: Éléments d'un cadre d'analyse de l'activité: Quelques conceptions essentielles de la psychologie soviétique (1979). Travail et Apprentissages **5**(1), 91–107 (2010). https://doi.org/10.3917/ta.005.0091). Cairn.info

Shapiro, S.S., Wilk, M.B.: An analysis of variance test for normality (complete samples). Biometrika **52**(3/4), 591–611 (1965). https://doi.org/10.2307/2333709

Steinhauer, H.M.: Correlation between a student's performance on the mental cutting test and their 3D parametric modeling ability. Eng. Des. Graph. J. **76**(3), 44–48 (2012)

Techsmith: Camtasia (2019.0.1) [Computer software]. Techsmith (2019). https://www.techsmith.fr/camtasia.html

# Mental Representations About Tangible Programming in Early Childhood Education

A. Misirli[✉] and V. Komis

Department of Educational Sciences and Early Childhood Education, University of Patras, Patras, Greece
amisirli@upatras.gr

**Abstract.** Representations of early childhood children and their development of computational thinking skills while tangibly programming a robot are presented in this study. Data was taken from multiple case studies in preschool settings. Research protocols recorded children's representations in individual basis semi-structured interviews in pre and post-sessions of an almost monthly educational intervention delivered by teachers in their classrooms. Some examples of depictions on children's drawings will also be presented, showing children's learning development in their computational thinking skills. The results show that while children attribute an animate identity to a robot, at the same time, they state and depict data for its properties and basic functional features.

**Keywords:** mental representations · educational robotics · computational thinking · early childhood education

## 1 Introduction

The term' educational robotics' refers to the teaching practice in which the teacher, by using robots, approaches learning either through them or about them. It appeared in the 1960's through the Logo educational movement. Educational robotics constitutes a teaching approach that recruits programmable methods and approaches learning through component projects (project-based learning). It is defined by the use of information technologies in the context of their ability to observe, analyse, modelisation and control various physical processes (Depover et al. 2007). It is about an approach which allows the trainee to familiarise themself with information technologies, with the broad sense of the term, and use them to determine a project, structure it and find a specific solution to the problem which is set by contrasting their opinion with others' (Denis & Baron 1993; Leroux et al. 2005). A specific category in educational robotics is programmable or tangible robots, a reproduction of Logo programming language that applies mainly to preschool and early school children. These are programmable robots that the user controls and are meant to move on the floor accordingly. The user lays out and determines the number of commands that the robot will input under specific circumstances. Following the principles of the Logo programming language favours the development of metacognitive ability, during which children rethink the procedures of thinking they have

J. Baratgin et al. (Eds.): HAR 2023, LNCS 14522, pp. 375–385, 2024.
https://doi.org/10.1007/978-3-031-55245-8_24

followed, improve problem-solving ability and promote spatial orientation (Clements & Sarama 2002). One of the most widespread programmable/tangible robots is the Bee-Bot, which embodies the Logo turtle philosophy and the programming principles of this specific language for its control. Thus, children can program a route/path on the floor in a themed mat or a free-flow activity. The tangible robot Bee-Bot has the shape and the colours of a bumblebee. Its programming interface is on the upper side and is based on seven (07) different coloured buttons or, in programming terms, commands. Four orange commands support the forward and backward motion and the left/proper rotation. The 'Go' command executes a program (green colour). Two blue commands ('Clear' and 'Pause') support different operations. The 'Clear' command is one of the most important commands and concepts around the development of algorithmic thinking and, thus, the present teaching intervention.

## 2 Mental Representations of Early Childhood Children

Integrating educational robotics in early childhood educational programs presumes an understanding of how children perceive and represent them. An early study by Scaife and Duuren (1995) refers to children's (5–11 years old) understanding of intelligent objects such as computers and robots, showing that five-year-olds were reluctant to attribute brains to objects compared to older children. Some years later, Greff (2005) studied the representations of eight pre-schoolers (5 years) about the tangible robot Roamer, showing how they depicted the programming procedure as it was taught. Levy and Mioduser (2008) investigated young children's (5–6) perspectives in explaining a self-regulating mobile robot as it learns to program its behaviours from rules. The findings revealed that the most straightforward tasks were approached from a technological perspective, while when the task became more complex, most children shifted to a psychological perspective.

Furthermore, Highfield and Milligan's study (2008) reports the representation of a preschool-aged child through the side movement of the programmable robot to attain circular movement. Other researchers in the last decade provided data on children's perceptions and interpretations of robots as well as their behaviour, but primarily for children aged five to sixteen years old without focusing on and providing evidence, especially for preschool children (Bhamjee et al. 2010; Beran et al. 2011; Bhamjee et al. 2010; Malinverni & Valero 2020). More recent research exploring preschool children's (3–6 years old) cognitive representations of robots showed that only some children mentioned some programming-related concepts (Monaco et al. 2018). In contrast, Mioduser and Kuperman (2020) argue that children have more difficulty explaining the behaviour of a robot than programming such behaviour. A study by Misirli, Nikolos, and Komis (2021) provided evidence for examining children's drawings alongside implementing an intervention on computational thinking concepts. The findings suggest that concepts such as thinking logically and algorithmically can be formed in children's representation after a developmentally appropriate teaching intervention.

## 3  Purpose and Questions of the Research

The present research aims to study representations of preschool-age children for the tangible robot Bee-Bot and how they differentiate after a scenario-based teaching intervention. The questions of the research are the following:

a. What are the children's prior representations of the tangible robot Bee-Bot?
b. What are the children's final representations of the tangible robot Bee-Bot?
c. Are there any changes between the two recordings (prior and final representations), and how are they interpreted?

## 4  Methodology

In the present study, we used multiple case studies to collect qualitative and quantitative developmental data (Kelly et al. 2008) applied to four preschool settings. An educational scenario was planned based on a scenario-based design of computer science concepts such as programming and robotics. Programming was primarily focused on algorithmic thinking and robotic technological features of the tangible robot Bee-Bot. The educators were trained to implement the educational scenario in their classes within this frame. The conceptualisation of a scenario-based design is based on constructionism pedagogical approaches such as project-based learning, child-centred learning, problem-solving learning environment, collaborative learning, and scaffolding and reflection process. It includes seven parts: 1. Identify the teaching subject, 2. Children's prior mental representations and subject knowledge, 3. Learning goals, 4. Teaching activities, 5. Artifacts and material, 6. Children's post-mental representations and subject knowledge, and 7. Documentation (Misirli & Komis 2014; Komis & Misirli 2015).

The educational material of the scenario-based approach was the tangible robot Bee-Bot for problem-solving, which was the educational material developed for this study. The data collection about children's representations was accomplished using a structured interview and an individual depiction (Appendix 1 shows the interview's plan). Both techniques were tools for educators to record and evaluate the learning process pre- and post. They were included in the detection and assessment activities of the educational scenario that precede or follow the teaching activities, respectively, reporting on mental representations about the robot's properties and programming concepts (subject knowledge). The interview included two strands: i) questions concerning exploring ideas about tangible robot's properties (What do you think it is? What do you think it does? How does Bee-Bot fly/move?) and ii) questions concerning its function and operation, for every command children formulated the idea that it made on them. Additionally, every interview included a corresponding depiction of the robot. Therefore, answers and drawings of the children's ideas were collected before the experimentation with the robot and after the completion of the teaching intervention. The interviews were used individually to assess (prior) and evaluate (post) the transformation of children's mental representations and consolidation of algorithmic thinking. The study sample consists of ninety-two (92) children (n = 42 boys, n = 50 girls) between the ages of four and six (M = 5,4). It comes from three urban public preschool settings and one from the outskirts of the region in Western Greece. The children were arranged in groups of four to six persons. The educational scenario was carried out in actual class conditions.

# 5  Data Analysis

The individual interviews and children's drawings were organised in a qualitative way and were classified in categories (seven variables with twenty – five values each) and were analysed with factor analysis of multiple correspondence a posteriori. The seven categorical variables that relate to the children's prior mental representations about the quality and operation of the tangible robot as shown by children's verbal representations and depictions are analysed in the table below (Table 1).

**Table 1.** Children's prior mental representations for the tangible robot Bee-Bot

| Description of keystone 1 for the variable values | | |
|---|---|---|
| Variable label | Title of variable values | Frequency |
| 1st group: Absence of ideas – representations | | |
| Prior Representation of Functional Definition | Absence of reply | 45 |
| Prior Representation of Operation/Function | Absence of reply | 45 |
| Prior Representation of Imaginary Content | Absence of reply | 45 |
| Sex | Absence of reply | 18 |
| What do you think Bee-Bot is? | Absence of reply | 18 |
| What do you think Bee-Bot does? | Absence of reply | 20 |
| How does Bee-Bot fly/move? | Absence of reply | 23 |
| Are all the keys/buttons the same? | Absence of reply | 29 |
| 2nd group: Confused ideas – representations | | |
| What do you think Bee-Bot does? | I_Imaginary Interpretation | 43 |
| Prior Representation of Operation/Function | I_Representation of Operation NO | 37 |
| Prior Representation of Functional Definition | I_Representartion of Functional Definition Confused | 25 |
| Prior Representation of Functional Definition | I_Representartion of Functional Definition Incomplete | 39 |
| What do you think Bee-Bot is? | I_Animal | 64 |
| Prior Representation of Operation/Function | I_Representation of Operation YES | 28 |
| Prior Representation of Imaginary Content | I_Representation Imaginary Content YES | 43 |
| Are all the keys-buttons the same? | I_Buttons NO | 75 |

*(continued)*

**Table 1.** (*continued*)

| Description of keystone 2 for the variable values | | |
|---|---|---|
| Variable label | Title of variable values | Frequency |
| 3<sup>rd</sup> group: Partly structured ideas – representations – Incomplete | | |
| Prior Representation of Operation | P_Representation of Operation YES | 28 |
| Prior Representation of Functional Definition | P_Representation of Operation Definition Confused | 25 |
| What do you think Bee-Bot does? | P_Action | 28 |
| What do you think Bee-Bot is? | P_Object_Animal | 21 |
| Age | 5–6 years old | 45 |
| Prior Representation Imaginary Content | P_Representation Imaginary Content NO | 22 |
| How does Bee-Bot fly/move? | P_Through Energy | 9 |
| How does Bee-Bot fly/move? | P_Through Movement | 21 |
| 2nd group: Confused ideas – representations | | |
| What do you think Bee-Bot is? | P_Animal | 64 |
| How does Bee-Bot fly/move? | P_Imaginary Interpretation | 26 |
| Prior Representation Imaginary Content | P_Representation Imaginary Content YES | 43 |
| What do you think Bee-Bot does? | P_Imaginary Interpretation | 43 |
| Prior Representation of Operation | P_Representation Operation NO | 37 |
| Prior Representation of Functional Definition | P_Representation Functional Definition Incomplete | 39 |
| Age | 4–5 years old | 26 |

As it resulted from Table 1, three groups of children's prior representations appear for the tangible robot Bee-Bot. The first group show lack of ideas in verbal formulations (from 18 to 29 children) and their depictions (45 children in total). In particular, regarding verbal formulation, the children's ideas about the quality of the tangible robot are absent (What do you think Bee-Bot is?, What do you think Bee-Bot does? How does Bee-Bot fly/move?, Are all the buttons/keys the same?), either because they did not come up with an idea or they tend to not replying negatively in unknown topics, which is a common behaviour for this age group. The picture of children's depictions in total is formed accordingly. Their drawings about the tangible robot are entirely absent. The second group contains the largest number of children (between 25 and 75 children), a big part of which belonging to the age group of 4–5 years old. In this group the prior children's representations are presented, that relate to the idea they formulate about the nature of the tangible robot and its corresponding depiction. As regards the quality of the tangible robot the prevalent idea among most children is that it is 'Animal-Bee' (What do you think Bee-Bot is?), by attributing its corresponding behaviour that is 'it will extract honey', 'it will fly' (What do you think Bee-Bot does?) as well as corresponding way of action, that is 'with its wings', 'with its sting' (How does Bee-Bot fly/move?). Associating the variable 'What do you think Bee-Bot does?' with the variable 'Prior Representation Imaginary Content' complete correspondence is observed. The majority of the children seem to recognise the existence of buttons on the top of the tangible robot for which they state differences about them such as in colour or semiotic representations. Nevertheless, there are a few cases of children having depicted partly (Prior Representation of Functional Definition Incomplete) but the majority of this group they do not depict at all the operation system and its connection to semiotic representations (buttons-symbols), as well as they do not use functional definitions in their descriptions, This specific variable associates directly with the children (between 37 and 39) which do not mention elements about the movement of the tangible robot in their verbal descriptions of their representations (Prior Representation of Operation_NO & Prior Representation of Functional Definition Incomplete). Proportionally, the operation system (buttons) and its semiotic representations (symbols) appear partly in the representations of a smaller portion of children, as well as the use of functional definitions in their description (Prior Representation of Functional Definition_Confused). From Table 1 it appears that there is direct correlation to the group of children whose verbal descriptions of their representations report that the tangible robot 'Moves Forward-Backwards-Left-Right' (Prior Representation of Operation_YES). In the third group the representations appear more structured than the previous. Most of the children belong to the age group of 5–6 years old and about the quality of the tangible robot, the idea that it is 'Object_Animal-Car/Bee' prevails (What do think Bee-Bot is?) by attributing to it corresponding behaviour, that is it 'goes on', 'it moves' (What do you think Bee-Bot does?, as well as corresponding way of action, that is 'with its wheels' (How does Bee-Bot fly/move?). A very small number of children refer to the necessity of batteries, as an indispensable element for the corresponding way of action of the tangible robot. Correlating the above variables to the variable of the children's representations, it arises that the children enter more elements that relate to the operation system (buttons) or the symbols that are located on the tangible robot, as well as they use functional definitions in their descriptions even in the circumstances they have partly depicted it (Prior Representation of Functional Definition_Confused). Therefore, consistency is attributed to the verbal description of their representation, by mentioning that the toy 'Moves' 'Forward/Backwards/Left/Right' (Prior Representation of Operation_YES). Subsequently, the children's final representations for the tangible robot are analysed in Table 2.

**Table 2.** Children's final representations for the tangible robot

| Description of keystone 1 for the variable values | | |
|---|---|---|
| Variable title | Title of variable values | Frequency |
| 1st group: Absence of ideas – representations | | |
| Final Representation of Functional Definition | Absence of reply | 29 |
| Final Representation of Imaginary Content | Absence of reply | 29 |
| Final Representation of Operation/Function | Absence of reply | 29 |
| What do you think Bee-Bot does? | Absence of reply | 31 |
| How does Bee-Bot fly/move? | Absence of reply | 31 |
| Are all the keys/buttons the same? | Absence of reply | 31 |
| 3rd group: Fully structured ideas – representations | | |
| Are all the keys-buttons the same? | F_Buttons NO | 76 |
| How does the Bee-Bot fly/move? | F_Through Operation | 62 |
| What do you think Bee-Bot does? | F_Interpretation/Description of Action | 55 |
| What do you think Bee-Bot is? | F_Animal | 39 |
| Final Representation of Imaginary Content | F_Representation Imaginary Content NO | 56 |
| Final Representation of Operation/Function | F_Representation of Operation YES | 42 |
| Age | 5–6 years old | 45 |
| Description of keystone 2 for the variable values | | |
| Variable title | Title of variable values | Frequency |
| 2nd group: Incomplete ideas – representations | | |
| Are all the keys-buttons the same? | F_Buttons YES | 3 |
| What do you think Bee-Bot is? | F_Animal | 39 |
| Final Representation Imaginary Content | F_Representation Imaginary Content | 25 |
| Final Representation of Functional Definition | F_Representation of Functional Definition Incomplete | 33 |
| Final Representation of Operation | F_Representation of Operation NO | 39 |
| 3rd group: Fully structured ideas – representations | | |
| Final Representation of Operation | F_Representation of Operation YES | 42 |
| What do you think Bee-Bot is? | F_Object | 19 |

(*continued*)

**Table 2.** (*continued*)

| Description of keystone 2 for the variable values | | |
|---|---|---|
| Variable title | Title of variable values | Frequency |
| Final Representation of Functional Definition | F_Representation of Functional Definition Complete | 13 |
| Final Representation of Functional Definition | F_Representation of Functional Definition Confused | 35 |
| Final Representation Imaginary Content | F_Representation Imaginary Content NO | 56 |
| What do you think Bee-Bot does? | F_Interepretation Description of Action | 5 |
| Age | 5–6 years old | 45 |

In the final evaluation the personal interview was repeated, with the same keystone of questions and a new individual depiction about the tangible robot Bee-Bot. In Table 2 three groups of the children's final representations are presented. The first group is about a small number of children (between 29 and 31 children), whose replies and representations are absent. For the second group the representations can be described as incomplete. The quality of animated features on the tangible robot, as well as the corresponding elements of imaginary content are attributed by a very small number of children. Additionally, although the largest the number of representations observes and formulates the diversity of the operation system/the operation key-buttons and the symbols depicted on them they seem to be absent from their corresponding depictions. An interpretation is that maybe they did not find the appropriate way for the depiction of their representation. In the third group of representations mostly children 5–6 years old are presented. Their representations relate to complete ideas about the tangible robot concerning its behaviour and action. The representations of its behaviour refer to the action and the operation of the tangible robot, whereas the representations of its action refer wholly to the operation mode. Consequently, elements of imaginary content are totally absent in the children's depictions whereas elements of the operation system/operation keys/buttons appear more, as well as their corresponding symbols, as they are depicted in them with consequent and more systematic use of functional definitions.

## 6   Discussion – Conclusion

The results of the present study provide data about the representations of preschool-age children concerning the quality and operation of a tangible robot. Specifically, they differentiate between the children's prior and post-mental representations of the tangible robot Bee-Bot. In the final representations, the number of answers that were absent from the prior representations decreases since these children seem to have formed an idea about the tangible robot. Although most of the children continue to attribute an animated quality to the robot (Rincon, Duenas, Torres, Bohorquez & Cruz 2021), they enter data that correlates to the programming procedure for its control and operation with the corresponding use of functional definitions more systematically. Thus, algorithmic thinking shapes cognitive models of different concepts involved. Children's depictions seem to give emphasis on the procedure required for the creation of the program without giving the relevant outline, as suggested in related research by Greff (2005), in order to reduce the emphasis on the procedure of programming depicting. The development of algorithmic thinking and programming concepts for children of preschool age with the use of the tangible robot Bee-Bot is facilitated through the designing and implementation of the appropriate educational scenarios (Misirli & Komis 2014; Komis & Misirli 2015; Komis et al. 2017). The findings showed that children were likely to demonstrate meaningful uses, although no curriculum outcomes were involved, provided explicit scaffolding and planned educational activities (Newhouse et al. 2017). The designing and implementation of educational scenarios suitable to the development of programming abilities in preschool-aged children appear from results to play a catalytic role in the creation of corresponding representations. Activities about controlling and operating the tangible robot were delivered at the beginning of the intervention, and we assume it was the factor giving children the tendency to attribute animate characteristics to the robot (Beran et al. 2011; Bhamjee et al. 2010). Consequently, the final drawings of the children of the third group and the corresponding formulations give elements about the control and operation procedure of the tangible robot through the educational activities they were taught, which is partially by the study of Sümeyye, Canan & Mustafa (2021). The present study may provide a framework for other researchers to combine data from interviews and depictions and use it to plan their teaching according to children's learning needs and social backgrounds.

## Appendix 1

**Assessment of prior & post subject-knowledge Individual interview (pre & post-test)**

| Date: | Duration: |
|---|---|
| Name of child/Age | |
| 1. What do you think the Bee-Bot is? | |
| 2. What do you think the Bee-Bot does? | |
| 3. How does the Bee-Bot fly/move…? | |
| 4. Are all buttons the same? <br> If the answer is 'Yes' continue with question: What they have different? | |
| 5. What you think you can do with buttons? | |
| 6. Button ↑: What do you think will happen if you press it? | |
| 7. Button ↓: What do you think will happen if you press it? | |
| 8. Button →: What do you think will happen if you press it? | |
| 9. Button ←: What do you think will happen if you press it? | |

(continued)

(*continued*)

| Date: | Duration: |
|---|---|
| 10. Button 'GO': What do you think will happen if you press it? | |
| 11. Button 'CLEAR': What do you think will happen if you press it? | |
| 12. What do you think is a robot? | |

# References

Beran, T.N., Ramirez-Serrano, A., Kuzyk, R., Fior, M., Nugent, S.: Understanding how children understand robots: perceived animism in child–robot interaction. Int. J. Hum. Comput. Stud. **69**(7–8), 539–550 (2011)

Bhamjee, S., Griffiths, F., Palmer, J.: Children's perception and interpretation of robots and robot behaviour. In: Lamers, M.H., Verbeek, F.J. (eds.) HRPR 2010. LNICSSITE, vol. 59, pp. 42–48. Springer, Heidelberg (2011). https://doi.org/10.1007/978-3-642-19385-9_6

Bhamjee, S., Griffiths, F., Palmer, J.: Children's perception and interpretation of robots and robot behavior. In: Proceedings of HRPR 2010 International Conference on HUMAN ROBOT PERSONAL RELATIONSHIPS, Leiden, The Netherlands, June 2010, pp. 42–48 (2010)

Bilotta, E., Gabriele, L., Servidio, R., Tavernise, A.: Investigating mental representations in children interacting with small mobile robots. In: Conference ICL 2007, 26–28 September 2007, pp. 15-pages. Kassel University Press (2007)

Clements, D.H., Sarama, J.: The role of technology in early childhood learning. Teach. Child. Math. **8**(6), 340–343 (2002)

Denis, B., Baron, G.L.: Regards sur la robotique pédagogique. Actes du quatrième colloque international sur la robotique pédagogique. INRP: Technologies nouvelles et éducation, Paris (1993)

Depover, C., Karsenti, T., Komis, V.: Enseigner avec les Technologies: Favoriser les apprentissages, développer des competences. Presses de l'Université du Quebec, Montréal (2007)

en éducation: Perspectives curriculaires et didactiques, pp. 209–226. Clermont-Ferrand: Presses Universitaires Blaise-Pascal

Greff, E.: Programme cognitique. Aux Actes de Colloque International «Noter pour penser». Université de Psychologie, Angers (France), 27–28 janvier 2005 (2005)

Highfield, K., Mulligan, J.: Young Children's engagement with technological tools: the impact on mathematics learning. In: Proceedings of International Congress in Mathematical Education 11, Monterrey, Mexico, 6–13 July 2008 (2008)

Kelly, A.E., Lesh, R.A., Baek, J.Y.: Handbook of Design Research Methods in Education: Innovations in Science, Technology, Engineering, and Mathematics Learning and Teaching. Routledge, New York (2008)

Komis, V., Misirli, A.: Robotique pédagogique et concepts préliminaires de la programmation à l'école maternelle: une étude de cas basée sur le jouet programmable Bee-Bot. Aux Actes DIDAPRO 4, Dida et STIC, Patras, Grèce, 24–26 octobre 2011, pp. 271–284 (2011)

Komis, V., Misirli, A.: Apprendre à programmer à l'école maternelle à l'aide de jouets programmables. In: Baron, G.-L., Bruillard, É., Drot-Delange, B. (eds.) Informatique en éducation: Perspectives curriculaires et didactiques, pp. 209–226. Presses Universitaires Blaise-Pascal, Clermont-Ferrand (2015)

Komis, V., Romero, M., Misirli, A.: A scenario-based approach for designing educational robotics activities for co-creative problem solving. In: Alimisis, D., Moro, M., Menegatti, E. (eds.) Edurobotics 2016 2016. AISC, vol. 560, pp. 158–169. Springer, Cham (2017). https://doi.org/10.1007/978-3-319-55553-9_12

Leroux, P., Nonnon, P., Ginestié, J.: Actes du 8ème colloque francophone de Robotique Pédagogique. IUFM, Aix-Marseille, Revue Skhôlé (2005)

Levy, S.T., Mioduser, D.: Does it "want" or "was it programmed to..."? Kindergarten children's explanations of an autonomous robot's adaptive functioning. Int. J. Technol. Des. Educ. **18**(4), 337–359 (2008)

Malinverni, L., Valero, C.: What is a robot? An artistic approach to understand children's imaginaries about robots. In: Proceedings of the Interaction Design and Children Conference, pp. 250–261 (2020)

Mertala, P.: Young children's conceptions of computers, code, and the Internet. Int. J. Child-Comput. Interact. **19**, 56–66 (2019)

Misirli, A., Komis, V.: Robotics and programming concepts in early childhood education: a conceptual framework for designing educational scenarios. In: Karagiannidis, C., Politis, P., Karasavvidis, I. (eds.) Research on e-Learning and ICT in Education, pp. 99–118. Springer, New York (2014). https://doi.org/10.1007/978-1-4614-6501-0_8

Misirli, A., Komis, V.: L'usage des jouets programmables à l'école maternelle : concevoir et utiliser des scenarios éducatifs de robotique pédagogique. Revue Scholé (2012, in press)

Misirli, A., Nikolos, D., Komis, V.: Investigating early childhood children's mental representations about the programmable floor robot Bee-Bot. Mediter. J. Educ. **1**(2), 223–231 (2021)

Monaco, C., Mich, O., Ceol, T., Potrich, A.: Investigating mental representations about robots in preschool children (2018)

Newhouse, C.P., Cooper, M., Cordery, Z.: Programmable toys and free play in early childhood classrooms (2017)

Pogadaeva, Zakharova, Melnikova, Yakovlevna, Sergeevna (n.d.) (2020). "RoboKids": Additional general developmental program technical focus

Scaife, M., van Duuren, M.: Do computers have brains? What children believe about intelligent artifacts. Br. J. Dev. Psychol. **13**(4), 367–377 (1995)

Seçim, E., Durmusoglu, M., Çiftçioglu, M.: Investigating pre-school children's perspectives of robots through their robot drawings. Int. J. Comput. Sci. Educ. Sch. **4**(4) (2021)

# Connecting Basic Proportional Thinking with Reasoning About Risks

Ulrich Hoffrage[1], Laura Martignon[2(✉)], Tim Erickson[3], and Joachim Engel[2]

[1] Faculty of Business and Economics (HEC Lausanne), University of Lausanne, Lausanne, Switzerland
[2] Institute of Mathematics, Ludwigsburg University of Education, Ludwigsburg, Germany
martignon@ph-ludwigsburg.de
[3] Eeps Media, Oakland, CA, USA

**Abstract.** Risk literacy requires basic elements of numeracy. It requires some ease with basics of probability theory. Yet, it is these basics which often cause difficulties, in particular when they are presented using abstract formalisms. This paper reviews a systematic framework for representations of information that eliminate difficulties and frequent fallacies in dealing with probabilities. A large subfamily of these representations is inspired by Otto Neurath's isotypes and consists of so-called icon arrays. Another subfamily contains trees, which are hierarchical noncyclic graphs. Yet another subfamily consists of double trees that foster intuitions for Bayesian inferences. An interactive webpage is presented that can be used by both adults and children, with buttons and sliders to set parameters and with 3 different levels of statistical literacy. Furthermore, trees are examined as structures for combining multiple cues in order to classify situations under risk and make decisions. Plugins for constructing such trees and for reckoning with risks are presented and discussed.

## 1 Introduction

Humans have dealt with risks since the beginnings of human history. Yet the rigorous, formal treatment of risks is a modern achievement. This mathematical treatment is essentially based on probability theory and was cemented during the early nineteenth century with the work of De Morgan on probability and life contingencies (1838). It attained full formal rigor only during the first decades of the last century, being embedded in the edifice of Mathematics. This edifice, solidly built on axioms and theorems proven by means of inferences based on classical logic, can appear daunting to untrained lay people. In fact, precisely this formal rigor hinders the natural approach to „doing" mathematics based on fruitful intuitions. The tension between rigorous, formal mathematics and mathematical intuitions remains a hot topic of mathematics education, especially at school level: *what* should be taught in school and *how* should it be taught so that school students acquire mathematical competencies beyond procedural techniques? The tension between the axiomatic treatment of Kolmogorov and the intuitive, quasi- empirical treatment of Pascal and Laplace, based mainly on numerical proportions, is so strong that it led Leo Breiman (1968) to state that "probability theory is condemned to having a right

© The Author(s), under exclusive license to Springer Nature Switzerland AG 2024
J. Baratgin et al. (Eds.): HAR 2023, LNCS 14522, pp. 386–406, 2024.
https://doi.org/10.1007/978-3-031-55245-8_25

and a left hand—the right hand being the measure-theoretical approach that guarantees mathematical rigor, and the left hand meaning 'intuitive probabilistic thinking'" (p. 7). The modern concept of a probability is defined by means of a real-valued function on a sigma algebra of subsets of a set that satisfies certain axioms. The enthusiasm for this definition in the first decades of the 20th century was enormous. It proved, once again, that set theory had become, as David Hilbert put it, "A paradise, from which no-one will throw us out." (Hilbert 1926).

In the sixties, mathematicians both in Europe and in the United States prompted the introduction of set theory in schools. This enthusiasm was enhanced by the possibility of representing sets by means of Venn diagrams. This representation of sets was at hand since John Venn had introduced his diagrams in the nineteenth century (Venn 1880). Venn diagrams represent sets, their intersections and their unions by circles or ovals that may overlap or be nested within each other. In the sixties and seventies, these Venn diagrams were introduced as representations of sets in primary and secondary schools in most European countries and in several other countries around the world. To the dismay of mathematicians, school students, their parents and even many of their teachers were irritated and frustrated, to say the least. We cite here a typical example: Stephanie Krug, who went to school in Baden Württemberg back in the seventies and whom we interviewed, recalls her reluctance to draw triangles and circles in different colors placed in those Venn diagrams (Fig. 1). "What for?" she asked.

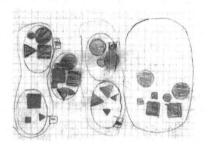

**Fig. 1.** From a notebook of Stephanie Krug in 1974

In fact, the reaction to set theory and Venn diagrams in schools represents one of those rare victories of teachers united with parents in many countries of the world. In Germany, the protests from teachers and parents were so strong that set theory was banned from primary school, and with it the Venn diagrams that had confused everyone.

Convinced that probability theory requires both set theory and the functions defined on sets, German mathematics educators were reluctant to introduce probabilities earlier than in advanced secondary school. This tendency changed somewhat during the nineties, and elements of probability theory and statistics were covered in short chapters in school math books. Sadly, these chapters were often left to be treated at the end of school years, with little emphasis and little enthusiasm. In this article, we focus on that realm of elementary stochastics, which examines the possible relationships between two bivariate variables and is essentially based on proportional reasoning. We posit that proportional reasoning, supported by dynamic visualizations of "natural frequencies" (see below for a definition of this term), can provide effective intuitions of probabilistic situations and risk literacy. We exhibit digital plugins for working with proportions related to risks, that have successfully been used at different institutions.

## 2   Proportions, Frequencies, and Probabilities

With the results of cognitive psychologists during the second half of the twentieth century on people's dealing with probabilistic inferences, things became even worse for the acceptance of probabilities in math education. Experiments by psychologists such as Amos Tversky and Daniel Kahneman discredited Breiman's left (intuitive) hand. Using many examples, Kahneman and Tversky (e.g., 1974) had shown that people can have enormous difficulties in dealing with various tasks that involved probability judgments. Note that in these tasks the necessary information was typically expressed in terms of probabilities. Likewise, they had shown that people have a hard time to correctly compare the probability of compound events with the probabilities of the constituents of these compounds. In the opinion of these two psychologists and many others who followed their line of research, people are unable to handle probabilities. More generally, the conclusion of these authors' heuristics-and-biases program was that people are not "rational".

This pessimistic view of human thinking did not go unchallenged. It provoked new empirical work as well as theoretical and methodological discussions, which were driven, to a considerable extent, by Gerd Gigerenzer and his students. In many experiments, these authors have demonstrated that so-called cognitive illusions can be made to disappear (e.g., Gigerenzer 1991, Gigerenzer, Hertwig, Hoffrage, & Sedlmeier 2008). They have argued that sometimes the "wrong" statistical (or logical) norm has been applied or that the stimulus materials used in experiments have not been representative of participants' natural environment to which they have adapted. Most important for the present article were, however, their argument and their demonstrations that *information needs representation* and that performance in judgment tasks can improve tremendously when information is presented in terms of frequencies instead of probabilities. Some of the studies by Gigerenzer and his students then led to a complete redesign of Venn diagrams, which made a big difference for children and adults. Ovals with small abstract figures such as triangles and squares are not helpful. What helps, however, are grids with representations of individuals, items of all sorts, mythical creatures, or animals that are easy to sort and count. It is easy to choose content that is appealing and motivating even for young children. With these representations, elementary probabilistic thinking and risk literacy boils down to reckoning with proportions, comparing them, and drawing conclusions from comparisons.

## 3   Classifications in Risky Situations

The main scope of this work is to address how elementary proportions are at the basis of risk literacy. Let us recall that one of the main theorems of probability theorem states that in any aleatory experiment which can be repeated „ad infinitum", the relative frequencies converge to the real probability. This result basically implies that relative frequencies, which are proportions of successful results divided by the total number of results, are the concrete, palpable approximations of probabilities. Mathematics educators often insist that working well with these approximations is a sufficient basis for risk literacy. Decisions in risky situations often depend on classifications of situations, which, on their turn, depend on the features or cues that characterize them. For instance, a doctor classifies a patient as "in high risk of heart attack" based on certain features extracted from the electrocardiogram of the patient and from behavioral cues, like the intake of certain medicines or chest pain. Once the patient is classified as "high risk" he/she is sent to the coronary care unit. If not, that is if he/she is not "at high risk" then the patient can be assigned a regular nursery bed. Medical situations are one of the great application fields in decision making, where risk literacy becomes fundamental.

Observe that in most medical situations, one feature alone is not enough for making a good decision. The immense progress in medicine and epidemiology is precisely the discovery of tests, symptoms, and behavioral traits that can fully characterize a patient's risk situation, so that accurate decisions tailored to the specific risk can be made.

The Covid pandemic that recently shocked the world is a wellknown example of medical decision making under high risk. Which are the relevant cues and how can their reliability be measured? We had some direct experience because we, at least one of the authors, had to deal with the disease in the region of the German city of Tübingen. It was interesting to consult doctors who worked together with the main Hospital of Tübingen and capture their simple strategy for decision making in case of symptomatic patients. Basically, the decision tree of doctors during the months of April and May 2020 looked like this:

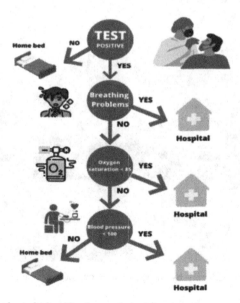

**Fig. 2.** A frequently adopted classification and decision tree for symptomatic patients at the beginning of the Covid pandemic

Observe that a positive test did not lead to immediate hospitalization. Breathing problems led to hospitalization, but, in its absence, other cues were checked. The order of those other cues was important here. In general, research has shown that a robust and accurate tree similar to the one shown in Fig. 2 can be constructed by ordering the cues according to their validity, that is, the proportion of correct classifications. However, the validity of the cues considered at the beginning of the Covid pandemic was not yet established by large empirical studies, so that the situation was characterized by uncertainty rather than risk (Mousavi & Gigerenzer 2014). Nevertheless, there were some preliminary numbers and estimates that could be used to construct such trees and so we witnessed the development of these decision trees as an ad hoc process, in the sense that "the science had to be developed along the way" without a large base of prior knowledge.

## 4   Scaffolding Risk Literacy

Being able to construct tools, like trees, for decisions under risk, is one component of risk literacy. Which are other components? In what follows, we briefly present our (Martignon & Hoffrage 2019) four-stage model of risk literacy. It consists of four components (Fig. 3):

I) Detecting risk and uncertainty
II) Analyzing and representing uncertain or risky situations
III) Comparing alternatives and dealing with trade-offs
IV) Making decisions and acting

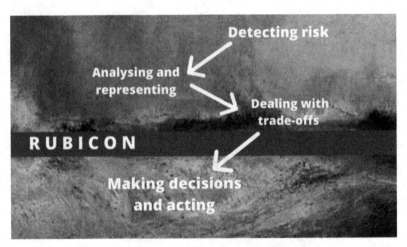

**Fig. 3.** Scaffolding Risk Literacy by means of four components (adapted from Martignon & Hoffrage 2019).

Of these components the first one – detecting and identifying risks and uncertainties in ordinary life – requires sufficient psychological disposition, either innate or acquired during childhood and youth. The second component – analysing and modelling risks – is not only psychological but, above all, adaptive. It requires basic skills but also basic education, especially in these times when we are confronted with a sheer amount of quantitative information on a daily basis – information that requires and provokes the acquisition of basic numeracy skills. The availability of information goes hand in hand with the availability of digital media and its tools, including those that can improve risk literacy. Concepts of risk and related tools need to be understood and trained in order to improve basic numeracy skills and risk literacy. To illustrate: if a person's Covid-19 test was negative and she was told that the false negative rate of the test was, say, 2%: is she now at risk? And what if the test was positive? Answers to this type of question became crucial to citizens all over the world during 2020. But it is, of course, not just the Covid 19 pandemic which prompts our need to understand the validity of features and how it can be computed. For instance, when a woman is told that regular screening reduces the risk of breast cancer by 50%, what should she do? Or if we are told that eating bacon sandwiches increases our risk of getting bowel cancer by 20%, how seriously should we try to avoid bacon sandwiches? One message of this paper is that there are simple tools and principles for analysing and modelling risks and uncertainties, so that these become amenable to being assessed and compared in terms of elementary proportions that can provide the basis for sound decisions, even of young students. The third component of risk literacy – namely comparing alternatives and dealing with trade-offs – builds on the first two, but goes beyond them and therefore represents an additional skill. This paper is concerned with the second, third and fourth components of risk literacy.

# 5    Proportional Thinking and Logical Principles at the Foundation of Classification

A child of 10 years can deal with the following situation: Consider 25 pupils in a classroom, some are boys, and some are girls. Some have short hair. Some are boys and have short hair. Some wear skirts, some wear trousers. Some wear skirts and have short hair. Would one bet that a child of our class who wears a skirt is a girl? Most probably we would. But would we bet that a child with short hair is a boy? Probably not! Features or characteristics, like „short hair" or „wears a skirt", are the essence of classification and inference in everyday life and we should early learn to deal with them. We extract features, items, and concepts out of situations with ease, we are able to classify based on features and tend to define situations, items, and concepts based on features.

In this example, three bi-variate variables were mentioned: Gender (boy/girl), Hair length (short/long), and Dress (trousers/skirt). The variables (here, three) and the individual objects (here, 25 pupils) can be conceived of as two poles. To start with the former: the variables can be used to define classes and to classify objects. They are, hence, on a population level and define what all have in common that belong to this class. Note that 25 is, in the present example, the entire population; but from a higher point of view this number can be seen as describing just one sample of a population that is, in principle, unlimited – and in fact, there are other classrooms of the same school, other schools in the same city, other cities in other countries, and other cohorts in coming years. The direction from "class to object", or from "population to individual" is deductive.

Conversely, one could also start from the other pole, the individuals. To make it concrete: A teacher could ask her 25 pupils to put all tables aside and to gather all in the middle of the room. She could then send some to the left side of the room – Peter, Thomas... and Rafael go here, and Sandra, Yvonne ..., and Kim go there – and let the pupils detect her organizing principle. While the teacher proceeds in a deductive manner and applies a general rule that determines whether a given individual goes left or right, the pupils whose task it is to detect the rule, need to make an inductive inference: They need to ask what does the one group of individuals have in common, what does the other have in common, and what discriminates between them? Having detected this principle, they should be able to predict where the teacher would place a new pupil (that is, to make an out-of-sample prediction if the door would open and pupil number 26 enters).

This little exercise, which can easily be implemented in the classroom, illustrates the difference between deductive reasoning (from population to sample, or top-down) and inductive reasoning (from sample to population, or bottom-up). When the teacher uses a variable to form groups, she uses deductive reasoning ("All girls should be on the left side – you are a girl – therefore you go left"). When she does not reveal her criterion but asks pupils to find it out, they engage in inductive reasoning. Note that this distinction is akin to Piaget's (1956) distinction between intensional and extensional reasoning about features in a given sample, which goes back to the Port Royal Logic. Features are intensional aspects of elements of sets, like „wearing a skirt". This is a variable that characterizes an unlimited number of objects and specifies what the members of this category (or set, or subpopulation) have in common. In contrast, if we start with a list of the names of all children wearing a skirt in our class, then we are performing an extensional operation rendering all children wearing skirts by listing them. The question whether extension and intension can be treated in one framework goes back a long way: The dichotomy can be found at the heart of what is considered the second epoch of logic initiated by Antoine Arnauld and Pierre Nicole in their book "Logic or the Art of Thinking" which was published in 1662. The dichotomies we just discussed (deductive vs. inductive; top-down vs. bottom-up, intensional vs. extensional) are intimately related to the two visual representations that we already mentioned in our introduction. Let us consider the deductive, top-down, and intensional viewpoint first. It is directional: from population to sample. It goes from features that characterize an, in principle, unlimited (and hence, not countable) population to (countable) individuals that possess these features. In other words, it goes from qualities (variables) to quantities (individuals). This viewpoint focusses on sets, but not just on how many individuals are in these sets. *Structure* (classes defined by features) comes first – *content* (individuals as carriers of features) second. A way to visualize sets and structure is to use adequate "good-old" Venn-diagrams. The regions or areas depicted in these diagrams specify what *all* have in common who are in a specific area, but these individuals (who, together, build the "all" in a specific area) are not individually identified.

The other viewpoint – inductive, bottom-up, and extensional – is also directional, but now from individual to population. The starting point is constituted by "countable individuals", say pupils in a classroom. By inspecting them closer and by comparing them, the question arises how they can be described, what they have in common, and, eventually, how they can be distinguished from each other, that is, which variables could be used to describe and to classify them. A way to visualize countable individuals is to use icons for representing each of them. A given icon depicts features of a given individual, but these features (along which individuals can be distinguished) are not displayed as a set.

Venn-diagrams and icons can be seen as the two polar representations that visualize the starting points of the two perspectives described above. Venn-diagrams visualize sets and icons visualize individuals. Sets as a starting point allow one to use a variable to classify an individual, and individuals as a starting point invites one to ask how they can be described and grouped. To illustrate, the abovementioned teacher in the class with 25 pupils starts with a variable, say gender that defines the set of boys and girls, then looks at her pupils, one after the other, and applies this variable as a classifier to determine, in a deductive manner, who should go to which side of the room. Conversely, those children who find themselves on one side of the room, without having been told anything about the organizing principle, first look at their ingroup and on those on the other side. They hence start from the individuals and then, in a second step, consider potential variables to test whether they could explain the assortment. Finding potential variables to scrutinize them requires inductive reasoning, testing whether a candidate variable can explain the observed grouping requires deductive reasoning.

For each of the two directions described above, we have seen that the natural next step was to leave the viewpoint (as defined by one's standpoint) and go in the direction of the other pole. The deductive view would hence apply intensional reasoning, starting with a classifier, and then look at a given individual in order classify it. This amounts to placing a certain individual into one of the areas in a Venn-diagram. Content is used to fill structure in a top-down manner. Conversely, the inductive view wonders how individuals, represented by icons, could be sorted. By shifting scattered icons around and grouping them according to defining characteristics, sets emerge in a bottom-up manner. Both directions from the two starting points meet in the same middle. No matter whether sets are filled with individuals, or whether individuals are identified as members of sets and sorted accordingly, at the end of each process there is structure with content, or, conversely, content with structure.

The discussion above can be supported by the panels displayed in Fig. 4. Panels A and F of Fig. 4 artificially separate objects and variables. Panel A focuses on objects but is mute about their features. Panel F focusses on features and possible relationships among them but does not contain any countable objects. Panels A and F can thus be seen as two poles: Objects without features and variables without objects, respectively. We already said that in our daily perceptions, these two poles are not separated, and it is hence only straightforward to explore the middle-ground between the two poles. Moving from Panel A towards Panel F leads us to Panel B which adds features to the objects. These features allow for sorting objects, which goes together with grouping them into classes or sets. The result of such grouping is shown in Panel C. Evidently, grouping facilitates counting. Note that sorting, classifying, and counting are elementary statistical operations even children at a very young age are capable of.

Starting at the other pole and moving from Panel F towards Panel A leads us to Panel E which fills the space of possibilities with objects. Compared to Panel F, Panel E leaves the world of pure structure and reminds one that the space consists of countable units – even if, in contrast to the present Panel E, these units are not yet counted, and even if these numbers may be infinite (e.g., repeated outcomes of a chance device like a roulette wheel).

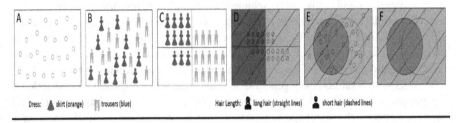

**Fig. 4.** Panel A: 25 distinguishable objects without any features. Panel B: The same objects, but now described with respect to two variables, dress and hair length. Panel C: The same objects, but now sorted according the two variables. Panel D: The same objects and assortment are now visualized on a more abstract level. Panel E: Visualization of possibilities that arise from combining two dichotomous variables, hair length and dress. For each of the possibilities, the number of objects from a given sample is visible. Panel F: Venn-diagram visualizing the possibilities that arise from combining two dichotomous variables.

Johnson-Laird's mental model theory is essentially based on exactly this step from Panel F to Panel E. When solving reasoning tasks that could be supported by Venn diagrams (e.g., "All P are Q" and "Some Q are R"; Is it true that "some R are not-P"?), mental model theory posits that people do not operate in an abstract variable space, but construct individual instances, thereby searching for examples that confirm or disconfirm the conclusion that they are asked to scrutinize. Arranging the objects of a finite sample that facilitates counting may lead to a representation such as the one depicted in Panel D.

Panels C and D hit the middle ground between the two poles. On the one hand (when coming from Panel A), description and structure are added to otherwise indistinguishable individual objects, and on the other hand (when coming from Panel F), an abstract structure and a space of possibilities is filled with concrete and countable cases. While Panel C maintains the analogous representation of individual cases that were already displayed in Panel B, Panel D inherits the level of abstraction that comes with a focus on possible features and their combinations displayed at the right end of the Figure.

The results on icon arrays, as presented in the next section, clearly indicate that the extensional approach fosters probabilistic intuitions of untrained people. We recall here that it was Otto Neurath who, during the first half of the twentieth century, used and introduced formats such those depicted in Panel C – he called those little icons *isotypes,* henceforth we will refer to arrays of such isotypes also as *icon grids* or *icon arrays.*

# 6  Icon Arrays for Risk Literacy: Following Otto Neurath

Icon arrays are a form of graphical representation that illustrates principles for the design of risk communications, inspired by Neurath's Isotypes (Trevena et al. 2013). An icon array is a form of pictograph or graphical representation that uses grids of matchstick figures, faces or other symbols to represent statistical information. An indicator of good quality risk communication is an adherence to the principle of *transparency*, which is definitely a characteristic of icon arrays: such representations define an appropriate reference class, and risks are presented in absolute rather than relative numbers (e.g., 1 out of 1,000 fewer women die from breast cancer with mammography screening as opposed to communicating the relative risk reduction of 20%). Icon arrays can be designed to communicate a variety of statistics transparently, including simple and conditional event frequencies (e.g., conditional probabilities). In medical risk communication, for instance, icons typically represent individuals who are affected by a risk, side-effect, or other outcome. Icon arrays are helpful for communicating risk information because they draw on people's natural tendency to count (Dehaene 1996), while also facilitating the visual comparison of proportions. For example, to represent a 3% risk of infection, icon arrays represent the proportion of individuals who end up with an infection, for instance, an icon array may just depict 100 icons, of which 3 are marked as „special". The one-to-one match between individual and icon has been proposed to invite identification with the individuals represented in the graphic to a greater extent than other graphical formats. Icon arrays are suitable for facilitating the understanding of risk information due to two characteristics: First, they arrange the icons systematically (Fig. 4C) rather than randomly (as in Fig. 4B). Second, and relatedly, they visualize a part-whole relationship.

Institutions devoted to fostering the intuitions of "responsible patients" like the Harding Center in Berlin are interested in propagating basics of risk literacy. Another institution, the AOK, which is one of the main insurance companies in Germany, communicates information to patients by means of fact boxes, as illustrated in Fig. 5.

What does the patient perceive here? The array on the left shows icons for 1,000 women who do not perform screening. Five of them die of breast cancer. The array on the right side shows 1,000 women who undergo screening regularly. Four of these women die of breast cancer, implying that the absolute risk reduction through regular screening is 1 per 1,000 (the reduction from 5 per 1,000 to 4 per 1,000 is 1 per 1,000). The so-called relative risk reduction in the breast cancer situation is 20% for women performing screening (1 per 1,000 whose lives could be saved compared to 5 per 1,000 who would lose their lives without screening; see Sect. 10 for dynamical representations of risk changes).

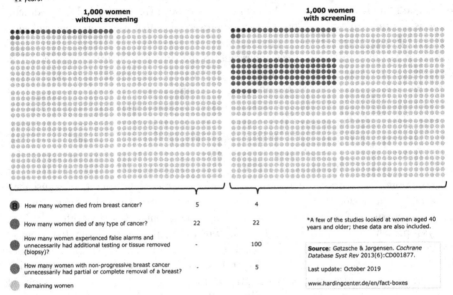

**Fig. 5.** A fact box designed by the Harding Center and used by the largest health insurance company in Germany, the AOK, for communicating information on the risk reduction caused by regular screening.

## 7   Beyond Neurath: Dynamic Icon Arrays

Even though icon arrays are already very helpful in facilitating understanding (Garcia-Retamero & Hoffrage 2013), there is still room for improvement: one can make them dynamic. In a dynamic webpage designed by Tim Erickson (https://www.eeps.com/projects/wwg/wwg-en.html), icon arrays can be sorted and organized by a simple click on a button (see, for instance, the HIV example, which is also briefly described below), so that relevant features can be quantified at a glance. *Sorting* is the first elementary statistical action we perform, sometimes just mentally (Martignon & Hoffrage 2019). Dynamic displays can thus become particularly useful for communicating about co-occurring or conditional events, which is relevant for understanding the meaning of features in the medical domain. For this purpose, sorting icon arrays becomes essential, as we illustrate with the example of 100 people who were tested as to whether they are HIV positive (Fig. 6).

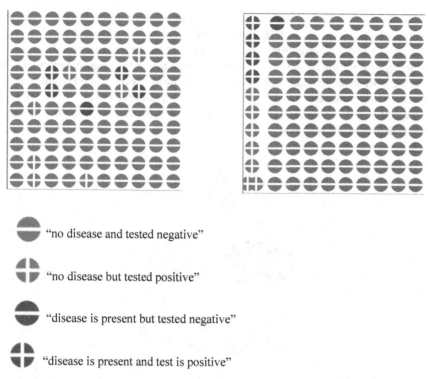

"no disease and tested negative"

"no disease but tested positive"

"disease is present but tested negative"

"disease is present and test is positive"

**Fig. 6.** This icon array, unsorted (left) and sorted (right), represents 100 people, diseased or not diseased, who are tested as to whether or not they are HIV positive.

## 8  Constructing Trees and Double Trees Starting from Icon Arrays

The validity or predictive value of a feature (e.g., a positive test, or breathing problems) can be computed by means of Bayes' Theorem. Consider, a physician receives new evidence (E) in form of a positive result. To infer whether a certain disease (D) is present or not, the physician should use her prior probability that the disease (D) is present, as well as the two likelihoods of a positive test result (if the disease is present and if the disease is not present, respectively) to calculate the so-called *posterior* probability that the disease is indeed present given the evidence, i.e., the positive test result. The corresponding formula is called *Bayes' Rule*, and was first formulated by the mathematician, philosopher and minister Thomas Bayes in the eighteenth century. Using Bayes' rule, the probability that the disease (D) is present once a new piece of evidence becomes known is calculated as follows:

$$P(D|E) = \frac{P(E|D)P(D)}{P(E|D)P(D) + P(E|\overline{D})P(\overline{D})}$$

Formula 1: Bayes' rule

The formula shows how to solve this evidential reasoning problem: In the medical setting, P(D|E) is the probability that the patient has the disease given that they tested positive on the test. People are notoriously bad at manipulating probabilities, as a plethora of empirical studies have shown (Eddy 1982; Gigerenzer & Hoffrage 1995).

The tree in Fig. 7 represents information about disease and test in a causal, sequential and hierarchical setting by means of a tree: The presence of the disease, represented by D+, is the label on the left node or leaf of the tree in the first level, while D- represents its absence. In this particular case the number 0.01 on the branch between the initial node and the node labeled with D+ in Fig. 7 represents the probability of the disease being present, also called its base rate.

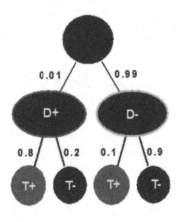

$P(D+|T+) = 7\%$

**Fig. 7.** A tree representing the binary categories "Disease" and "Test result" (D+ means that the disease is present, while D- denotes absence of the disease; T+ and T- denote a positive and negative test result, respectively).

Gigerenzer and Hoffrage (1995) have proposed a didactical simplification or reduction of the initial probabilistic treatment, which deserves the name of a heuristic: the systematic use of so-called natural frequencies. They argued that the kind of reasoning needed to make assessments on the diagnosticity of a test or symptom (or, in general, of a feature characterizing a category) can be facilitated by changing the format of information representation. In the same article (Gigerenzer & Hoffrage 1995) have empirically shown that diagnostic assessments based on new evidence could be substantially improved when the statistical information was provided by means of natural frequencies compared to representation in terms of probabilities.

## 9   Dynamic Trees of Natural Frequencies

Natural frequencies are the frequencies that naturally result if a sample is taken from a population. In case of one hypothesis or cause, like a disease, and of a piece of binary evidence, like the positive or negative result of a test, natural frequencies are the result of counting members of a given sample in each category. Translating probabilities into natural frequencies is always possible and becomes an ecologically rational heuristic that facilitates reasoning. In medicine, physicians' diagnostic inferences have been shown to improve considerably when natural frequencies are used instead of probabilities (Gigerenzer 1996; Hoffrage, Lindsey, Hertwig, & Gigerenzer 2000). Figure 8 illustrates both approaches to the causal tree: one by means of probabilities and one by means of natural frequencies.

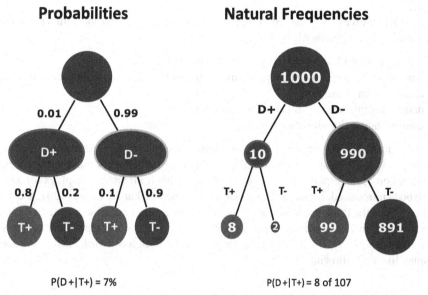

**Fig. 8.** Two trees, one labeled with probabilities, the other labeled with natural frequencies, representing the knowledge of the physician on a certain patient concerning breast cancer.

Trees with natural frequencies as labels can be constructed in the causal direction, i.e., from cause to evidence, as in the Figures above. The beneficial effect of natural frequencies could also be used as a basis to design tutorials that teach students to better cope with probability representations. Instead of teaching them how to plug probabilities into Bayes' rule, they have been taught how to translate these probabilities into natural frequencies and subsequently derive the solution for there. In an online-tutorial, Sedlmeier & Gigerenzer 2001) could show that such frequency-tree representations were superior to probability trainings (long-term performance of over 90% compared to 20%, respectively). Likewise, representation training also proved to be superior over rule training in a classical classroom setting with medical students, using medical problems as content (Kurzenhäuser & Hoffrage 2002).

Natural frequencies go hand in hand with icon arrays as illustrated by the dynamic web page the reader can reach by means of the QR Code below (https://www.eeps.com/projects/wwg/wwg-en.html)

This page can help the public to become "informed" and "competent" when dealing both with the sensitivity or specificity of a test and with its positive/negative predictive value.

These resources are designed to support instruction of children and adults to become informed and competent when

- dealing both with the sensitivity or specificity of a test,
- dealing with positive/negative predictive values of tests and dependence on base rates,
- understanding base rates,
- understanding relative and absolute risks, and
- understanding the subtleties of features' conjunctions.

To summarise, the resource is designed to make the teaching and training of risk literacy easy and transparent, by offering multiple complementary and interactive perspectives on the interplay between key parameters. Such interactive displays for adults have been introduced, for instance, by Garcia-Retamero, Okan, & Cokely (2012). Clicking on any of the three sections leads to pages where a variety of contexts are presented. For instance, in *The explanatory power of features* one can choose between contexts – one is *Pets and bells*, which is quite appropriate for children of fourth class – and see a display like the following (Fig. 9):

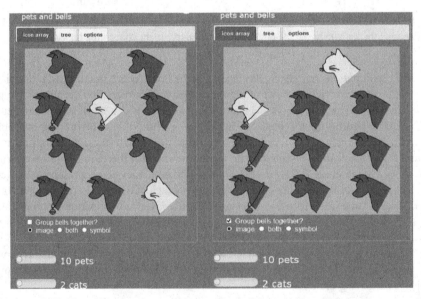

**Fig. 9.** Representations of 10 pets, cats and dogs, random and sorted.

The natural question is: "If a pet is wearing a bell, is it likely to be a cat?" The task is to judge the validity, or predictive value of this feature for the category *Cats*. The button "group bells together", at the left side under the picture, sorts the pets, so that it becomes easy to visualize pets wearing a bell.

The role of base rates is illustrated through the use of the sliders placed under the array. Maintaining the total number of pets equal to 10 one can enhance the base rate of "cats", while keeping the sensitivity of "bell" constant, as illustrated in Fig. 10.

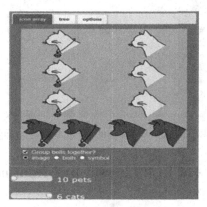

**Fig. 10.** In this display the number of pets remains 10 but the base rate of cats is now "6 out of 10"

Wearing a bell now becomes moderately predictive for the category "cats". The next instructional step is to construct trees. A button at the top left of the grid in Fig. 10 leads to the corresponding double tree, illustrated in Fig. 11.

The double tree in Fig. 11 exhibits two inference directions: one is causal the other is diagnostic. The double tree is a simple and transparent way of approaching Bayesian reasoning. Studies by Wassner (2004) clearly demonstrated the effectiveness of such double trees for fostering successful Bayesian reasoning in the classroom. He worked with ninth grade students in Germany.

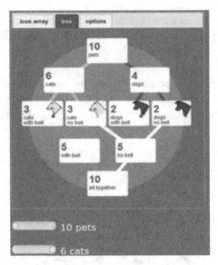

**Fig. 11.** The double tree from the icon array in Fig. 10.

## 10 Other Elements of Risk Literacy by Means of Dynamical Representations

The Webpage "Worth the risk?", also illustrates the subtleties connected with risk reductions and increases in transparent ways that are easy to grasp. Figure 12 shows 20 boys who have had a bike accident, ten of which were wearing a helmet. The faces with a pad and a black eye represent boys with severe injury caused by the bike accident.

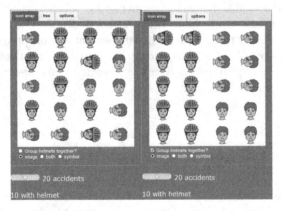

**Fig. 12.** Icon array exhibiting 20 boys having a bike accident, ten of them wearing helmets

Simply sorting the icon array by grouping helmets together allows an easy grasp of the risk reduction provided by helmets.

While all the dynamical resources described so far are devoted to the second component of Risk Literacy, as described in Fig. 3, the third author of this paper has also designed and produced plugins for the third and fourth component. Opening https://codap.xyz/ the reader finds, among many plugins, also ARBOR. This plugin is designed for the construction of trees for classification and decision, based on data concerning features for classification. Decision trees constructed by means of clever algorithms which can be used in complex medical situations exist and have been developed by eminent statisticians, such as CART (Breiman 1996); implementations of those algorithms are available in common software platforms such as R (Erickson & Engel 2023). The trees proposed in ARBOR are utterly simple and intuitive and young students can easily understand them. The basic idea is that the display starts with an initial node that specifies a response variable—the binary criterion or category the tree is designed to predict. Then the user can drag any attribute to any node in order to make (or replace) a branch based on that attribute. Finally, to make a prediction (in the case of a classification tree), the user has to add a "diagnosis leaf" to the end of every branch to indicate what conclusion you should come to if a case arrives at that branch. Let's see what that looks like in practice using a famous dataset about heart patients from Green and Mehr (1999). It has 89 cases with four attributes: MI (whether the patient had a myocardial infarction, a heart attack); pain (whether the patient complained of chest pain); STelev (whether the "ST" segment on an EKG was elevated); and oneOf (whether the patient showed any of four other symptoms). First, the user must set up a *response variable*. This involves deciding which attribute the tree will predict, and which value the tree will orient towards. In our situation, we are trying to predict whether the patient will have a heart attack ( MI ) using the other three attributes. We also need to orient our tree: will we look at the proportion of patients that do get a heart attack ( MI = yes ) or the proportion that do not ( MI = no )? In Arbor, the response variable and its orientation appear in the "root" node of the tree, represented by a box as shown in Fig. 13.

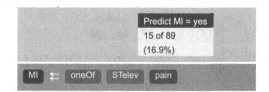

**Fig. 13.** The positive value of the criterion (MI = Infarction)

*Growing the Tree.* Now we want to "grow" our tree. We will do so by dragging attributes ( oneOf, STelev, or pain) from the palette and dropping them on a node. CODAP's built-in graphing helps us explore the data before constructing a tree. In the case of this question, we can make a graph that shows how the value of MI is associated with pain. The figure shows that, indeed, a patient is more likely to have an MI if they complain of chest pain—but to a student, the relationship probably does not look as they expected. Even with chest pain, a large majority of patients do not get a heart attack (Fig. 14).

The plugin allows then to construct individual trees for each cue and then compose trees with three cues like the one shown in Fig. 15.

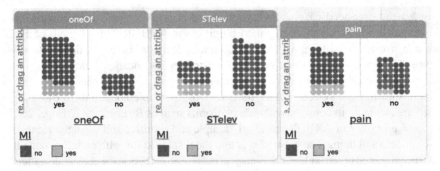

**Fig. 14.** User-friendly display created by the plugin for the statistics of three cues

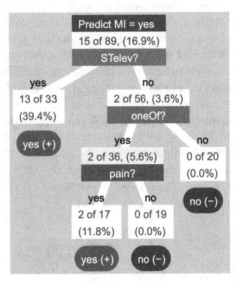

**Fig. 15.** This tree has been constructed: One can drag and place cues one under the other and construct trees.

This tree could be constructed also with other orderings and we would be able to compare their performances and choose the „best" one. The plugin is thus a facilitating tool for doing the necessary steps for good classification in a simple user-friendly way. There are many other useful plugins in the webpage https://codap.xyz. Another plugin is, for instance, Lotti, which is specifically constructed for component 4 of risk literacy. Here the plugin presents the user with „doors" and alternatives with different outcomes. The plugin provides the typical „lottery" situation, with fixed gain versus situations of risk versus benefits (Fig. 16).

Concluding, we simply stress the benefits of the dynamical, interactive tools for classifications based on dynamic, extensional representations of information, fostering intuitions on risk and provide tools of risk literacy.

**Fig. 16.** A game for acquiring an understanding of risk through clicking either on plan A, with a fixed allowance, or B with a varying allowance, with larger expected value.

# References

Breiman, L.: Probability. SIAM, New York (1968)

De Morgan, A.: An Essay on Probabilities, and Their Application to Life Contingencies and Insurance Offices. Longmans, London (1838)

Dehaene, S.: The organization of brain activations in number comparison: event-related potentials and the additive-factors method. J. Cogn. Neurosci. **8**(1), 47–68 (1996). https://doi.org/10.1162/jocn.1996.8.1.47

Eddy, D.: Probabilistic reasoning in clinical medicine: problems and opportunities. In: Slovic, P., Tversky, A. (eds.) Judgment under Uncertainty: Heuristics and Biases, pp. 249–267. Cambridge University Press (1982)

Erickson, T., Engel, J.: What goes before the CART? Introducing classification trees with Arbor and CODAP. Teach. Stat. **45**, S104–S113 (2023)

Garcia-Retamero, R., et al.: Using visual aids to improve communication of risks about health: a review. Sci. World J. 562637 (2012). https://doi.org/10.1100/2012/562637

Garcia-Retamero, R., Hoffrage, U.: Visual representation of statistical information improves diagnostic inferences in doctors and their patients. Soc Sci Med **83**, 27–33 (2013). https://doi.org/10.1016/j.socscimed.2013.01.034

Gigerenzer, G.: How to make cognitive illusions disappear: beyond "heuristics and biases." Eur. Rev. Soc. Psychol. **2**(1), 83–115 (1991)

Gigerenzer, G., Hoffrage, U.: How to improve Bayesian reasoning without instruction: frequency formats. Psychol. Rev. **102**, 684–704 (1995)

Gigerenzer, G., Hertwig, R., Hoffrage, U., Sedlmeier, P.: Cognitive illusions reconsidered. In: Handbook of Experimental Economics Results, vol. 1, pp. 1018–1034 (2008)

Hilbert, D.: Über das Unendliche. Mathematische Annalen **95**, 170 (1926)

Hoffrage, U., et al.: Communicating statistical information. Science **290**(5500), 2261–2262 (2000). https://doi.org/10.1126/science.290.5500.2261

Kurzenhäuser, S., Hoffrage, U.: Teaching Bayesian reasoning: an evaluation of a classroom tutorial for medical students. Med. Teach. **24**(5), 516–521 (2002)

Martignon, L., Hoffrage, U.: Wer wagt, gewinnt? Wie Sie die Risikokompetenz von Kindern und Jugendlichen fördern können. Hogrefe, Göttingen (2019)

Mousavi, S., Gigerenzer, G.: Risk, uncertainty, and heuristics. J. Bus. Res. **67**(8), 1671–1678 (2014)

Sedlmeier, P., Gigerenzer, G.: Teaching Bayesian reasoning in less than two hours. J. Exp. Psychol. Gen. **130**(3), 380 (2001)

Trevena, L.J., et al.: Presenting quantitative information about decision outcomes: a risk communication primer for patient decision aid developers. BMC Med. Inform. Decis. Mak. **13**(2), 7 (2013)

Venn, J.: On the employment of geometrical diagrams for the sensible representation of logical propositions. Proc. Camb. Philos. Soc. **4**, 47–54 (1880)

Wassner, C.: Förderung Bayesianischen Denkens — Kognitionspsychologische Grundlagen und didaktische Analysen. Franzbecker, Hildesheim (2004)

# Author Index

J. Baratgin et al. (Eds.): HAR 2023, LNCS 14522, pp. 407–408, 2024.
https://doi.org/10.1007/978-3-031-55245-8

Printed in the United States
by Baker & Taylor Publisher Services

Printed in the United States
by Baker & Taylor Publisher Services